Analysis of Global Expansion Methods: Weakly Asymptotically Diagonal Systems

MATH STAT LIBRARY

COMPUTATIONAL MATHEMATICS AND APPLICATIONS

Series Editor
J. R. WHITEMAN

Institute of Computational Mathematics, Brunel University, England.

E. HINTON and D. R. J. OWEN: Finite Element Programming

M. A. JASWON and G. T. SYMM: Integral Equation Methods in Potential Theory and Elastostatics

J. R. CASH: Stable Recursions: with applications to the numerical solution of stiff systems

H. ENGELS: Numerical Quadrature and Cubature

L. M. DELVES and T. L. FREEMAN: Analysis of Global Expansion Methods: Weakly Asymptotically Diagonal Systems

Analysis of Global Expansion Methods: Weakly Asymptotically Diagonal Systems

L. M. DELVES

Department of Computational and Statistical Science
University of Liverpool
Liverpool, UK

and

T. L. FREEMAN

Department of Mathematics
University of Manchester
Manchester, UK

1981

London New York Toronto Sydney San Francisco

MATH. STAT. LIBRARY

6381-0256
MATH

SEP/AE
MATH

XD81
2008
MATH

ACADEMIC PRESS INC. (LONDON) LTD
24–28 Oval Road
London NW1

US edition published by
ACADEMIC PRESS INC.
111 Fifth Avenue
New York, New York 10003

Copyright © 1981 by
ACADEMIC PRESS INC. (LONDON) LTD

All Rights Reserved
No part of this book may be reproduced in any form, by photostat, microfilm or any other means, without written permission from the publishers

British Library Cataloguing in Publication Data

Delves, L. M.
Analysis of global expansion methods. – (Computational mathematics and applications).
1. Differential equations – Asymptotic theory
2. Integral equations
3. Matrices
I. Title II. Freeman, T. L. III. Series
515.3′5 QA371

ISBN 0-12-208880-8
LCCCN 80-42084

QA371
D39
1981
MATH

Preface

The study of *Weakly Asymptotically Diagonal* systems originated in an attempt to analyse the convergence of expansion methods for differential and integral equations of "global" type; that is, methods which employ expansions of the form

$$f(x) = \sum_{i=1}^{N} a_i^{(N)} h_i(x),$$

where the $h_i(x)$ have global rather than local support and are typically chosen to be orthogonal polynomials in an appropriate set of variables. For a linear problem, such methods lead to a set of linear equations for the coefficients $a_i^{(N)}$, $i = 1, \ldots, N$, and it is possible to treat both the convergence and the stability properties of the method by analysing the structure of the matrix and right-hand side of these equations. Such an approach has the advantage that these properties are then characterised directly in terms of quantities which are available without additional cost during the course of the calculations: the error estimates which result are cheap to compute.

For the analysis to be of general use, it is necessary to abstract the essential structure of the equations, and to analyse the class of matrices having this structure. The definition of a *Weakly Asymptotically Diagonal* (WAD) matrix arises directly from this necessity. Historically, two subclasses of WAD matrices were introduced first: matrices of type A, which closely model the equations that arise from Fredholm integral equations, and type B, which model differential equations.

These subclasses yield a rather straightforward analysis of the convergence and stability properties of global expansion methods for one-dimensional problems, together with cheaply computable and very effective error estimates for such methods. Systems of type A in particular yield a rather simple theory, which is discussed in detail in Chapter 2.

8266

The more abstract general WAD definitions arise from a natural wish to broaden the class of problems covered by the analysis as much as possible, and to formalise the essential results contained in the rather specific theorems of type A and type B systems. It appears that this can be achieved in quite a natural manner, and a considerable body of analysis has grown up for WAD matrices. Although it is by no means complete, it seems worthwhile to bring together the available results together with some of their applications to expansion methods; the result is this book, which splits naturally into three sections.

The theory and numerical analysis of WAD systems is discussed in Chapters 1–8; we hope that the ordering chosen, and the examples given, illustrate sufficiently the analysis to motivate this detailed study. The third section, Chapters 12–14, discusses applications of the theory to the solution of integral and differential equations. Space considerations limit the detail that can be given in these chapters. However, it is our hope that they, and also Chapter 8 on the numerical analysis of WAD systems, form a practical justification for Chapters 1–7: namely, that the study of WAD systems, and the insight which these systems give into the structure of global expansion methods, has led directly to the development of improved algorithms and to cheap and effective error estimates for a wide class of problems. The error estimates depend on the parameters of the WAD systems, and these parameters can themselves be directly estimated during the calculations. They can, however, also be related to the analytic properties (the smoothness) of the coefficients in the equation being solved. Section 2, Chapters 9–11, is devoted to the analysis of the convergence properties of orthogonal expansions, and discusses this relationship in detail. These chapters thus form a bridge between the abstract WAD theory and its applications in Chapters 12–14. They also form a bridge between the basis—dependent analysis of this book, and the more common basis—independent analysis of the convergence of variational and Galerkin methods in which the assumed smoothness properties of the equations appear directly.

This book contains a number of new and extended results, but its main purpose is to draw together previously published results. We are grateful to all of our colleagues who have worked in this field, and particularly to Drs M. Bain, K. O. Mead and F. A. Musa, for permission to draw upon their theses and published work, as listed in the references, as well as for many discussions. Without their collaboration and cooperation, this book could not have been written. Responsibility for any errors which remain in it, however, rests with us. We would also like to acknowledge with gratitude the patience and forbearance of Miss K. Anderson and Mrs G. M. Eyres, in seeing us through successive drafts.

Liverpool & Manchester
March 1981

L. M. Delves
T. L. Freeman

Contents

Section 2: Orthogonal Expansions

Section 3: Applications

Section 1

Theory of WAD Systems

1

Introduction and Motivation

1.1 Introduction

This book is primarily concerned with the analysis of a class of infinite matrices: the class of *Weakly Asymptotically Diagonal* (WAD) matrices of the title. This analysis is interesting in its own right; however, the major interest lies (for the authors at least) in the application of the theory to problems ir the general field of approximation theory. WAD matrices arise naturally when global or global element expansion methods are used to solve numerically a wide class of problems which involve integral and differential equations in one or more dimensions; the WAD assumptions were in fact introduced in order to model the structure of the matrices that arise from these problems. In this chapter we therefore try to motivate the analysis of Chapters 2–7 by giving a brief discussion of such methods. We demonstrate by example their main source of interest: the possibility of obtaining very rapid convergence; convergence which in "suitable" cases is much more rapid than that attainable with a conventional finite difference or finite element approach. However, rapid convergence is of little use in practice unless it can be recognised: what is needed is a computable error estimate which reflects the actual error as faithfully as possible while adding as little as possible to the cost of the calculations. The provision of such error estimates is one of the main achievements of the WAD theory developed here. The estimates depend on an analysis of the structure of the defining equations for the expansion method; this structure is displayed in a simple case in the example of Section 1.5. Once the structure is understood, surprisingly simple and effective error estimates follow. In addition, the structure suggests efficient ways of both setting up and of solving the defining equations; see Chapters 8 and 12–14 for a discussion of the savings which can be made.

1.2 Galerkin methods for linear operator equations

1.2.1 The Galerkin formalism

Let R be a Hilbert space with inner product $(\cdot, \cdot)$ and $\mathscr{L}: R \to R$ a linear operator defined on R (or perhaps a subspace of R). Let g be a known element of R, and f the solution (assumed to exist and to be unique in R) of the linear operator equation

$$\mathscr{L}f = g. \tag{1.2.1}$$

An *expansion method* introduces a complete set $\{h_i\}$, $i = 1, 2, \ldots,$ of elements in R, and expansions for the exact solution of (1.2.1) and the truncated form f_N which approximates it. Thus

$$f = \sum_{i=1}^{\infty} b_i h_i \tag{1.2.2}$$

$$f_N = \sum_{i=1}^{N} a_i^{(N)} h_i. \tag{1.2.3}$$

The method also provides an algorithm for computing the coefficients $a_i^{(N)}$ in the approximate solution. Many algorithms can be constructed; we consider here only the *unsymmetric Galerkin method* or *method of moments*, which introduces a second set of elements $\{\hat{h}_i\}$ and computes $\mathbf{a}^{(N)} = (a_i^{(N)})$ as the solution of the $N \times N$ linear system

$$\mathbf{L}\mathbf{a}^{(N)} = \mathbf{g} \tag{1.2.4}$$

where the matrix $\mathbf{L}$ and vector $\mathbf{g}$ are defined by

$$L_{ij} = (\hat{h}_i, \mathscr{L}h_j), \quad i, j = 1, \ldots, N, \tag{1.2.5}$$

$$g_i = (\hat{h}_i, g), \quad i = 1, \ldots, N. \tag{1.2.6}$$

In practice, the sets $\{\hat{h}_i\}$, $\{h_i\}$ may be related; for example, we may choose $\hat{h}_i = \mathscr{L}h_i$ (method of least squares) or $\hat{h}_i = wh_i$ (weighted Galerkin). The choice $\hat{h}_i = h_i$ yields the symmetric Galerkin technique; we shall use the generic term Galerkin method, the precise choice of $\hat{h}_i$ being evident from the context.

The use of a Galerkin technique for numerical calculations raises a number of interesting and interrelated questions:

(i) *Posing the problem*

How do we pose a given type of problem in the form (1.2.1)? In particular, for differential equations, this question usually reduces to: How do we treat the boundary conditions?

(ii) *Choosing the expansion set*
How do we choose the basis set $\{h_i\}$ in (1.2.2), (1.2.3), and the companion set $\{\hat{h}_i\}$ in (1.2.5), (1.2.6)? How does this choice affect the accuracy of the calculation? The cost of the calculation? The stability against numerical or round-off errors? The convenience?

(iii) *Computing the solution*
Setting up the matrix problem (1.2.4) usually involves providing numerical approximations to the inner products involved; the choice of approximation can crucially affect both the accuracy obtained and the time taken. For the large systems which result from multi-dimensional problems, the methods used to solve (1.2.4) are also important since the solution time can be as great as or greater than the time taken to form the equations.

(iv) *Analysing the errors*
Three classes of numerical error can be distinguished in a Galerkin calculation:

(a) *Truncation errors*
These stem from the truncation of (1.2.2) after N terms.
(b) *Discretisation errors*
These stem from the differences $b_i - a_i^{(N)}$, $i = 1, 2, \ldots, N$.
(c) *Quadrature errors*
Given that we approximate the inner products involved (using quadrature rules) we in fact solve not (1.2.4) but the perturbed system

$$(\mathbf{L} + \delta\mathbf{L})(\mathbf{a}^{(N)} + \delta\mathbf{a}^{(N)}) = \mathbf{g} + \delta\mathbf{g}, \qquad (1.2.7)$$

these perturbations yielding an additional source of error, $\delta\mathbf{a}^{(N)}$.

1.2.2 An example

We provide a partial answer to questions (i) and (ii) above by means of an example. Consider the real, linear Fredholm integral equation of the second kind

$$f(x) = g(x) + \lambda \int_a^b K(x, y) f(y)\, dy. \qquad (1.2.8)$$

Under suitable assumptions (on the kernel λK, and driving term g), this equation has a unique solution†. If we choose an inner product on the

† For example: if $g(x)$ is continuous in $[a, b]$, $K(x, y)$ is continuous in $[a, b] \times [a, b]$, and λ is not a characteristic value of (1.2.8), then there exists a unique continuous solution $f(x)$.

interval $[a, b]$:

$$(f_1, f_2) \equiv \int_a^b f_1(x) f_2(x) w(x) \, dx \tag{1.2.9}$$

and set

$$\hat{h}_i(x) = h_i(x), \quad i = 1, 2, \ldots \tag{1.2.10}$$

then (1.2.5), (1.2.6) takes the form

$$L_{ij} = \int_a^b w(x) h_i(x) h_j(x) \, dx - \lambda \int_a^b w(x) h_i(x) \int_a^b K(x, y) h_j(y) \, dy \, dx \tag{1.2.11}$$

$$g_i = \int_a^b w(x) h_i(x) g(x) \, dx. \tag{1.2.12}$$

Equations (1.2.4), (1.2.11), (1.2.12) are the formal defining equations for the Galerkin method applied to the problem (1.2.8). The same equations result if we choose the *unweighted inner product*

$$(f_1, f_2) = \int_a^b f_1(x) f_2(x) \, dx \tag{1.2.12a}$$

and set

$$\hat{h}_i(x) = w(x) h_i(x), \quad i = 1, 2, \ldots. \tag{1.2.12b}$$

A *numerical method* results from these defining equations once we:

(a) specify the basis set $\{h_i\}$;
(b) specify how the integrals are to be performed.

We are not concerned here with (b) (but see Chapter 12). The choice (a) lies primarily between *local* and *global* bases; we can informally describe these as follows:

Local basis
We split the region $a \leqslant x \leqslant b$ into intervals of width h. On each subinterval $[a + mh, a + (m + 1)h]$ the approximate solution $f_N(x)$ is taken to be a polynomial of fixed degree p. Continuity conditions may be imposed across subinterval boundaries (for example, a *spline basis* results if $f_N(x)$ is constrained to have $p - 1$ continuous derivatives on $[a, b]$). Convergence is obtained by letting $h \to 0$ for fixed p.

This description of $f_N(x)$ is not in terms of an expansion of the form (1.2.3), but for any choice of p and of continuity constraints, it can be put in that form. For example, if $p = 1$ (piecewise linear approximation) and $f_N(x)$ is constrained to be continuous on $[a, b]$, the description can be rephrased as follows:

Define the *hat function* $\mathrm{Hat}_i(x)$:

$$\begin{aligned}
\mathrm{Hat}_i(x) &= 0, & x &\leqslant a + (i-1)h \\
&= \frac{x - a - (i-1)h}{h}, & a + (i-1)h &< x \leqslant a + ih \\
&= \frac{a + (i+1)h - x}{h}, & a + ih &< x \leqslant a + (i+1)h \\
&= 0, & x &> a + (i+1)h
\end{aligned}$$

$$h = (b - a)/N. \tag{1.2.13}$$

Thus, $\mathrm{Hat}_i(x)$ is zero except on the interval $[a + (i-1)h, a + (i+1)h]$ and is linear over the two halves of this interval, with discontinuous first derivative at $x = a + ih$, and at $a + (i-1)h$, $a + (i+1)h$. Now choose as basis

$$h_i(x) = \mathrm{Hat}_i(x), \quad i = 1, \ldots, N. \tag{1.2.14}$$

Global basis

The local basis described above has the feature that each of the *basis functions* $\mathrm{Hat}_i(x)$ depends explicitly on N, and has *local support*; that is, is zero everywhere except over a small subinterval of $[a, b]$. A *global basis* is one in which the basis functions have support over the whole region; in practice, they are also chosen to be independent of N. Within these restrictions, many choices are available. The most common choice is to take $f_N(x)$ to be a *polynomial of degree* $N - 1$ *in* x.

We can clearly achieve this in a number of ways; for example, the choice

$$h_i(x) = x^{i-1}, \quad i = 1, 2, \ldots, \tag{1.2.15}$$

and

$$h_i(x) = T_{i-1}\left(\frac{2x - (a+b)}{b - a}\right), \quad i = 1, 2, \ldots, \tag{1.2.16}$$

where $T_k(z)$, $-1 \leqslant z \leqslant 1$, is a Chebyshev polynomial of degree k, and each defines a polynomial approximating function $f_N(x)$; we show in Section 1.3 that these choices are in fact *equivalent* in the sense that, if quadrature and round-off errors are ignored, they will yield the *same solution* $f_N(x)$.

The choice between *local* and *global* basis is crucial; the two types lead to quite different numerical techniques both for setting up and for solving the Galerkin equations, and to quite different types of error analysis. They also perform quite differently in practice, as the following example demonstrates.

In (1.2.8), we take

$$\lambda K(x, y) = e^{xy}$$
$$g(x) = e^{x} - (e^{x+1} - 1)/(x + 1) \qquad (1.2.17)$$
$$a = 0, \quad b = 1.$$

Then the exact solution of (1.2.8) is

$$f(x) = e^{x}. \qquad (1.2.18)$$

Table 1.1 shows the maximum errors, $\|f - f_N\|_\infty$, obtained using expansion methods with the expansion sets (1.2.14) and (1.2.16).

Looking first at the results using a local expansion, we remark that the errors reduce only relatively slowly as N increases. The results given are in fact well fitted by the form

$$\|f - f_N\| \sim CN^{-p}, \quad p = 2,$$

and it is a simple matter to predict this value for the exponent p in advance: it is related to the continuity of the expansion set $\{\mathrm{Hat}_i(x)\}$ used, and not at all to the problem being solved, provided that this is "smooth enough". It is more difficult to predict the amplitude C; still, it is certainly an advantage of local methods that the behaviour of the error can be predicted, and hence checked.

Looking next at the global expansion results, we see that very rapid convergence is obtained; for this problem, there is no doubt that the global method is preferable. This rapid convergence is typical of that achieved with a global polynomial basis for problems which have "smooth" solutions. Of course, not all problems are of this type. When the solution is "non-smooth", a global basis may converge no faster than a local basis, and may well be more expensive overall. Further, techniques exist for improving the performance of both types of expansion. For example, extrapolation procedures such as the "deferred approach to the limit" may be applied to increase the convergence rate of a local calculation, while we may choose to subdivide the region $[a, b]$ into two or more subintervals and apply a global expansion over each subinterval, to improve the performance of a global method. It is not the purpose of this book to argue for or against the use of global as opposed to local bases. Instead, we note the interestingly fast convergence which can be achieved, and ask the question: can this error be predicted in practice for a global method, as it can for a local method? In particular, we would like to provide error estimates for a global expansion method which are *cheaply computable* and *realistic*. It transpires that such an analysis of some generality can be given, provided that we limit attention to orthogonal bases as typified by (1.2.16); fortunately, the use of such bases,

TABLE 1.1

Solution of the integral equation (1.2.8) with parameters (1.2.17), using local and global bases. The "predicted error" for the global basis is that given by the error estimates discussed in Chapter 12, Section 12.2.

Max error \ N	3	5	7	9	15	31	63	127	255
Local (1.2.14)	$3{\cdot}8 \times 10^{-1}$	9·8, −2	—	2·4, −2	6·1, −3	1·5, −3	3·8, −4	9·5, −5	2·4, −5
Global (1.2.16)	$4{\cdot}3 \times 10^{-2}$	1·3, −4	2·7, −7	2·7, −10	—	—	—	—	—
Predicted error (1.2.16)	—	7·5, −3	1·5, −5	1·6, −8	—	—	—	—	—

rather than monomials such as (1.2.15), is dictated also by stability considerations. The error estimates which result are exemplified by those given in Table 1.1. In this example, these estimates are rather pessimistic; but as N increases, the error estimates reduce at about the same rate as the actual error. Thus, as for local expansions, it turns out to be easier to predict *rates of convergence* than absolute error amplitudes. The relationship between (1.2.15), (1.2.16) and the special role played by orthogonal expansions, is discussed in the next two sections.

1.3 Convergence and error analysis of Galerkin methods

For simplicity, let us first ignore any quadrature errors (that is, set $\delta\mathbf{L} = \delta\mathbf{g} = \mathbf{0}$ in (1.2.7)) and consider a symmetric Galerkin calculation. The error $e_N = f - f_N$ then depends only on the choice of the expansion set $S_N = \{h_i, i = 1, \ldots, N\}$. However, this dependence is not on the detailed choice of the h_i but rather on the subspace (of R) spanned by S_N, as Theorem 1.1 shows:

Theorem 1.1. *Let $S_h = \{h_i, i = 1, \ldots, N\}$ and $S_k = \{k_i, i = 1, \ldots, N\}$ span the same subspace, and let $f_N(h)$ and $f_N(k)$ represent the solutions of* (1.2.1) *obtained with S_h and S_k in a symmetric Galerkin calculation. Then for each N:*

$$f_N(h) = f_N(k). \tag{1.3.1}$$

Proof. We may write

$$f_N(h) = \sum_{i=1}^{N} \alpha_i h_i$$

$$f_N(k) = \sum_{i=1}^{N} \beta_i k_i$$

where

$$\begin{aligned} \mathbf{L}(h)\boldsymbol{\alpha} &= \mathbf{g}(h) \\ \mathbf{L}(k)\boldsymbol{\beta} &= \mathbf{g}(k) \end{aligned} \tag{1.3.2}$$

with

$$\left.\begin{aligned} \mathrm{L}(p)_{ij} &= (p_i, \mathscr{L}p_j) \\ g_i(p) &= (p_i, g) \end{aligned}\right\} p = k, h.$$

But since S_h, S_k span the same space there exists a non-singular $N \times N$

matrix $\mathbf{\Gamma}$ such that

$$k_i = \sum_{j=1}^{N} \Gamma_{ij} h_j \tag{1.3.3}$$

whence it follows that

$$\begin{aligned} \mathbf{L}(k) &= \mathbf{\Gamma}\mathbf{L}(h)\mathbf{\Gamma}^t \\ \mathbf{g}(k) &= \mathbf{\Gamma}\mathbf{g}(h) \end{aligned} \tag{1.3.4}$$

and hence from (1.3.2)

$$\boldsymbol{\alpha} = \mathbf{\Gamma}^t \boldsymbol{\beta}.$$

That is,

$$f_N(h) = \sum_{i=1}^{N} \alpha_i h_i = \sum_{i,\, j=1}^{N} \Gamma_{ji} \beta_j h_i = \sum_{j=1}^{N} \beta_j k_j = f_N(k).$$

It is clear from Theorem 1.1 that the convergence properties of a Galerkin method are *basis independent*; the proof applies also to the weighted Galerkin and least squares methods outlined above. As an example, we consider a one-dimensional problem defined on the region $-1 \leqslant x \leqslant 1$. Then the following bases are equivalent in the sense of Theorem 1.1:

(a) $\{h_i\} = \{1, x, x^2, \ldots, x^{N-1}\}$

(b) $\{h_i\} = \{P_{i-1}(x), \quad i = 1, \ldots, N\}$

(c) $\{h_i\} = \{T_{i-1}(x), \quad i = 1, \ldots, N\}$

(d) $\{h_i\} = \{1, x, (1 - x^2)T_j(x), \quad j = 0, 1, \ldots, N - 3\}$

where $P_i(x)$, $T_i(x)$ are Legendre and Chebyshev polynomials, respectively. Theorem 1.1 suggests that basis-independent methods should be used to analyse the convergence of a Galerkin method, and this is usually the case (see, for example, Schultz, 1969; Strang and Fix, 1973).

However, a basis-independent approach, although elegant, has two limitations. First, the error estimates which it generates are also basis independent, depending typically on continuity parameters of the solution and of the coefficients in the equation being solved, such as bounds on the values attained by high order partial derivatives or on generalised Lipschitz constants. These parameters are generally not available numerically, or at best are themselves costly to estimate, so that the price of an error estimate may be higher than that of the calculation itself. Second, the *stability* of a calculation, that is its sensitivity to quadrature errors, is *not* basis independent at all. Of the four bases (a)–(d) above, (b) and (c) will (normally) yield stable calculations, (a) is likely to be grossly unstable, and (d) "almost" stable.

We develop in this book a convergence and stability analysis of Galerkin calculations which depends on a direct analysis of the matrix equations (1.2.4), and hence is certainly basis dependent. The analysis depends on an application of the theory of WAD systems given in Chapters 2–7, and hence on the assumption that the basis yields a WAD system. Sufficient conditions for this to be so are given in Chapters 9–11; roughly speaking, the methods are applicable to orthogonal polynomial bases such as (b), (c) and (d) above, in one or more dimensions, but not to bases such as (a). The analysis therefore covers the Fast Galerkin method (Chapter 12) for integral equations, and the global or Global Element method (Chapters 13 and 14) for ordinary and partial differential equations. It does not cover local (spline or finite element) bases, for which alternative approaches are required.

In return for this basis dependence, we find that all the parameters which enter the theory involve the matrix **L** and vector **g** (see (1.2.4)) and hence are available at no extra cost during the course of the calculation. The error estimates which result are therefore not only computable but free; in addition, we can deal directly with the *stability* of the calculation.

1.4 The special role of orthogonal expansions

The special role played by orthogonal bases can be demonstrated by a simple example.

Let $\mathscr{L} = \mathscr{I}$, the identity operator; then (1.2.1) reduces to the identity $f = g$, and the problem to that of approximating g by the truncated expansion (1.2.3). Let $\{h_i\}$ be an orthonormal set; then (1.2.5), (1.2.6) reduce to

$$L_{ij} = (h_i, h_j) = \delta_{ij}$$

$$g_i = (h_i, g)$$

and the solution of (1.2.4) is, trivially,

$$a_i^{(N)} = (h_i, g) = b_i. \tag{1.4.1}$$

For such a problem the discretisation error is zero and only the truncation error remains. This in turn can be estimated very simply. We have, with $e_N = f - f_N$:

$$\|e_N\|^2 = (e_N, e_N) = \sum_{i=N+1}^{\infty} b_i^2 = \sum_{i=N+1}^{\infty} g_i^2. \tag{1.4.2}$$

Now, provided only that (g, g) exists, this last sum is finite; that is, certainly $g_n = o(n^{-1/2})$. In fact, it is shown in Chapter 9 that typically we may expect for some c_g independent of n

$$|g_n| \leqslant C_g n^{-p}, \tag{1.4.3}$$

where p depends on the analyticity properties of g. If we assume $p > \frac{1}{2}$ we may bound (1.4.2) as follows:

$$\|e_N\|^2 \leqslant C_g^2 \sum_{i=N+1}^{\infty} i^{-2p} < \frac{C_g^2}{2p-1} N^{-2p+1}. \tag{1.4.4}$$

Both C_g and p may be estimated cheaply by inspecting the behaviour of the g_i, $i = 1, \ldots, N$, which are needed during the calculation; thus (1.4.4) represents a computable error estimate for this problem.

Of course, in general $\mathscr{L} \neq \mathscr{I}$ and we must work rather harder. With an orthonormal basis, we obtain in place of (1.4.2) the identity

$$\begin{aligned} \|e_N\|^2 &= \sum_{i=1}^{N} (a_i^{(N)} - b_i)^2 + \sum_{i=N+1}^{\infty} b_i^2 \\ &= S_1 + S_2. \end{aligned} \tag{1.4.5}$$

The estimation of S_1 and S_2 involves the behaviour of the coefficients $a_i^{(N)}$ as a function of both i and N, and of b_i as a function of i. We shall find it convenient to talk of the *horizontal*, *vertical* and *diagonal* convergence of a given calculation, and we define these terms as follows:

(a) *Vertical convergence*:

$$\lim_{i \to \infty} b_i = 0.$$

(b) *Horizontal convergence*:

$$\lim_{N \to \infty} (a_i^{(N)} - b_i) = 0, \quad i = 1, 2, \ldots.$$

(c) *Diagonal convergence*:

$$\lim_{i \to \infty} (a_i^{(i)} - b_i) = 0.$$

Clearly, if $\lim_{i \to \infty} b_i = 0$, diagonal convergence is obtained only if $\lim_{i \to \infty} a_i^{(i)} = 0$.

In practice, we are interested in predicting not only *convergence* (of $\|e_N\|$ to zero) but also the *rate* of convergence. In general, the rate of decrease of $\|e_N\|$ is approximately given by the slowest of the rates of vertical, horizontal and diagonal convergence. For example, consider the (hypothetical) case when the following bounds are available:

$$\begin{aligned} &|b_i| \leqslant k i^{-r}, \\ &|a_i^{(N)} - b_i| \leqslant K N^{-q} i^{-p+q}, \quad p, q, r > 1. \end{aligned} \tag{1.4.6}$$

The form of these bounds typifies those discussed in Chapters 2–4.

For such a system we have the following three *convergence rates*:

$$\begin{aligned} &\text{vertical:} && |b_i| = \mathcal{O}(i^{-r}) \\ &\text{horizontal:} && |a_i^{(N)} - b_i| = \mathcal{O}(N^{-q}), \quad i = 1, 2, \ldots \\ &\text{diagonal:} && |a_i^{(i)} - b_i| = \mathcal{O}(i^{-p}). \end{aligned}$$

Then from (1.4.5) we find

$$S_2 = \mathcal{O}(N^{-2r+1})$$

while

$$\begin{aligned} S_1 &\leqslant \sum_{i=1}^{N} K^2 i^{2(q-p)} N^{-2q} \\ &= \mathcal{O}(N^{-2q} + N^{-2p+1}) \end{aligned}$$

where the order estimates follow on bounding the series.

Thus we find

$$\|e_N\| = \mathcal{O}(N^{-s}) \qquad \textit{or} \qquad \mathcal{O}(N^{-s+1/2})$$

where $s = \min(p, q, r)$.

The concept of "rate of convergence" is fundamental to much of the material in this book, and for brevity and to avoid ambiguities we introduce the following definition.

Definition 1.1. *The function $f(i, \boldsymbol{\alpha})$ has i-convergence rate q if there exists a positive constant C such that for all positive i and fixed $\boldsymbol{\alpha}$:*

$$|f(i, \boldsymbol{\alpha})| \leqslant Ci^{-q}.$$

With this notation b_i, equation (1.4.6), has i-convergence rate r (vertical convergence); $b_i - a_i^{(N)}$ has N-convergence rate q (horizontal convergence); and $b_i - a_i^{(i)}$ has i-convergence rate p (diagonal convergence). From these, it then follows that $\|e_N\|$ has N-convergence rate at least $s - \frac{1}{2}$, where $s = \min(p, q, r)$.

In succeeding chapters, we show that estimates of the general form (1.4.6) can be derived for the coefficients $a_i^{(N)}$, b_i under suitable assumptions on the structure of the matrix $\mathbf{L}$ and right-hand side of equation (1.2.4). These assumptions can be shown to hold for a wide variety of expansion methods using orthonormal or "near orthonormal" bases (see Chapters 12–14). They do *not* hold (necessarily) for the monomial basis (1.2.15). For this basis, it is common to find that neither vertical nor horizontal convergence is achieved; that is, to find that

$$\lim_{i\to\infty} b_i \neq 0$$

$$\lim_{N\to\infty} (b_i - a_i^{(N)}) \neq 0.$$

This lack of convergence of the expansion coefficients does *not* (necessarily) imply a lack of convergence of f_N to f, since equation (1.4.5) is only valid for an orthonormal basis. It does, however, imply that inspection of the expansion coefficients is not sufficient to yield a reasonable error estimate in such cases. Conversely, we shall show that, given an orthonormal expansion, inspection of the expansion coefficients, and of the matrix $\mathbf{L}$ and vector $\mathbf{g}$, *is* sufficient to yield a reasonable error estimate; this fact forms one strong argument for the choice of an orthonormal basis.

1.5 An example

As an example of the matrix structure which results from the use of an orthogonal basis, we consider the boundary value problem

$$\mathscr{L}(f(x)) = [(\mathrm{d}^2/\mathrm{d}x^2) - \lambda x]f(x) = g(x) = x^4 - (1 + \pi)x^3 + \pi x^2 - 6x + 2(1 + \pi), \quad x \in [0, \pi], \tag{1.5.1}$$

with $\lambda > 0$ and homogeneous boundary conditions

$$f(0) = f(\pi) = 0. \tag{1.5.2}$$

We take the orthonormal basis

$$h_i(x) = (2/\pi)^{1/2} \sin ix, \quad i = 1, 2, \ldots, \tag{1.5.3}$$

which satisfies the boundary conditions term by term; then the Galerkin equations (1.2.4) have coefficients

$$\begin{aligned} L_{ij} = L_{ji} = \int_0^\pi h_i(x)\mathscr{L}h_j(x)\,\mathrm{d}x &= -i^2 - \tfrac{1}{2}\lambda\pi, \quad i = j, \\ &= \frac{8\lambda ij}{\pi(j^2 - i^2)^2}, \quad i + j \textit{ odd}, \\ &= 0, \quad \textit{otherwise}; \end{aligned} \tag{1.5.4}$$

$$\begin{aligned} g_i = \int_0^\pi h_i(x)g(x)\,\mathrm{d}x &= \left(\frac{2}{\pi}\right)^{1/2}\left\{\frac{48}{i^5} + \frac{2\pi(1 - 3\pi)}{i^3} - \frac{2(\pi - 2)}{i}\right\}, \quad i \textit{ odd}, \\ &= \left(\frac{2}{\pi}\right)^{1/2}\left\{\frac{6\pi(\pi - 1)}{i^3} + \frac{6\pi}{i}\right\}, \quad i \textit{ even}. \end{aligned} \tag{1.5.5}$$

TABLE 1.2

The computed variational parameters $a_i^{(N)}$ and exact coefficients b_i for $\lambda = 1$.

$N = 6$	$N = 12$	$N = 18$	$N = 24$	$N = 30$	b_i
1·8217994	1·8217222	1·8217189	1·8217185	1·8217184	1·821718
−1·8799758	−1·8799711	−1·8799712	−1·8799712	−1·8799712	−1·879971
6·75456 × 10^{-2}	6·74741 × 10^{-2}	6·74715 × 10^{-2}	6·74712 × 10^{-2}	6·74711 × 10^{-2}	6·747105 × 10^{-2}
−2·350145 × 10^{-1}	−2·349967 × 10^{-1}	−2·349964 × 10^{-1}	−2·349964 × 10^{-1}	−2·349964 × 10^{-1}	−2·349964 × 10^{-1}
1·46638 × 10^{-2}	1·45761 × 10^{-2}	1·45741 × 10^{-2}	1·45738 × 10^{-2}	1·45738 × 10^{-2}	1·457375 × 10^{-2}
−6·972215 × 10^{-2}	−6·96289 × 10^{-2}	−6·96286 × 10^{-2}	−6·96286 × 10^{-2}	−6·96286 × 10^{-2}	−6·962856 × 10^{-2}
	5·3133 × 10^{-3}	5·3114 × 10^{-3}	5·3112 × 10^{-3}	5·3112 × 10^{-3}	5·311132 × 10^{-3}
	−2·9375 × 10^{-2}	−2·93746 × 10^{-2}	−2·93746 × 10^{-2}	−2·93746 × 10^{-2}	−2·937455 × 10^{-2}
	2·5015 × 10^{-3}	2·4992 × 10^{-3}	2·499 × 10^{-3}	2·4989 × 10^{-3}	2.498928 × 10^{-3}
	−1·50405 × 10^{-2}	−1·50398 × 10^{-2}	−1·50398 × 10^{-2}	−1·50398 × 10^{-2}	−1·503977 × 10^{-2}
	1·373 × 10^{-3}	1·3689 × 10^{-3}	1·3687 × 10^{-3}	1·3687 × 10^{-3}	1·368684 × 10^{-3}
	−8·7075 × 10^{-3}	−8·7036 × 10^{-3}	−8·7036 × 10^{-3}	−8·7036 × 10^{-3}	−8·703570 × 10^{-3}
		8·295 × 10^{-4}	8·292 × 10^{-4}	8·292 × 10^{-4}	8·29184 × 10^{-4}
		−5·481 × 10^{-3}	−5·481 × 10^{-3}	−5·481 × 10^{-3}	−5·48097 × 10^{-3}
		5·401 × 10^{-4}	5·398 × 10^{-4}	5·398 × 10^{-4}	5·39768 × 10^{-4}
		−3·6719 × 10^{-3}	−3·6718 × 10^{-3}	−3·6718 × 10^{-3}	−3·67182 × 10^{-3}
		3·715 × 10^{-4}	3·708 × 10^{-4}	3·708 × 10^{-4}	3·70796 × 10^{-4}
		−2·5794 × 10^{-3}	−2·5788 × 10^{-3}	−2·5788 × 10^{-3}	−2·57884 × 10^{-3}
			2·657 × 10^{-4}	2·656 × 10^{-4}	2·65595 × 10^{-4}
			−1·88 × 10^{-3}	−1·88 × 10^{-3}	−1·87997 × 10^{-3}
			1·968 × 10^{-4}	1·967 × 10^{-4}	1·96709 × 10^{-4}
			−1·4125 × 10^{-3}	−1·4125 × 10^{-3}	−1·41245 × 10^{-3}
			1·499 × 10^{-4}	1·497 × 10^{-4}	1·49726 × 10^{-4}
			−1·0881 × 10^{-3}	−1·0879 × 10^{-3}	−1·08795 × 10^{-3}
				1·166 × 10^{-4}	1·1659 × 10^{-4}
				−8·557 × 10^{-4}	−8·5570 × 10^{-4}
				9·26 × 10^{-5}	9·2553 × 10^{-5}
				−6·851 × 10^{-4}	−6·8512 × 10^{-4}
				7·48 × 10^{-5}	7·4694 × 10^{-5}
				−5·571 × 10^{-4}	−5·5703 × 10^{-4}

Table 1.2 lists the computed coefficients $a_i^{(N)}$ for this problem with $\lambda = 1$, when the exact solution is:

$$f(x) = x(x-1)(\pi - x). \tag{1.5.6}$$

It is clear from this table that vertical, horizontal and diagonal convergence are all attained and that convergence is quite rapid; a numerical fit to the results yields the estimates

$$\begin{aligned} &\text{vertical:} && |b_i| \leqslant C_v i^{-p}, \quad p \sim 3 \\ &\text{horizontal:} && |b_i - a_i^{(N)}| \leqslant C_h N^{-q}, \quad q \sim 4{\cdot}7\text{–}5{\cdot}0 \\ &\text{diagonal:} && |b_i - a_i^{(i)}| \leqslant C_d i^{-r}, \quad r \sim 4{\cdot}6. \end{aligned} \tag{1.5.7}$$

The analysis of Chapters 2–4 is designed to predict these results *a priori*, from a knowledge of the overall structure of the matrix $\mathbf{L}$ and vector $\mathbf{g}$. To achieve this, it is necessary to extract the essential features from the detail associated with the particular problem being solved. We therefore return to (1.5.4), (1.5.5) and make the following observations:

(a) g_i satisfies a bound of the form (1.4.3) with $p = 1$.

(b) The elements of $\mathbf{L}$ satisfy the following bounds

$$\begin{aligned} |L_{ii}| &\geqslant i^2 \\ |L_{ij}| &\leqslant \frac{8\lambda}{\pi} ij^{-1}(j-i)^{-2}, \quad j > i \\ &\leqslant \frac{8\lambda}{\pi} ji^{-1}(i-j)^{-2}, \quad i > j, \end{aligned}$$

where the latter bounds follow from the inequality $j + i > j$.

(c) The apparent structure of $\mathbf{L}$, $\mathbf{g}$, and certainly the form of the bounds, are not invariant under a scaling of the expansion functions h_i. The normalisation change

$$h_i \to \gamma_i h_i$$

leads to a renormalised set of Galerkin equations

$$\bar{\mathbf{L}}\bar{\mathbf{a}} = \bar{\mathbf{g}} \tag{1.5.8}$$

with $\bar{\mathbf{L}} = \mathbf{\Gamma L \Gamma}$, $\bar{\mathbf{g}} = \mathbf{\Gamma g}$, $\mathbf{\Gamma} = \mathbf{diag}(\gamma_i)$ and solution $\bar{\mathbf{a}} = \mathbf{\Gamma}^{-1}\mathbf{a}$.

This scale change leaves the computed solution f_N invariant (and does not affect the numerical stability of the calculation; see Chapter 5). We may therefore choose any convenient normalisation in which to carry out the analysis, and we choose in practice to normalise so that $|\bar{L}_{ii}| = 1$, $\forall i$, or equivalently to look at the normalisation-invariant quantities:

$$\frac{|L_{ij}|}{|L_{ii}|^{1/2}|L_{jj}|^{1/2}} \leqslant \frac{8\lambda}{\pi} j^{-2}(j-i)^{-2}, \quad j > i, \tag{1.5.9}$$

$$|g_i|/|L_{ii}|^{1/2} \leqslant Ci^{-2}. \tag{1.5.10}$$

We now pose the question: given bounds of the form (1.5.9), (1.5.10), can we predict the convergence rates (1.5.7)? Chapters 2–4 are devoted to answering this question and some of its generalisations.

2

Asymptotically Diagonal Systems of Type A

2.1 Introduction

We now turn to the study of WAD systems, and in this chapter illustrate the methods used by considering the simplest WAD systems: those characterised by the inequalities of Definition 2.1 below.

Following the notation of Chapter 1, we consider the infinite system of equations

$$\mathbf{Lb} = \mathbf{g}, \tag{2.1.1}$$

and the related $N \times N$ system

$$\mathbf{L}^{(N)}\mathbf{a}^{(N)} = \mathbf{g}^{(N)}, \tag{2.1.2}$$

where $\mathbf{L}^{(N)}$ is the $N \times N$ leading submatrix of $\mathbf{L}$, and $\mathbf{g}^{(N)}$ is the leading N-vector of $\mathbf{g}$. We assume that $\mathbf{L}$ is symmetric, and that it is also asymptotically diagonal (AD) of type $A(p; C)$:

Definition 2.1. *A symmetric matrix* $\mathbf{L}$ *is said to be "AD of type* $A(p; C)$*" if constants* $p > 0$ *and* $C > 0$ *exist such that, for all* $i > j$,

$$F_{ij} \equiv \frac{|L_{ij}|}{\{|L_{ii}|\,|L_{jj}|\}^{1/2}} \leqslant Ci^{-p}. \tag{2.1.3}$$

Bounds for the case $i < j$ follow from the symmetry. Additionally, $\mathbf{L}$ is said to be normalised, if for each i,

$$|L_{ii}| = 1.$$

We recall from Chapter 1 that property (2.1.3) is invariant under a diagonal transformation of **L**. If $\bar{\mathbf{L}} = \mathbf{DLD}^t$, where **D** is the diagonal matrix, $\mathbf{diag}(d_i)$, then

$$\frac{|\bar{L}_{ij}|}{\{|\bar{L}_{ii}|\,|\bar{L}_{jj}|\}^{1/2}} = \frac{|d_i L_{ij} d_j|}{\{|d_i L_{ii} d_i|\,|d_j L_{jj} d_j|\}^{1/2}} = \frac{|L_{ij}|}{\{|L_{ii}|\,|L_{jj}|\}^{1/2}}.$$

Without loss of generality, we may therefore consider only normalised matrices, and we assume henceforth that **L** is in fact normalised.

Every symmetric $N \times N$ matrix with non-zero diagonal elements satisfies (2.1.3) for sufficiently large C and any positive p. However, the theorems which follow impose limitations on C for a given p, and we shall assume that these limitations are satisfied for every finite N.

2.2 Triangular decomposition of symmetric AD matrices of type A

The central Theorem 2.1 of this chapter relates to the triangular decomposition of the matrix $\mathbf{L}^{(N)}$ in the form

$$\mathbf{T}^{(N)}\mathbf{L}^{(N)}\mathbf{T}^{(N)t} = \mathbf{diag}^{(N)}(L_{ii}), \qquad (2.2.1)$$

where $\mathbf{T}^{(N)}$ is an $N \times N$ lower triangular matrix, and $\mathbf{diag}^{(N)}(L_{ii})$ is an $N \times N$ diagonal matrix, with diagonal elements L_{ii}. In Theorem 2.1, we will show that, under suitable restrictions on the parameters C and p, the elements of $\mathbf{T}^{(N)}$ satisfy inequalities similar to those satisfied by the elements of $\mathbf{L}^{(N)}$. Further, the existence of a decomposition (2.2.1) is guaranteed by the constructive nature of the proof of this theorem.

Theorem 2.1. *Let the infinite Hermitian matrix* **L** *be normalised so that its diagonal elements satisfy the condition* $|L_{ii}| = 1$ *and be AD of type* $A(p; C)$. *If the inequality*

$$p > 1 \qquad (2.2.2)$$

holds, C *is a constant satisfying the inequalities*

$$C \leqslant 2^p/12 \quad \textit{and} \quad 2C < 1 + \sqrt{(1 - 12C/2^p)} \qquad (2.2.3)$$

and K' *is the constant*

$$K' = K\left[1 - K2^{-p} - \frac{Kp}{(p-1)}\right]^{-1} > 0, \qquad (2.2.4)$$

where

$$K = \frac{2C}{1 + \sqrt{(1 - 12C/2^p)}}, \qquad (2.2.5)$$

then $\mathbf{L}^{(N)}$, *the leading* $N \times N$ *submatrix of* $\mathbf{L}$, *may be factorised in the form* (2.2.1) *where* $\mathbf{T}^{(N)}$ *is a lower triangular matrix whose elements* T_{ij} *do not depend on* N *and satisfy the inequalities*

$$|T_{ij}| \leqslant K' i^{-p} \quad \text{for } i > j \tag{2.2.6}$$

and

$$|T_{ii} - 1| \leqslant K' i^{-p}. \tag{2.2.7}$$

Proof. We prove this theorem in two stages. We write $\mathbf{T}^{-1} = \mathbf{Y} = \mathbf{I} - \mathbf{X}$ in (2.2.1), and consider first the associated reduction of $\mathbf{L}$

$$\mathbf{L}^{(N)} = \mathbf{Y}^{(N)} \mathbf{diag}^{(N)} (L_{ii}) \mathbf{Y}^{(N)^t}, \tag{2.2.1a}$$

where $\mathbf{Y}^{(N)}$ is also lower triangular.

We show that under the conditions of the theorem

$$|X_{ij}| \leqslant K i^{-p}, \quad i \geqslant j. \tag{2.2.8}$$

The result is trivial for $N = 1$, since in this case $Y_{11} = 1$. Now assume inductively that the result is true for all orders up to N. Then we may factorise $\mathbf{L}^{(N-1)}$, the leading $(N-1) \times (N-1)$ submatrix of $\mathbf{L}^{(N)}$, in the required way and $\mathbf{Y}^{(N-1)}$ is suitable as the leading submatrix of $\mathbf{Y}^{(N)}$. It remains to show that the last row of $\mathbf{Y}^{(N)}$ may be constructed and has elements which satisfy inequalities (2.2.8). For $i = 1,2,\ldots,N-1$ we choose Y_{Nj} by the formula

$$Y_{Nj} = \left(L_{Nj} - \sum_{k=1}^{j-1} Y_{Nk} L_{kk} Y_{jk} \right) \Big/ (L_{jj} Y_{jj}), \tag{2.2.9}$$

so that the (N,j)-th component of (2.2.1a) is satisfied. Since $Y_{11} = 1$ we deduce that $|Y_{N1}| \leqslant CN^{-p}$. But, from equations (2.2.3), (2.2.5) we find $C \leqslant K$ and deduce that (2.2.8) holds for Y_{N1}. Now assume inductively that it holds for Y_{Ni}, $i = 1,2,\ldots,j-1$, where $1 < j < N$; from equation (2.2.9), since $\mathbf{L}$ is AD type $A(p; C)$ and $p > 1$, $K < 1$, we find

$$\begin{aligned} |Y_{Nj}| &\leqslant \left(|L_{Nj}| + \sum_{k=1}^{j-1} |Y_{Nk}|\, |L_{kk}|\, |Y_{jk}| \right) \Big/ (|L_{jj}|\, |Y_{jj}|) \\ &\leqslant \left(CN^{-p} + \sum_{k=1}^{j-1} KN^{-p} K j^{-p} \right) \Big/ (1 - Kj^{-p}) \\ &\leqslant (CN^{-p} + K^2 N^{-p} j^{-p+1})/(1 - Kj^{-p}) \\ &\leqslant N^{-p}(C + K^2 2^{-p+1})/(1 - K2^{-p}). \end{aligned} \tag{2.2.10}$$

The condition on C and the choice of K ensure that this last expression is just KN^{-p} so that we have proved the required result (2.2.8). For the diagonal term we wish to satisfy the equation

$$Y_{NN}^2 = \left(L_{NN} - \sum_{k=1}^{N-1} Y_{Nk}^2 L_{kk}\right) \Big/ L_{NN} = 1 - \sum_{k=1}^{N-1} Y_{Nk}^2 L_{kk}/L_{NN}. \tag{2.2.11}$$

By our inductive hypothesis and inequalities $p > 1$, $K < 1$ we find the inequalities

$$\left|\sum_{k=1}^{N-1} Y_{Nk}^2 L_{kk}/L_{NN}\right| \leqslant \sum_{k=1}^{N-1} |Y_{Nk}|^2 |L_{kk}|/|L_{NN}| \leqslant \sum_{k=1}^{N-1} K^2 N^{-2p} \leqslant K^2 N^{-2p+1} \leqslant KN^{-p} < 1 \tag{2.2.12}$$

hold and therefore Y_{NN} (see (2.2.11)) exists. It is readily verified that if $|\varepsilon| < 1$ then $|1 - \sqrt{(1-\varepsilon)}| \leqslant |\varepsilon|$; it therefore follows from (2.2.11) and (2.2.12) that Y_{NN} satisfies (2.2.8). This completes the first stage of the proof of Theorem 2.1.

For the second stage, we consider the inversion of $\mathbf{Y}$ in the following Lemma 2.1.

Lemma 2.1. *Let* $\mathbf{Y} = \mathbf{I} - \mathbf{X}$ *be a lower triangular matrix with elements satisfying*

$$|X_{ij}| \leqslant Ki^{-p}, \quad \text{with } p > 1, \tag{2.2.13}$$

then $\mathbf{T} = \mathbf{Y}^{-1} = \mathbf{I} + \mathbf{F}$, *which is also lower triangular, has elements satisfying*

$$|F_{ij}| \leqslant K'i^{-p}, \tag{2.2.14}$$

with

$$K' = K\left[1 - K2^{-p} - \frac{Kp}{p-1}\right]^{-1} \tag{2.2.15}$$

provided $K' > 0$.

Proof. Using the (i,j)-th component of the identity $\mathbf{T}^{-1}\mathbf{T} = \mathbf{I}$, we find for $i \geqslant j$

$$F_{ij} = X_{ij} + \sum_{k=j}^{i} X_{ik} F_{kj}.$$

Hence,

$$|F_{ij}| \leqslant \left\{|X_{ij}| + \sum_{k=j}^{i-1} |X_{ik}|\,|F_{kj}|\right\} \Big/ (1 - |X_{ii}|). \tag{2.2.16}$$

Now,

$$|F_{11}| \leqslant \frac{|X_{11}|}{1 - |X_{11}|} \leqslant \frac{K}{1 - K} \leqslant K'.$$

Hence, (2.2.14) is true for F_{1j}, since $j \leqslant i$.

Now, assume inductively that (2.2.14) is true for F_{kj}, $k = 2, \ldots, i - 1$. Then

$$\begin{aligned}
|F_{ij}| &\leqslant \left(Ki^{-p} + \sum_{k=j}^{i-1} Ki^{-p} K' k^{-p} \right) \Big/ (1 - Ki^{-p}) \\
&\leqslant Ki^{-p} \left[1 + \frac{K' pj^{-p+1}}{p - 1} \right] \Big/ (1 - Ki^{-p}) \\
&\leqslant Ki^{-p} \left[1 + \frac{K' p}{p - 1} \right] \Big/ (1 - K2^{-p}) \quad \textit{since } i \geqslant 2, j \geqslant 1, \\
&\leqslant K' i^{-p}.
\end{aligned}$$

Hence F_{ij} satisfies (2.2.14) by choice of K', and the lemma is proved.

Now Theorem 2.1 follows immediately.

Corollary 2.1. Since the results of Theorem 2.1 are independent of N, it follows that the infinite matrix $\mathbf{L}$ can be formally decomposed as

$$\mathbf{TLT}^t = \mathbf{diag}(L_{ii}), \tag{2.2.17}$$

where the elements of the infinite, lower triangular, matrix $\mathbf{T}$ satisfy (2.2.6), (2.2.7), and $\mathbf{diag}(L_{ii})$ is an infinite diagonal matrix.

Comment.

Justification of the limiting procedure involved in this corollary is given in Section 3.4.

2.3 Behaviour of $\mathbf{a}^{(N)}$ and $\mathbf{b}$

We can now use the results of Theorem 2.1 to categorise the i-convergence rate of b_i, the components of the solution of (2.1.1), in terms of the behaviour of the matrix $\mathbf{L}$, given by Definition 2.1, and the i-convergence rate of g_i, the components of the right-hand side vector $\mathbf{g}$. In a similar way we can categorise the N-convergence rate of $b_i - a_i^{(N)}$.

We assume that the components g_i of $\mathbf{g}$ satisfy the inequality,

$$|g_i| \leqslant \mathscr{C} i^{-s}. \tag{2.3.1}$$

The following theorem then gives a bound on the components b_i of the solution of (2.1.1).

Theorem 2.2. *Let the infinite matrix* **L** *satisfy the conditions of Theorem* 2.1 *and let the infinite vector* **g** *satisfy* (2.3.1), *with* $s > 1$. *Then, for all i and some* A_1,

$$|b_i| \leqslant A_1 i^{-q}, \tag{2.3.2}$$

where

$$q = \min\{p, s\}. \tag{2.3.3}$$

Proof. The vector **b** satisfies the infinite system of equations (2.1.1).

If we write $\mathbf{J} = \mathbf{diag}(L_{ii})$, set $\mathbf{b} = \mathbf{T}^t\mathbf{c}$ and premultiply (2.1.1) by **T**, we obtain

$$\mathbf{TLT}^t\mathbf{c} = \mathbf{Tg}$$

i.e.

$$\mathbf{c} = \mathbf{JTg}, \text{ since } J_{ii} = \pm 1.$$

Hence,

$$\mathbf{b} = \mathbf{T}^t\mathbf{JTg},$$

and

$$b_i = \sum_{j=1}^{\infty} (T^t JT)_{ij} g_j$$

$$= \sum_{j=i}^{\infty} \sum_{k=j}^{\infty} T_{ki} (JT)_{kj} g_j + \sum_{j=1}^{i-1} \sum_{k=i}^{\infty} T_{ki} (JT)_{kj} g_j,$$

since **T** is lower triangular. Writing $\mathbf{T} = \mathbf{I} + \mathbf{F}$, we find

$$|b_i| \leqslant \sum_{j=i}^{\infty} \sum_{k=j}^{\infty} |\delta_{ki} + F_{ki}| \, |\delta_{kj} + F_{kj}| \, |g_j| + \sum_{j=1}^{i-1} \sum_{k=i}^{\infty} |\delta_{ki} + F_{ki}| \, |\delta_{kj} + F_{kj}| \, |g_j|$$

$$\leqslant |g_i| + |F_{ii}| \, |g_i| + \sum_{j=i}^{\infty} |F_{ji}| \, |g_j| + \sum_{j=i}^{\infty} \sum_{k=j}^{\infty} |F_{ki}| \, |F_{kj}| \, |g_j|$$

$$+ \sum_{j=1}^{i-1} |F_{ij}| \, |g_j| + \sum_{j=1}^{i-1} \sum_{k=i}^{\infty} |F_{ki}| \, |F_{kj}| \, |g_j|.$$

These sums may be bounded as before by inserting the bounds (2.2.14), (2.3.1) and then bounding the resulting series. We find

$$
\begin{aligned}
|b_i| &\leqslant \mathscr{C} i^{-s} + \mathscr{C} K' i^{-(p+s)} + \frac{\mathscr{C} K'(p+s)}{(p+s-1)} i^{-(p+s)+1} \\
&\quad + \frac{\mathscr{C} K'^2 \, 2p(2p+s-1)}{(2p-1)(2p+s-2)} i^{-(2p+s)+2} + \frac{\mathscr{C} K' s}{(s-1)} i^{-p} \\
&\quad + \frac{\mathscr{C} K'^2 \, 2ps}{(2p-1)(s-1)} i^{-2p+1} \\
&\leqslant A_1 i^{-q},
\end{aligned}
$$

where $q = \min\{p, s\}$ and

$$
\begin{aligned}
A_1 = \mathscr{C}\Bigg[1 &+ K'\left(1 + \frac{(p+s)}{(p+s-1)} + \frac{s}{(s-1)}\right) \\
&+ \frac{K'^2 \, 2p}{(2p-1)}\left(\frac{(2p+s-1)}{(2p+s-2)} + \frac{s}{(s-1)}\right)\Bigg].
\end{aligned}
$$

This theorem shows that the i-convergence rate of b_i is bounded by the behaviour of the matrix $\mathbf{L}$, Definition 2.1, and the i-convergence rate of g_i, (2.3.1).

Also, under the assumptions of Theorem 2.2, we can investigate the N-convergence rate of $b_i - a_i^{(N)}$.

Theorem 2.3. *Under the conditions of Theorem 2.2, it follows that, for all N, $i = 1,2,\ldots,N$, and some A_2,*

$$|b_i - a_i^{(N)}| \leqslant A_2 N^{-(p+q)+1}, \tag{2.3.4}$$

where q is given by (2.3.3).

Proof. From Theorem 2.2,

$$b_i = \sum_{j=i}^{\infty} \sum_{k=j}^{\infty} T_{ki}(JT)_{kj} g_j + \sum_{j=1}^{i-1} \sum_{k=i}^{\infty} T_{ki}(JT)_{kj} g_j. \tag{2.3.5}$$

In a similar manner, we consider the $N \times N$ system of equations (2.1.2) and, noting that $\mathbf{T}^{(N)}$ is the $N \times N$ leading submatrix of $\mathbf{T}$, we obtain

$$a_i^{(N)} = \sum_{j=i}^{N} \sum_{k=j}^{N} T_{ki}(JT)_{kj} g_j + \sum_{j=1}^{i-1} \sum_{k=i}^{N} T_{ki}(JT)_{kj} g_j. \tag{2.3.6}$$

Subtracting (2.3.6) from (2.3.5) gives

$$
\begin{aligned}
b_i - a_i^{(N)} &= \left\{ \sum_{j=i}^{\infty} \sum_{k=j}^{\infty} + \sum_{j=1}^{i-1} \sum_{k=i}^{\infty} - \sum_{j=i}^{N} \sum_{k=j}^{N} - \sum_{j=1}^{i-1} \sum_{k=i}^{N} \right\} T_{ki}(JT)_{kj} g_j \\
&= \left\{ \sum_{j=1}^{N} \sum_{k=N+1}^{\infty} + \sum_{j=N+1}^{\infty} \sum_{k=j}^{\infty} \right\} T_{ki}(JT)_{kj} g_j.
\end{aligned}
$$

Substituting $\mathbf{T} = \mathbf{I} + \mathbf{F}$, we find

$$
\begin{aligned}
|b_i - a_i^{(N)}| &\leqslant \left\{ \sum_{j=1}^{N} \sum_{k=N+1}^{\infty} + \sum_{j=N+1}^{\infty} \sum_{k=j}^{\infty} \right\} |(\delta_{ki} + F_{ki})(\delta_{kj} + F_{kj})\, g_j| \\
&\leqslant \sum_{j=N+1}^{\infty} |F_{ji}|\,|g_j| + \left\{ \sum_{j=1}^{N} \sum_{k=N+1}^{\infty} + \sum_{j=N+1}^{\infty} \sum_{k=j}^{\infty} \right\} |F_{ki}|\,|F_{kj}|\,|g_j| \\
&\leqslant \frac{\mathscr{C} K' N^{-(p+s)+1}}{(p+s-1)} + \frac{\mathscr{C} K'^2 s N^{-2p+1}}{(2p-1)(s-1)} + \frac{\mathscr{C} K'^2\, 2p N^{-(2p+s)+2}}{(2p-1)(2p+s-2)} \\
&\leqslant A_2 N^{-(p+q)+1},
\end{aligned}
$$

where q is given by (2.3.3), and

$$
A_2 = \mathscr{C}\left[\frac{K'}{(p+s-1)} + \frac{K'^2}{(2p-1)} \left(\frac{s}{(s-1)} + \frac{2p}{(2p+s-2)} \right) \right].
$$

The uniform nature of the bound (2.3.4), with respect to i, implies that the (bound on the) diagonal rate of convergence is the same as the (bound on the) horizontal rate of convergence. This bound again depends on the behaviour of both the matrix $\mathbf{L}$ and the vector $\mathbf{g}$.

2.4 Convergence of variational or Galerkin calculations

We now consider the application of these theorems when the matrix problem arises from a variational or Galerkin calculation. In this case we may relate the vertical, horizontal and diagonal convergence rates to the N-convergence rate of e_N, where $e_N = f - f_N$ is the error in the N-th approximate solution f_N, (1.2.3); the following theorems explore this relationship.

The theorems of this section, with the exception of Theorem 2.7, require that the operator $\mathscr{L}$ of (1.2.1) be positive definite. Therefore, throughout we shall assume that this is valid except where we explicitly state otherwise.

In these theorems, we provide bounds on the energy norm and natural norm of the error $e_N = f - f_N$ in the N-th approximate solution. The energy ($\mathscr{L}$-)norm $\|u\|_{\mathscr{L}}$ of an element u of the domain of a positive definite operator $\mathscr{L}$ is defined by

$$
\|u\|_{\mathscr{L}} = (u, \mathscr{L}u)^{1/2}. \tag{2.4.1}
$$

It is trivial to show that $\|u\|_{\mathscr{L}}$ satisfies the norm axioms.

To simplify the statement of the theorems of this section, we define the following class of systems:

Definition 2.2. *Let $\mathscr{L}f = g$ be an inhomogeneous operator equation, and let* $\mathbf{Lb} = \mathbf{g}$ *be the corresponding infinite linear system, with $\{h_i, i = 1, 2, \ldots\}$ as expansion set. We refer to this system as a "nice system of type A", if for the given expansion set, every submatrix $\mathbf{L}^{(N)}$ of $\mathbf{L}$ satisfies the conditions of Theorem 2.1.*

Theorem 2.4. *For a "nice system of type A", the error in the N-th approximate solution, defined by*

$$e_N = f - f_N = f - \sum_{i=1}^{N} a_i^{(N)} h_i$$

satisfies the bounding inequality

$$\|e_N\|_{\mathscr{L}} \leqslant k \sum_{i=N+1}^{\infty} |a_i^{(i)}|, \quad \text{where } 0 < k \leqslant k(p, C).$$

Proof. Let us orthogonalise the first N expansion functions with respect to the operator $\mathscr{L}$, using a Gram-Schmidt process. That is, we define

$$\bar{h}_i = \sum_{j=1}^{i} T_{ij} h_j, \quad i = 1, \ldots, N,$$

where $\mathbf{T}$ is a lower triangular matrix such that

$$(\bar{h}_i, \mathscr{L}\bar{h}_j) = \delta_{ij}.$$

The transformed operator matrix is

$$\bar{\mathbf{L}}^{(N)} = \mathbf{T}\mathbf{L}^{(N)}\mathbf{T}^t = \mathbf{I}^{(N)}, \quad \textit{the } N \times N \textit{ identity matrix.}$$

The N-th approximate solution is invariant under this orthogonalisation and may be written

$$f_N = \sum_{i=1}^{N} \alpha_i^{(N)} \bar{h}_i = \boldsymbol{\alpha}^{(N)} \cdot \bar{\mathbf{h}} \quad \text{(in an obvious notation)},$$

where $\boldsymbol{\alpha}^{(N)}$ is the (trivial) solution of

$$\mathbf{I}^{(N)}\boldsymbol{\alpha}^{(N)} = \boldsymbol{\beta}^{(N)} \qquad \text{and} \qquad \beta_i^{(N)} = (\bar{h}_i, g).$$

Let us now add h_{N+1} to the expansion set. The operator matrix becomes

$$\bar{\mathbf{L}}^{(N+1)} = \begin{bmatrix} \mathbf{T} & \mathbf{0} \\ \mathbf{0}^t & 1 \end{bmatrix} \begin{bmatrix} \mathbf{L}^{(N)} & \gamma^{(N)} \\ \gamma^{(N)t} & 1 \end{bmatrix} \begin{bmatrix} \mathbf{T}^t & \mathbf{0} \\ \mathbf{0}^t & 1 \end{bmatrix} = \begin{bmatrix} \mathbf{I}^{(N)} & \mathbf{T}\gamma^{(N)} \\ \gamma^{(N)t}\mathbf{T}^t & 1 \end{bmatrix},$$

where $\gamma_i^{(N)} = (h_i, \mathscr{L}h_{N+1})$ and hence the system

$$\begin{bmatrix} \mathbf{I}^{(N)} & \mathbf{T}\gamma^{(N)} \\ \gamma^{(N)t}\mathbf{T}^t & 1 \end{bmatrix} \begin{bmatrix} \boldsymbol{\alpha}^{(N+1)} \\ a_{N+1}^{(N+1)} \end{bmatrix} = \begin{bmatrix} \boldsymbol{\beta}^{(N)} \\ g_{N+1} \end{bmatrix}.$$

This implies

$$\boldsymbol{\alpha}^{(N+1)} + a_{N+1}^{(N+1)}\,\mathbf{T}\boldsymbol{\gamma}^{(N)} = \boldsymbol{\beta}^{(N)} \tag{2.4.2}$$

and

$$\boldsymbol{\gamma}^{(N)t}\mathbf{T}^t\,\boldsymbol{\alpha}^{(N+1)} + a_{N+1}^{(N+1)} = g_{N+1}. \tag{2.4.3}$$

The $(N + 1)$-th approximate solution may be written:

$$f_{N+1} = \sum_{i=1}^{N} \alpha_i^{(N+1)}\,\bar{h}_i + a_{N+1}^{(N+1)}\,h_{N+1}.$$

Hence

$$\begin{aligned} e_N - e_{N+1} &= f_{N+1} - f_N \\ &= \boldsymbol{\alpha}^{(N+1)}\cdot\bar{\mathbf{h}} + a_{N+1}^{(N+1)}\,h_{N+1} - \boldsymbol{\alpha}^{(N)}\cdot\bar{\mathbf{h}} \\ &= \boldsymbol{\beta}^{(N)}\cdot\bar{\mathbf{h}} - a_{N+1}^{(N+1)}(\mathbf{T}\boldsymbol{\gamma}^{(N)})\cdot\bar{\mathbf{h}} + a_{N+1}^{(N+1)}\,h_{N+1} - \boldsymbol{\beta}^{(N)}\cdot\bar{\mathbf{h}} \\ &= a_{N+1}^{(N+1)}\left[h_{N+1} - (\mathbf{T}\boldsymbol{\gamma}^{(N)})\cdot\bar{\mathbf{h}}\right]. \end{aligned}$$

Therefore,

$$\|e_N - e_{N+1}\| \leqslant |a_{N+1}^{(N+1)}|\left\{\|h_{N+1}\| + \sum_{i=1}^{N} |(\mathbf{T}\boldsymbol{\gamma}^{(N)})_i|\;\|\bar{h}_i\|\right\}.$$

Using the $\mathscr{L}$ norm, $\|h_{N+1}\| = \|\bar{h}_i\| = 1$, and hence, since $\mathscr{L}$ is positive definite,

$$\|e_N\|_{\mathscr{L}} - \|e_{N+1}\|_{\mathscr{L}} \leqslant \|e_N - e_{N+1}\|_{\mathscr{L}} \leqslant |a_{N+1}^{(N+1)}|\{1 + \|\mathbf{T}\boldsymbol{\gamma}^{(N)}\|_1\}.$$

But

$$\begin{aligned} \|\mathbf{T}\boldsymbol{\gamma}^{(N)}\|_1 &\leqslant \|\boldsymbol{\gamma}^{(N)}\|_1\{\|\mathbf{I}\|_1 + \|\mathbf{F}\|_1\} \quad \textit{using notation of Lemma 2.1}, \\ &\leqslant CN^{-p+1}\left\{1 + \max_{j=1}^{N}\sum_{i=j}^{N}|F_{ij}|\right\}. \end{aligned}$$

From Theorem 2.1, $|F_{ij}| \leqslant K'i^{-p}$, so

$$\sum_{i=j}^{N}|F_{ij}| \leqslant \frac{p}{p-1}K'j^{-p+1}$$

is maximised by $j = 1$. Thus

$$\|\mathbf{T}\boldsymbol{\gamma}^{(N)}\|_1 \leqslant C + (p/(p-1))CK' \qquad \text{and} \qquad \|e_N\|_{\mathscr{L}} - \|e_{N+1}\|_{\mathscr{L}} \leqslant k|a_{N+1}^{(N+1)}|$$

with $k \leqslant 1 + C + (p/(p-1))CK'$. For $M > N$,

$$\|e_N\|_{\mathscr{L}} - \|e_M\|_{\mathscr{L}} = \sum_{i=N+1}^{M}\{\|e_{i-1}\|_{\mathscr{L}} - \|e_i\|_{\mathscr{L}}\} \leqslant \sum_{i=N+1}^{M} k|a_i^{(i)}|.$$

As $M \to \infty$, $\|e_M\|_{\mathscr{L}} \to 0$, and we obtain the inequality

$$\|e_N\|_{\mathscr{L}} \leqslant k \sum_{i=N+1}^{\infty} |a_i^{(i)}|.$$

We have now shown that, for the class of problems under consideration, the N-convergence rate of the error norm is characterised by the diagonal convergence rate of the system. However, the diagonal elements $|a_i^{(i)}|$ are not readily computable, so we proceed by relating them to quantities which are. The next theorem shows a connection with the column of free terms $\mathbf{g}$.

Theorem 2.5. *For a "nice system of type A" having*

$$g_i = (h_i, g) \leqslant \mathscr{C} i^{-s} \quad \textit{with } s > 1,$$

and

$$\left[1 + \left(\frac{p}{p-1}\right)K'\right](1 + K')C \leqslant K, \tag{2.4.4}$$

it follows that

$$|a_{N+1}^{(N+1)}| \leqslant A_3 N^{-q},$$

and hence that

$$\|e_N\|_{\mathscr{L}} \leqslant A_4 N^{-q+1},$$

where A_3, A_4 are positive constants, and q is given by (2.3.3).

Proof. From the proof of Theorem 2.4, equation (2.4.2) gives

$$\boldsymbol{\alpha}^{(N+1)} = \boldsymbol{\beta}^{(N)} - a_{N+1}^{(N+1)} \mathbf{T}\boldsymbol{\gamma}^{(N)}.$$

Substituting into equation (2.4.3),

$$(\mathbf{T}\boldsymbol{\gamma}^{(N)})^t \boldsymbol{\beta}^{(N)} - (\mathbf{T}\boldsymbol{\gamma}^{(N)})^t a_{N+1}^{(N+1)} \mathbf{T}\boldsymbol{\gamma}^{(N)} + a_{N+1}^{(N+1)} = g_{N+1}.$$

Hence

$$a_{N+1}^{(N+1)} = \frac{g_{N+1} - (\mathbf{T}\boldsymbol{\gamma}^{(N)})^t \boldsymbol{\beta}^{(N)}}{1 - (\mathbf{T}\boldsymbol{\gamma}^{(N)})^t \mathbf{T}\boldsymbol{\gamma}^{(N)}}$$

$$= \frac{g_{N+1} - (\mathbf{T}\boldsymbol{\gamma}^{(N)})^t (\mathbf{T}\mathbf{g}^{(N)})}{1 - (\mathbf{T}\boldsymbol{\gamma}^{(N)})^t (\mathbf{T}\boldsymbol{\gamma}^{(N)})}$$

since

$$\boldsymbol{\beta}^{(N)} = (\bar{\mathbf{h}}, g) = (\mathbf{T}\mathbf{h}, g) = \mathbf{T}\mathbf{g}^{(N)}.$$

Therefore

$$\begin{aligned}
|a_{N+1}^{(N+1)}| &\leqslant \frac{|g_{N+1}| + |(\mathbf{T}\gamma^{(N)})^t (\mathbf{T}\mathbf{g}^{(N)})|}{1 - |(\mathbf{T}\gamma^{(N)})^t (\mathbf{T}\gamma^{(N)})|} \\
&\leqslant \frac{|g_{N+1}| + \|\mathbf{T}\gamma^{(N)}\|_\infty \|\mathbf{T}\mathbf{g}^{(N)}\|_1}{1 - \|\mathbf{T}\gamma^{(N)}\|_\infty \|\mathbf{T}\gamma^{(N)}\|_1} \\
&\leqslant \frac{|g_{N+1}| + \|\mathbf{T}\|_1 \|\mathbf{T}\|_\infty \|\gamma^{(N)}\|_\infty \|\mathbf{g}^{(N)}\|_1}{1 - \|\mathbf{T}\|_1 \|\mathbf{T}\|_\infty \|\gamma^{(N)}\|_\infty \|\gamma^{(N)}\|_1},
\end{aligned} \tag{2.4.5}$$

provided the denominator of (2.4.5) is positive.

We now bound the norms as follows:

$$\begin{aligned}
\|\mathbf{T}\|_1 &\leqslant 1 + \|\mathbf{F}\|_1 \leqslant 1 + \left(\frac{p}{p-1}\right) K', \\
\|\mathbf{T}\|_\infty &\leqslant 1 + \|\mathbf{F}\|_\infty \leqslant 1 + K', \\
\|\mathbf{g}^{(N)}\|_1 &\leqslant \mathscr{C} \sum_{i=1}^{N} i^{-s} \leqslant \mathscr{C}\left(\frac{s}{s-1}\right), \\
\|\gamma^{(N)}\|_\infty &\leqslant C(N+1)^{-p} \leqslant CN^{-p}, \\
\|\gamma^{(N)}\|_1 &\leqslant C(N+1)^{-p+1} \leqslant CN^{-p+1}.
\end{aligned}$$

Hence, the denominator of (2.4.5)

$$\begin{aligned}
&\geqslant 1 - \left[1 + \left(\frac{p}{p-1}\right) K'\right]\left[1 + K'\right] C^2 N^{-2p+1} \\
&\geqslant 1 - KCN^{-2p+1} \\
&\geqslant 1 - KC > 0.
\end{aligned}$$

Therefore,

$$|a_{N+1}^{(N+1)}| \leqslant \frac{\mathscr{C}}{(1-KC)}\left[N^{-s} + K\left(\frac{s}{s-1}\right) N^{-p}\right] \leqslant A_3 N^{-q},$$

where

$$A_3 \leqslant \frac{\mathscr{C}}{(1-KC)}\left[1 + \frac{Ks}{s-1}\right].$$

Using Theorem 2.4,

$$\|e_N\|_{\mathscr{L}} \leqslant k \sum_{i=N}^{\infty} A_3 i^{-q}$$

$$\leqslant \frac{kA_3 q}{q-1} N^{-q+1}$$

$$= A_4 N^{-q+1}.$$

This result enables us to predict the N-convergence rate of $\|e_N\|_{\mathscr{L}}$, provided we know the i-convergence rate of the terms $g_i = (h_i, g)$, i.e. the convergence rate of the generalised Fourier coefficients of the function g. It is frequently unnecessary, however, to compute these coefficients in order to determine their convergence rate. Given certain qualitative information about a function (such as its continuity, its differentiability, and its boundary behaviour) the convergence rate of generalised Fourier coefficients, with respect to a given expansion set, may frequently be predicted *a priori*. This problem is considered further in Chapters 9 and 10.

We now consider the case in which, either through physical considerations or otherwise, we have qualitative information concerning the true solution of our equation and are thus in a position to predict the i-convergence rate of the coefficients b_i. The next theorem relates this "vertical" convergence rate to the N-convergence rate of the error norm $\|e_N\|_{\mathscr{L}}$. Having determined the error convergence rate in terms of properties of a known function g, it may well seem unnecessary to obtain a similar result in terms of properties of the unknown solution f.

However, the step is of importance in considering the extension of these results to homogeneous systems (e.g. eigenproblems) for which no right-hand side function g exists. (Homogeneous systems are considered in detail in Chapter 6.) For an inhomogeneous system, the i-convergence rates of the b_i and g_i are related, as shown in Theorem 2.2.

Theorem 2.6. *For a "nice system of type A" with*

$$b_i = (h_i, f) \leqslant \kappa i^{-q}, \qquad \text{with } q > 1,$$

and with C satisfying (2.4.4), *it follows that* $|g_N| \leqslant A_5 N^{-q} + A_6 N^{-p}$ *and hence that* $\|e_N\|_{\mathscr{L}} \leqslant A_7 N^{-q+1} + A_8 N^{-p+1}$, *where* A_5, A_6, A_7, A_8 *are positive constants.*

Proof.

$$g_N = (h_N, g) = (h_N, \mathscr{L}f)$$
$$= \left(h_N, \mathscr{L} \sum_{i=1}^{\infty} b_i h_i\right) = \sum_{i=1}^{\infty} b_i (h_N, \mathscr{L} h_i)$$
$$= \sum_{i=1}^{\infty} b_i L_{Ni} = b_N + \sum_{i=1}^{N-1} b_i L_{Ni} + \sum_{i=N+1}^{\infty} b_i L_{Ni}.$$

Thus,

$$|g_N| \leqslant \kappa N^{-q} + \left(\frac{q}{q-1}\right) C\kappa N^{-p} + \frac{C\kappa}{(p+q-1)} N^{-p-q+1}$$
$$\leqslant \kappa\left(1 + \frac{C}{(p+q-1)}\right) N^{-q} + \left(\frac{q}{q-1}\right) C\kappa N^{-p}$$
$$= A_5 N^{-q} + A_6 N^{-p}.$$

From the proof of Theorem 2.5,

$$|a_{N+1}^{(N+1)}| \leqslant \frac{|g_N|}{(1-KC)} + \frac{K\mathscr{C}s}{(1-KC)(s-1)} N^{-p}.$$

Hence

$$|a_{N+1}^{(N+1)}| \leqslant \frac{A_5}{(1-KC)} N^{-q} + \left[\frac{A_6}{1-KC} + \frac{K\mathscr{C}s}{(1-KC)(s-1)}\right] N^{-p}$$

and so

$$\|e_N\|_{\mathscr{L}} \leqslant k \sum_{i=N+1}^{\infty} |a_i^{(i)}|$$
$$\leqslant \frac{kA_5 q N^{-q+1}}{(1-KC)(q-1)} + \frac{kpN^{-p+1}}{p-1}\left[\frac{A_6}{1-KC} + \frac{K\mathscr{C}s}{(1-KC)(s-1)}\right]$$
$$\leqslant A_7 N^{-q+1} + A_8 N^{-p+1}.$$

An important conclusion from the above theorem is that the vertical convergence rate will dominate, provided $p > q$.

It is also possible to bound the norm of the error e_N by using the bounds on $|b_i|$ and $|b_i - a_i^{(N)}|$ given in Theorems 2.2 and 2.3, respectively.

Theorem 2.7. *For a "nice system of type A", having*

$$|g_i| \leqslant \mathscr{C} i^{-s}, \quad \text{with } s > 1,$$

it follows that

$$\|e_N\|_{\mathscr{L}}^2 \leqslant A_9 N^{-2q+1},$$

where q is given by (2.3.3).

Proof.

$$\|e_N\|_{\mathscr{L}}^2 = (e_N, \mathscr{L} e_N)$$

where

$$e_N = \sum_{i=1}^{N} (b_i - a_i^{(N)})\, h_i + \sum_{i=N+1}^{\infty} b_i h_i.$$

Hence,

$$\begin{aligned}
\|e_N\|_{\mathscr{L}}^2 &\leqslant \sum_{i=1}^{N} \sum_{j=1}^{N} |b_i - a_i^{(N)}|\,|b_j - a_j^{(N)}|\,|L_{ij}| + \sum_{i=1}^{N} \sum_{j=N+1}^{\infty} |b_i - a_i^{(N)}|\,|b_j|\,|L_{ij}| \\
&\quad + \sum_{i=N+1}^{\infty} \sum_{j=1}^{N} |b_i|\,|b_j - a_j^{(N)}|\,|L_{ij}| + \sum_{i=N+1}^{\infty} \sum_{j=N+1}^{\infty} |b_i|\,|b_j|\,|L_{ij}| \qquad (2.4.6) \\
&\leqslant \sum_{i=1}^{N} |b_i - a_i^{(N)}|^2 + 2 \sum_{i=1}^{N} \sum_{j=i+1}^{N} |b_i - a_i^{(N)}|\,|b_j - a_j^{(N)}|\,|L_{ij}| \\
&\quad + 2 \sum_{i=1}^{N} \sum_{j=N+1}^{\infty} |b_i - a_i^{(N)}|\,|b_j|\,|L_{ij}| + \sum_{i=N+1}^{\infty} |b_i|^2 \\
&\quad + 2 \sum_{i=N+1}^{\infty} \sum_{j=i+1}^{\infty} |b_i|\,|b_j|\,|L_{ij}| \\
&\leqslant A_2^2 N^{-2(p+q)+3} + \frac{2CA_2^2}{(p-1)} N^{-2(p+q)+3} \\
&\quad + \frac{2CA_1 A_2}{(p+q-1)} N^{-2(p+q)+3} + \frac{A_1^2}{(2q-1)} N^{-2q+1} \\
&\quad + \frac{2CA_1^2}{(p+q-1)(p+2q-2)} N^{-(p+2q)+2} \\
&\leqslant A_9 N^{-2q+1},
\end{aligned}$$

where

$$A_9 = A_2^2 + \frac{2CA_2^2}{p-1} + \frac{2CA_1A_2}{p+q-1} + \frac{A_1^2}{2q-1} + \frac{2CA_1^2}{(p+q-1)(p+2q-2)}.$$

It should be noted that this theorem gives a somewhat stronger result than the second part of Theorem 2.5. Additionally, the constant C is no longer restricted to satisfy (2.4.4).

Theorem 2.7 does not explicitly assume that the operator $\mathscr{L}$ is positive definite; when $\mathscr{L}$ is indefinite we can define the energy ($\mathscr{L}$-) pseudonorm of e_N by

$$\|e_N\|_{\mathscr{L}} = |(e_N, \mathscr{L}e_N)|^{1/2}, \tag{2.4.7}$$

and the theorem then provides a bound on this pseudonorm.

We can also use the results of Theorems 2.2 and 2.3 to bound the error e_N in norms other than the energy norm. Consider, for example, the natural norm

$$\|e_N\| = (e_N, e_N)^{1/2}$$

under the assumption that the expansion set $\{\bar{h}_i, i = 1,2,\ldots\}$ is orthonormal. Then

$$\begin{aligned}\|e_N\|^2 &= \left(\left(\sum_{i=1}^{N} (\bar{b}_i - \bar{a}_i^{(N)})\,\bar{h}_i + \sum_{i=N+1}^{\infty} \bar{b}_i\bar{h}_i\right), \left(\sum_{j=1}^{N} (\bar{b}_j - \bar{a}_j^{(N)})\,\bar{h}_j + \sum_{j=N+1}^{\infty} \bar{b}_j\bar{h}_j\right)\right)\\ &= \sum_{i=1}^{N} (\bar{b}_i - \bar{a}_i^{(N)})^2 + \sum_{i=N+1}^{\infty} \bar{b}_i^2.\end{aligned} \tag{2.4.8}$$

Note, however, that this normalisation of the expansion set is not that used in Theorems 2.1–2.1. We therefore define a renormalised set of functions h_i:

$$\bar{h}_i = \gamma_i h_i, \qquad (\bar{h}_i, \bar{h}_j) = \delta_{ij},$$

where the functions h_i have the normalisation of Theorems 2.1–2.7, and therefore γ_i satisfies

$$(\bar{h}_i, \mathscr{L}\bar{h}_i) = \gamma_i^2. \tag{2.4.9}$$

In terms of the coefficients $a_i^{(N)}$, b_i appropriate to the expansion set $\{h_i, i = 1,2,\ldots\}$, (2.4.8) becomes

$$\|e_N\|^2 \leqslant \sum_{i=1}^{N} \gamma_i^{-2}\,|b_i - a_i^{(N)}|^2 + \sum_{i=N+1}^{\infty} \gamma_i^{-2}\,|b_i|^2. \tag{2.4.10}$$

Whence we obtain the following result.

Theorem 2.8. *For a "nice system of type A", having*

$$|g_i| \leqslant \mathscr{C}i^{-s}, \quad \text{with } s > 1,$$

and

$$\gamma_i^{-2} \leqslant \Gamma i^{\gamma} \quad \text{with } \gamma < 2q - 1,$$

it follows that

$$\|e_N\|^2 \leqslant A_{10}\, N^{-2(p+q)+\gamma+3} + A_{11}\, N^{-2q+\gamma+1},$$

where A_{10}, A_{11} *are positive constants and* q *is given by* (2.3.3).

Proof. From (2.4.10)

$$\|e_N\|^2 \leqslant \sum_{i=1}^{N} \Gamma i^{\gamma} A_2^2\, N^{-2(p+q)+2} + \sum_{i=N+1}^{\infty} \Gamma i^{\gamma} A_1^2 i^{-2q}$$

$$\leqslant \Gamma A_2^2 N^{-2(p+q)+\gamma+3} + \Gamma A_1^2 \frac{N^{-2q+\gamma+1}}{2q-\gamma-1}.$$

Comment. The two terms correspond to S_1, S_2 of (1.4.5); hence we conclude that for "nice systems of type A", S_2 will dominate the N-convergence rate of the error in the natural norm, $\|e_N\|$.

2.5 A theorem based on a variational principle

The theorems of the previous section make no reference to the way in which the matrix problem is derived from the underlying operator equation. Hence they are applicable to any of the expansion method solutions of (1.2.1) introduced in Chapter 1.

We now assume that there exists a functional $F[\omega]$, which depends on the operator $\mathscr{L}$ and the right-hand side g of (1.2.1), and which is minimised over the space of functions from the domain of $\mathscr{L}$ by that function which is the solution of (1.2.1), i.e.

$$\min_{\omega\in\mathscr{R}} F[\omega] = F[f], \tag{2.5.1}$$

where $\mathscr{R}$ is the domain of $\mathscr{L}$ and f is the solution of (1.2.1). The relationship (2.5.1) is referred to as a variational principle for (1.2.1).

In the case when $\mathscr{L}$ is positive definite and Hermitian, it is easy to verify that

$$F[\omega] = (\omega, \mathscr{L}\,\omega) - (\omega, g) - (g, \omega) \tag{2.5.2}$$

satisfies (2.5.1). A variational method solution of (1.2.1) then proceeds by replacing ω in (2.5.2) by the truncated expansion

$$f_N = \sum_{i=1}^{N} a_i^{(N)}\, h_i,$$

and minimising the resulting quadratic functional with respect to the parameters $a_i^{(N)}$, $i = 1,2,\ldots,N$.

We now derive a theorem based directly on the functional (2.5.2).

Theorem 2.9. *$\mathscr{L}$ is a positive definite, Hermitian operator, and the solution of (1.2.1) is given by (2.5.1). In addition, for the suitably normalised expansion set $\{h_i, i = 1,2,\ldots\}$, the matrix $\mathbf{L}$ is AD of type $A(p; C)$, with $L_{ii} = 1$ and $p > \frac{1}{2}$, and*

$$|b_i| \leqslant \kappa i^{-q}, \quad \textit{with } q > \tfrac{1}{2}.$$

If, in addition $p + 2q > 2$, then

$$\|e_N\|_{\mathscr{L}}^2 \leqslant A_{12} N^{-2q+1} + A_{13} N^{-(p+2q)+2},$$

where A_{12}, A_{13} are positive constants.

Proof. For any element $f_N = f - e_N$,

$$F[f_N] = F[f] + (e_N, \mathscr{L} e_N)$$
$$= F[f] + \|e_N\|_{\mathscr{L}}^2.$$

The minimum of $F[f_N]$ and hence of $\|e_N\|_{\mathscr{L}}^2$ is therefore no greater than that given by any choice of the coefficients $a_i^{(N)}$. We choose

$$a_i^{(N)} = b_i, \quad i = 1,2,\ldots,N,$$

and hence find,

$$\|e_N\|_{\mathscr{L}}^2 \leqslant \sum_{i,j=N+1}^{\infty} |b_i|\,|b_j|\,|L_{ij}|$$
$$\leqslant \sum_{i=N+1}^{\infty} |b_i|^2 + 2 \sum_{i=N+1}^{\infty} \sum_{j=i+1}^{\infty} |b_i|\,|b_j|\,|L_{ij}|$$
$$\leqslant \frac{\kappa^2 N^{-2q+1}}{2q-1} + \frac{2C\kappa^2 N^{-(p+2q)+2}}{(p+q-1)(p+2q-2)}.$$

This result agrees with that given by Theorem 2.7, although the restriction on the constant C is somewhat weaker.

2.6 Numerical examples

We illustrate the theorems given above with two numerical examples for which we can compare the observed numerical performance with the predictions of these theorems.

As a first example, we consider the Fredholm integral equation

$$f(x) = x^2 + \lambda \int_0^1 (x^2 + y^2) f(y)\, dy, \tag{2.6.1}$$

which has the exact solution

$$f(x) = \frac{(45 - 15\lambda)\, x^2 + 9\lambda}{45 - 30\,\lambda - 4\,\lambda^2}. \tag{2.6.2}$$

We use the orthonormal expansion set $\{h_i(x)\}$, where

$$h_i(x) = \begin{cases} 1, & i = 1, \\ \sqrt{2}\cos((i-1)\,\pi x), & i \geqslant 2. \end{cases}$$

The right-hand side vector $\mathbf{g}$ and the matrix $\mathbf{L}$ of the Galerkin equations (1.2.4) can be computed analytically. They have components

$$g_i = \begin{cases} \frac{1}{3}, & i = 1, \\ \dfrac{\sqrt{8}(-1)^{i-1}}{\pi^2 (i-1)^2}, & i \geqslant 2, \end{cases}$$

and $L_{ij} = \delta_{ij} - K_{ij}$, where $K_{ij} = \lambda(g_i\, \delta_{j1} + g_j\, \delta_{i1})$. Table 2.1 lists the computed coefficients $a_i^{(N)}$, for various values of N, for this problem with $\lambda = 1$. A numerical fit to these results yields estimates of the vertical, horizontal and diagonal convergence rates:

$$\text{vertical:} \quad |b_i| \sim C_v\, i^{-2.1}, \tag{2.6.3}$$
$$\text{horizontal:} \quad |b_3 - a_3^{(N)}| \sim C_h\, N^{-3.1}, \tag{2.6.4}$$
$$\text{diagonal:} \quad |b_i - a_i^{(i)}| \sim C_d\, i^{-5.4}. \tag{2.6.5}$$

Now, for $0 < \lambda < \frac{3}{2}$,

$$|L_{11}| \geqslant 1 - (2\lambda/3),$$

and

$$|L_{ii}| = 1, \quad i \geqslant 2,$$

and hence, for $i > j \geqslant 1$,

$$\begin{aligned} \frac{|L_{ij}|}{\{|L_{ii}|\,|L_{jj}|\}^{1/2}} &\leqslant \frac{|\lambda|\,|g_i|}{\{1 - (2\lambda/3)\}^{1/2}} \\ &\leqslant \frac{\lambda\sqrt{24}}{\pi^2\sqrt{(3 - 2\lambda)}\,(i-1)^2} \\ &\leqslant \frac{8\lambda\sqrt{6}}{\pi^2\sqrt{(3 - 2\lambda)} i^2}. \end{aligned}$$

TABLE 2.1

$a_i^{(N)}$

$N = 3$	$N = 6$	$N = 11$	$N = 16$	$N = 21$	b_i
1·7092288	1·7254676	1·7270101	1·7271909	1·7272373	1·7272727
$-7·7640967 \times 10^{-1}$	$-7·8106337 \times 10^{-1}$	$-7·8150542 \times 10^{-1}$	$-7·8155722 \times 10^{-1}$	$-7·8157054 \times 10^{-1}$	$-7·8158068 \times 10^{-1}$
$1·9410242 \times 10^{-1}$	$1·9526584 \times 10^{-1}$	$1·9537635 \times 10^{-1}$	$1·9538931 \times 10^{-1}$	$1·9539263 \times 10^{-1}$	$1·9539517 \times 10^{-1}$
	$-8·6784819 \times 10^{-2}$	$-8·6833935 \times 10^{-2}$	$-8·6839692 \times 10^{-2}$	$-8·6841171 \times 10^{-2}$	$-8·6842298 \times 10^{-2}$
	$4·8816461 \times 10^{-2}$	$4·8844089 \times 10^{-2}$	$4·8847327 \times 10^{-2}$	$4·8848158 \times 10^{-2}$	$4·8848793 \times 10^{-2}$
	$-3·1242535 \times 10^{-2}$	$-3·1260217 \times 10^{-2}$	$-3·1262289 \times 10^{-2}$	$-3·1262821 \times 10^{-2}$	$-3·1263227 \times 10^{-2}$
		$2·1708484 \times 10^{-2}$	$2·1709923 \times 10^{-2}$	$2·1710293 \times 10^{-2}$	$2·1710575 \times 10^{-2}$
		$-1·5949090 \times 10^{-2}$	$-1·5950148 \times 10^{-2}$	$-1·5950419 \times 10^{-2}$	$-1·5950626 \times 10^{-2}$
		$1·2211022 \times 10^{-2}$	$1·2211832 \times 10^{-2}$	$1·2212040 \times 10^{-2}$	$1·2212198 \times 10^{-2}$
		$-9·6482151 \times 10^{-3}$	$-9·6488547 \times 10^{-3}$	$-9·6490190 \times 10^{-3}$	$-9·6491442 \times 10^{-3}$
		$7·8150542 \times 10^{-3}$	$7·8155723 \times 10^{-3}$	$7·8157053 \times 10^{-3}$	$7·8158068 \times 10^{-3}$
			$-6·4591506 \times 10^{-3}$	$-6·4592606 \times 10^{-3}$	$-6·4593445 \times 10^{-3}$
			$5·4274807 \times 10^{-3}$	$5·4275732 \times 10^{-3}$	$5·4276436 \times 10^{-3}$
			$-4·6245990 \times 10^{-3}$	$-4·6246777 \times 10^{-3}$	$-4·6247378 \times 10^{-3}$
			$3·9875369 \times 10^{-3}$	$3·9876048 \times 10^{-3}$	$3·9876566 \times 10^{-3}$
			$-3·4735877 \times 10^{-3}$	$-3·4736468 \times 10^{-3}$	$-3·4736919 \times 10^{-3}$
				$3·0530099 \times 10^{-3}$	$3·0530495 \times 10^{-3}$
				$-2·7043963 \times 10^{-3}$	$-2·7044314 \times 10^{-3}$
				$2·4122547 \times 10^{-3}$	$2·4122861 \times 10^{-3}$
				$-2·1650153 \times 10^{-3}$	$-2·1650434 \times 10^{-3}$
				$1·9539263 \times 10^{-3}$	$1·9539517 \times 10^{-3}$

Hence, the symmetric matrix $\mathbf{L}$ is AD of type A $(2; 8\lambda(6/(3-2\lambda))^{1/2}/\pi^2)$. Also, in the normalisation of this chapter

$$|g_i| \leqslant \mathscr{C} i^{-2}, \quad i \geqslant 2,$$

where

$$\mathscr{C} = \frac{8\sqrt{2}}{\pi^2}.$$

Theorems 2.2 and 2.3 then provide the bounds

$$|b_i| \leqslant A_1\, i^{-2} \tag{2.6.6}$$

$$|b_i - a_i^{(N)}| \leqslant A_2\, N^{-3}. \tag{2.6.7}$$

We thus find that (2.6.6) and (2.6.7) provide estimates of the vertical and horizontal convergence rates which are very close to the observed numerical convergence rates. Since the bound (2.6.7) is valid uniformly in i, $i \leqslant N$, it is applicable also to the diagonal convergence rate. However, the observed diagonal rate of convergence (2.6.5) is rather faster than that predicted by (2.6.7).

As a second example we consider the boundary value problem introduced in Section 1.5. We see, from (1.5.9), that the matrix $\mathbf{L}$ is AD of type A $(2; 8\lambda/\pi)$. We also see, from (1.5.10), that in the normalisation assumed in this chapter,

$$|g_i| \leqslant \mathscr{C}\, i^{-2}.$$

Hence Theorems 2.2 and 2.3 again provide convergence rate bounds (for λ sufficiently small), and in the normalisation used in the solution of the problem in Section 1.5, we obtain

$$|\bar{b}_i| \leqslant \bar{A}_1\, i^{-3}, \tag{2.6.8}$$

$$|\bar{b}_i - \bar{a}_i^{(N)}| \leqslant \bar{A}_2\, N^{-3} i^{-1}. \tag{2.6.9}$$

We may compare these bounds with those obtained numerically in Section 1.5, see (1.5.7). We see that (2.6.8) and (2.6.9) certainly provide bounds on the observed convergence rates, although in the cases of horizontal and diagonal convergence these bounds are rather weak. Now, strictly speaking, we need not be surprised: the value $\lambda = 1$ corresponds to $C \approx 2.5$ in (2.1.3), which certainly does not satisfy restriction (2.2.3). Some restriction on C is inevitable in a theorem such as 2.1 (which guarantees that $\mathbf{L}^{(N)}$ is non-singular for all N); however, if we are prepared to assume that $\mathbf{L}^{(N)}$ is non-singular for sufficiently large N, inequality (2.2.3) can be weakened; we return to this point in Chapter 7. For now, we note that numerical results indicate that the convergence rates (1.5.7) apply independent of λ for a very wide range of λ, and clearly reflect the matrix structure rather than any technical limitation on λ. The weakness of the bound (2.6.9) is due to the incomplete way in which this structure is reflected by the bounding assumption (2.1.3). This bound takes account of the j^{-2} term in (1.5.9), but cannot take account of the $(j - i)^{-2}$ term. To obtain better estimates of the convergence rates we need to take account of this structure. In the next chapter, we generalise the assumptions of this chapter and the theorems which result; in Chapter 4 we discuss particular classes of matrices which satisfy these weaker assumptions, and which reflect quite closely the structure of practical problems. In particular, we use a special case of these generalisations to produce sharp convergence rate estimates for example 2.

3

Weakly Asymptotically Diagonal Systems

3.1 Introduction

In Chapter 2, we showed that by assuming that the infinite matrix **L** is AD of type A, we can closely characterise the convergence properties of the infinite system of equations

$$\mathbf{L}\mathbf{b} = \mathbf{g} \tag{3.1.1}$$

and the truncated $N \times N$ system of equations

$$\mathbf{L}^{(N)}\mathbf{a}^{(N)} = \mathbf{g}^{(N)}. \tag{3.1.2}$$

For symmetric matrices which are AD of type A, we also showed that if these sets of equations arise from an expansion method solution of the operator equation (1.2.1), then these characterisations can be used to consider the convergence of the N-th approximate solution f_N to the true solution f.

However, the assumption that **L** is AD of type A is rather strong, as is illustrated by the rather weak convergence rate estimates obtained for the second numerical example of Section 2.6. We now generalise the results of Chapter 2, by making the assumptions on **L** as weak as possible consistent with retaining theorems of the same character. Not only are such generalisations interesting mathematically, but, as we shall see in later chapters, they allow us to use the theory of AD matrices on a wider class of problems.

The results of Chapter 2 for type A systems can be recognised as having three stages:

Stage 1: We showed that under suitable conditions, the matrix $\mathbf{L}^{(N)}$ had a triangular decomposition

$$\mathbf{L}^{(N)} = \mathbf{YJY}^t, \quad \text{where } \mathbf{J} = \mathbf{diag}(L_{ii}),$$

for every N, and hence was non-singular.

Stage 2: Again under suitable restrictions, we could characterise the inverse of the triangular matrix $\mathbf{Y}$, and hence characterise the inverse of $\mathbf{L}$.

Stage 3: This information was used to bound the solution of linear equations with matrix $\mathbf{L}$ or $\mathbf{L}^{(N)}$.

We follow the same stages in this chapter, at each stage weakening the assumptions as much as seems practicable. In particular, we no longer assume that $\mathbf{L}$ is symmetric.

The analysis is considerably simplified by adopting the following notation: for arbitrary $N \times N$ matrices $\mathbf{A} = \{A_{ij}\}$, $\mathbf{B} = \{B_{ij}\}$ we write

$$|\mathbf{A}| = \{|A_{ij}|\}$$

and

$$\mathbf{A} \leqslant C\mathbf{B} \text{ when } A_{ij} \leqslant CB_{ij}, \quad i,j = 1,2,\ldots,N.$$

3.2 A class of non-singular matrices

Definition 3.1. *Let $\mathbf{L}^{(N)}$ be an $N \times N$ matrix with non-zero diagonal elements, and let $\mathbf{F}$,*

$$F_{ij} = \frac{|L_{ij}|}{\{|L_{ii}|\,|L_{jj}|\}^{1/2}}, \tag{3.2.1}$$

satisfy the following condition,

$$|\mathbf{F} - \mathbf{I}| \leqslant \mathbf{A} + \mathbf{B}, \tag{3.2.2}$$

where $\mathbf{A}$, $\mathbf{B}^t$ are strictly lower triangular matrices which satisfy

$$\mathbf{AB} \leqslant C(\mathbf{A} + \mathbf{D} + \mathbf{B}), \tag{3.2.3}$$

and $\mathbf{D}$ is a diagonal matrix which satisfies

$$\mathbf{D} \leqslant C_1\mathbf{I} \tag{3.2.4}$$

and C, C_1 are positive constants. $\mathbf{L}^{(N)}$ is then said to be weakly diagonal (WD). Additionally, we say that $\mathbf{L}^{(N)}$ is normalised if

$$|L_{ii}| = 1, \quad i = 1,2,\ldots,N. \tag{3.2.5}$$

Every $N \times N$ matrix with non-zero diagonal elements is WD for sufficiently large constants C, C_1, and any $\mathbf{A}$, $\mathbf{B}$ satisfying (3.2.2). However, as in Chapter 2, we shall impose restrictions on the constants and assume that these restrictions are satisfied for every value of N considered.

The major result of this section concerns the triangular decomposition of a WD matrix $\mathbf{L}^{(N)}$ in the form,

$$\mathbf{L}^{(N)} = \mathbf{UV}, \tag{3.2.6}$$

where $\mathbf{U}$, $\mathbf{V}^t$ are $N \times N$ lower triangular matrices. We show that, provided C, C_1 are suitably restricted, the decomposition (3.2.6) is essentially unique and the elements of $\mathbf{U}$ and $\mathbf{V}$ are bounded.

As in Section 2.2, we assume that $\mathbf{L}^{(N)}$ is normalised to satisfy (3.2.5), since the WD property is also invariant under a diagonal transformation.

We now extend the results of the first part of Theorem 2.1:

Theorem 3.1. *Let the normalised matrix* $\mathbf{L}^{(N)}$ *be WD with constants* C, C_1 *satisfying*

$$4C(1 + C_1) \leqslant 1. \tag{3.2.7}$$

Then $\mathbf{L}^{(N)}$ *may be uniquely factorised in the form* (3.2.6), *with* $|\mathbf{diag}(\mathbf{V})| = \mathbf{I}$. *Furthermore, the matrices* $\mathbf{U}$, $\mathbf{V}$ *satisfy*

$$|\mathbf{U} - \mathbf{diag}(\mathbf{U})| \leqslant K_1\mathbf{A} \tag{3.2.8}$$

$$|\mathbf{V} - \mathbf{diag}(\mathbf{V})| \leqslant K_2\mathbf{B} \tag{3.2.9}$$

$$|\mathbf{diag}(\mathbf{U})| \geqslant \mathbf{I} - K_1K_2C\mathbf{D} \geqslant \tfrac{1}{2}\mathbf{I}, \tag{3.2.10}$$

where the constants K_1, K_2 *satisfy* $0 < K_1, K_2 \leqslant 2$.

Proof. Let $\mathbf{u}_i^t$ denote the i-th row of $\mathbf{U}$ and let $\mathbf{v}_i$ denote the i-th column of $\mathbf{V}$.

It is trivial to show that $\mathbf{u}_1^t$ satisfies (3.2.10), since

$$|L_{11}| = |U_{11}\,V_{11}| = |U_{11}| = 1.$$

Now assume inductively that $\mathbf{u}_i^t$ satisfies (3.2.8), (3.2.10) and $\mathbf{v}_i$ satisfies (3.2.9) for $i = 1,2,\ldots,m - 1 < N$. It then remains to show that $\mathbf{u}_m^t$ satisfies (3.2.8), (3.2.10) and $\mathbf{v}_m$ satisfies (3.2.9). Now,

$$U_{ij}V_{jj} = L_{ij} - \sum_{k=1}^{j-1} U_{ik}V_{kj}, \qquad i \geqslant j \tag{3.2.11}$$

$$V_{ij} = \frac{1}{U_{ii}}\left\{L_{ij} - \sum_{k=1}^{i-1} U_{ik}V_{kj}\right\}, \quad i < j,\, U_{ii} \neq 0. \tag{3.2.12}$$

Hence,

$$|U_{m1}| = |L_{m1}|, |V_{1m}| = |L_{1m}|, \quad \text{since } |U_{11}| = 1.$$

Therefore U_{m1} satisfies (3.2.8) and V_{1m} satisfies (3.2.9). Assume inductively that

$$|U_{mr}| \leqslant K_1 A_{mr} \quad \text{and} \quad |V_{rm}| \leqslant K_2 B_{rm},$$

for $r = 1,2,\ldots,j-1 < m-1$; then we have, by (3.2.11), (3.2.12),

$$\begin{aligned} |U_{mj}| &\leqslant |L_{mj}| + \sum_{k=1}^{j-1} |U_{mk}||V_{kj}| \\ &\leqslant A_{mj} + \sum_{k=1}^{j-1} K_1 A_{mk} K_2 B_{kj} \\ &\leqslant (1 + K_1 K_2 C) A_{mj} \leqslant K_1 A_{mj} \end{aligned} \tag{3.2.13}$$

and

$$\begin{aligned} |V_{jm}| &\leqslant \frac{1}{1 - K_1 K_2 C D_{jj}} \left\{ |L_{jm}| + \sum_{k=1}^{j-1} |U_{jk}||V_{km}| \right\} \\ &\leqslant \frac{1}{1 - K_1 K_2 C D_{jj}} \left\{ B_{jm} + \sum_{k=1}^{j-1} K_1 A_{jk} K_2 B_{km} \right\} \\ &\leqslant \frac{(1 + K_1 K_2 C)}{(1 - K_1 K_2 C D_{jj})} B_{jm} \leqslant K_2 B_{jm}. \end{aligned} \tag{3.2.14}$$

Also, we have, from (3.2.11)

$$U_{mm} V_{mm} = L_{mm} - \sum_{k=1}^{m-1} U_{mk} V_{km}.$$

Hence

$$\begin{aligned} |U_{mm}| &\geqslant 1 - \sum_{k=1}^{m-1} K_1 A_{mk} K_2 B_{km} \\ &\geqslant 1 - K_1 K_2 C D_{mm} > 0. \end{aligned} \tag{3.2.15}$$

It now remains only to show that there exist real K_1 and K_2 that satisfy (3.2.13), (3.2.14) and (3.2.15).

From (3.2.13) we may take

$$K_1 = \frac{1}{1 - K_2 C}. \tag{3.2.16}$$

Substituting in (3.2.14), we have

$$K_2\left(1 - \frac{K_2CC_1}{1 - K_2C}\right) \geq 1 + \frac{K_2C}{1 - K_2C}$$

thus

$$K_2^2(C + CC_1) - K_2 + 1 \leq 0, \quad \text{since } 1 - K_2C > 0.$$

This inequality is satisfied by all K_2 which lie between the real roots of the equation

$$C(1 + C_1)\gamma^2 - \gamma + 1 = 0.$$

We then take K_2 to be the smaller of these roots,

$$K_2 = \frac{1 - \delta}{2C(1 + C_1)} \leq 2,$$

where $\delta^2 = 1 - 4C(1 + C_1)$. Now, from (3.2.16)

$$K_1 = \frac{2(1 + C_1)}{1 + 2C_1 + \delta} \leq 2.$$

Also,

$$K_1K_2CC_1 = \frac{(1 - \delta)\,C_1}{(1 + 2C_1 + \delta)} \leq \tfrac{1}{2},$$

and hence, from (3.2.15),

$$|U_{mm}| \geq 1 - K_1K_2CC_1 \geq \tfrac{1}{2}.$$

This completes the proof of the theorem.

Theorem 3.1 thus provides bounds on the triangular factors of $\mathbf{L}^{(N)}$ in terms of bounds on the matrix $\mathbf{L}^{(N)}$ itself. It can also be shown directly from this theorem that $\mathbf{L}^{(N)}$ is non-singular.

Corollary 3.1. Under the conditions of Theorem 3.1, $\mathbf{L}^{(N)}$ is non-singular.

Proof.

$$\det \mathbf{L}^{(N)} = \det \mathbf{U}$$

$$= \prod_{i=1}^{N} U_{ii} \neq 0.$$

It is well known that results which concern the non-singularity of $\mathbf{L}^{(N)}$ can be proved under the assumption that $\mathbf{L}^{(N)}$ is diagonally dominant. However, it should be noted that the assumptions of Theorem 3.1 do not necessarily imply that $\mathbf{L}^{(N)}$ is diagonally dominant. Hence Corollary 3.1 guarantees the non-singularity of a class of non-diagonally dominant matrices.

Consider, for example, the $N \times N$ matrix $\mathbf{L}^{(N)}$ defined by

$$\left.\begin{aligned} L_{ij} &= -1, \quad i = j \\ L_{ij} &= \frac{1}{8ij}, \quad i \neq j \end{aligned}\right\} \quad i,j = 1,2,\ldots,N. \tag{3.2.17}$$

It is easy to show that $\mathbf{L}^{(N)}$ satisfies the conditions of Theorem 3.1, with matrices $\mathbf{A}$ and $\mathbf{B}$:

$$A_{ij} = B_{ji} = \frac{1}{8ij}, \quad j = 1,2,\ldots,N-1, \quad i = j+1,\ldots,N,$$

and constants C, C_1:

$$C = \tfrac{7}{32}, \; C_1 = \tfrac{1}{32}.$$

For (3.2.2) is clearly satisfied, and

$$(AB)_{ij} = \sum_{k=1}^{\min\{i-1,j-1\}} \frac{1}{64ijk^2} \leqslant \tfrac{7}{32} \cdot \frac{1}{8ij},$$

$$\begin{aligned} (AB)_{ii} &= 0, \qquad i = 1, \\ &\leqslant \tfrac{7}{32} \cdot \tfrac{1}{32}, \quad i \geqslant 2. \end{aligned}$$

It is also clear that, for large N, $\mathbf{L}^{(N)}$ is not diagonally dominant; however, we can invoke Corollary 3.1 to prove that $\mathbf{L}^{(N)}$ is non-singular for every finite N.

3.3 Bounds on the inverses of certain matrices

The WD hypothesis enables us to characterise the triangular factors, (3.2.6), of $\mathbf{L}^{(N)}$, but it is too weak to give any information on the nature of the inverse $\mathbf{L}^{(N)-1}$. In this section we introduce further restrictions on $\mathbf{L}^{(N)}$, which firstly allow us partly and then completely to characterise $\mathbf{L}^{(N)-1}$.

We note that, if $\mathbf{L}^{(N)}$ can be decomposed as (3.2.6), then $\mathbf{L}^{(N)-1} = \mathbf{V}^{-1}\mathbf{U}^{-1}$, and $\mathbf{V}^{-1}$, $\mathbf{U}^{-1}$ are triangular matrices. We consider first the inversion of the matrices $\mathbf{U}$ and $\mathbf{V}$.

Definition 3.2. *Let* $\mathbf{A}$ *be a strictly lower (upper) triangular, positive,* $N \times N$ *matrix. Then* $\mathbf{A}$ *satisfies:*

(i) *Hypothesis weak H1: if there exists a strictly lower (upper) triangular matrix* $\overline{\mathbf{A}}$ *(the image matrix) and a positive constant* C_2 *such that*

(*a*) $\mathbf{A} \leqslant \overline{\mathbf{A}}$; (3.3.1)

(*b*) $\mathbf{A}\overline{\mathbf{A}} \leqslant C_2\overline{\mathbf{A}}$ *and* $\overline{\mathbf{A}}\mathbf{A} \leqslant C_2\overline{\mathbf{A}}$. (3.3.2)

(*ii*) *Hypothesis strong H*1: *if* $\mathbf{A}$ *satisfies weak H*1 *with* $\overline{\mathbf{A}} = \mathbf{A}$.

Theorem 3.2. *Let* $\mathbf{T}$ *be a lower (upper) triangular,* $N \times N$ *matrix with diagonal* $\mathbf{D}$, *and for some* $k > 0$, *let*

$$|\mathbf{T} - \mathbf{D}| \leqslant k\mathbf{A},$$

where $\mathbf{A}$ *satisfies weak H*1 *with image* $\overline{\mathbf{A}}$ *and constant* C_2. *Further, for some* $\alpha > 0$, *let*

$$|D_{ii}|^{-1} \leqslant \alpha.$$

Then

$$|\mathbf{T}^{-1}| \leqslant \alpha\mathbf{I} + \alpha^2 k\beta_N(\alpha C_2 k)\,\overline{\mathbf{A}},$$

where

$$\beta_N(x) = \frac{(1 - x^{N-1})}{(1 - x)},$$

and we assume $\alpha C_2 k \neq 1$.

Proof. We write

$$\mathbf{T} = \mathbf{D} - \mathbf{R}, \quad \text{and} \quad \mathbf{S} = \mathbf{D}^{-1}\mathbf{R},$$

and note that $\mathbf{T}^{-1}$ has the representation

$$\mathbf{T}^{-1} = [\mathbf{I} + \mathbf{S} + \mathbf{S}^2 + \ldots + \mathbf{S}^{N-1}]\,\mathbf{D}^{-1},$$

since $S_{ij} = 0$ for $i \leqslant j$ and hence $\mathbf{S}^N \equiv \mathbf{0}$. Moreover,

$$|\mathbf{S}^m| \leqslant (\alpha k)^m \mathbf{A}^m,$$

and hence, from (3.3.1), (3.3.2)

$$|\mathbf{S}^m| \leqslant (\alpha k)^m C_2^{m-1}\,\overline{\mathbf{A}}.$$

Hence,

$$\begin{aligned}
|\mathbf{T}^{-1}| &\leqslant \{\mathbf{I} + (\alpha k)\overline{\mathbf{A}} + (\alpha k)^2 C_2\overline{\mathbf{A}} + \ldots + (\alpha k)^{N-1} C_2^{N-2}\overline{\mathbf{A}}\}\,\alpha\mathbf{I} \\
&\leqslant \alpha\mathbf{I} + \alpha\{\alpha k + (\alpha k)^2 C_2 + \ldots + (\alpha k)^{N-1} C_2^{N-2}\}\overline{\mathbf{A}} \\
&\leqslant \alpha\mathbf{I} + \alpha^2 k\frac{(1 - (\alpha k C_2)^{N-1})}{(1 - \alpha k C_2)}\overline{\mathbf{A}}.
\end{aligned}$$

Corollary 3.2. Let $\mathbf{T} = \mathbf{D} - \mathbf{R}$ and $\mathbf{A}$ be lower and strictly lower, triangular, infinite matrices respectively, with $|\mathbf{R}| < k\mathbf{A}$; for each N, let the $N \times N$ leading submatrices of these matrices satisfy the conditions of Theorem 3.2, with uniform positive constants α, C_2. Then, if

$$\alpha k C_2 \leqslant \beta < 1,$$

the infinite matrix $\mathbf{T}^{-1}$ satisfies

$$|\mathbf{T}^{-1}| \leqslant \alpha \mathbf{I} + \frac{\alpha^2 k}{(1-\beta)} \overline{\mathbf{A}}.$$

We now use the results of Theorem 3.2 to impose further conditions on the WD matrix $\mathbf{L}^{(N)}$ to allow a characterisation of the inverse $\mathbf{L}^{(N)-1}$.

Definition 3.3. *Let $\mathbf{L}^{(N)}$ be WD, with $\mathbf{F}$, $\mathbf{A}$, $\mathbf{B}$, $\mathbf{D}$, C, C_1 as in Definition 3.1. Further, let*
(a) $\mathbf{A}$ satisfy weak H1, with image $\overline{\mathbf{A}}$ and constant C_A,
(b) $\mathbf{B}$ satisfy weak H1, with image $\overline{\mathbf{B}}$ and constant C_B,
(c) the images $\overline{\mathbf{A}}$, $\overline{\mathbf{B}}$ satisfy

$$\overline{\mathbf{B}}\overline{\mathbf{A}} \leqslant \Omega(\overline{\mathbf{A}} + \Omega_1 \mathbf{I} + \overline{\mathbf{B}}), \tag{3.3.3}$$

where Ω, Ω_1 are positive constants. Then $\mathbf{L}^{(N)}$ is said to be weakly asymptotically diagonal (WAD).

Definition 3.4. $\mathbf{L}^{(N)}$ *is strongly asymptotically diagonal (SAD) if in addition $\overline{\mathbf{A}} = \mathbf{A}$, $\overline{\mathbf{B}} = \mathbf{B}$.*

Theorem 3.3. *Let the normalised matrix $\mathbf{L}^{(N)}$ be WAD with C, C_1 satisfying (3.2.7). Then*

$$|\mathbf{L}^{(N)-1}| \leqslant K'_0 \mathbf{I} + K'_1 \overline{\mathbf{A}} + K'_2 \overline{\mathbf{B}},$$

where

$$K'_0 = (\alpha + \gamma_A(N)\gamma_B(N)\Omega\Omega_1)$$
$$K'_1 = \gamma_A(N)(1 + \gamma_B(N)\Omega)$$
$$K'_2 = \gamma_B(N)(\alpha + \gamma_A(N)\Omega)$$

$$\alpha = \frac{1}{1 - K_1 K_2 C C_1},$$

$$\gamma_A(N) = \alpha^2 K_1 \beta_N(\alpha K_1 C_A), \qquad \alpha K_1 C_A \neq 1,$$

and

$$\gamma_B(N) = K_2 \beta_N(K_2 C_B), \qquad K_2 C_B \neq 1.$$

Proof. Under the conditions of the theorem, Theorem 3.1 yields

$$\mathbf{L}^{(N)} = \mathbf{UV},$$

and Theorem 3.2 yields

$$|\mathbf{U}^{-1}| \leqslant \alpha\mathbf{I} + \gamma_A(N)\bar{\mathbf{A}},$$

and

$$|\mathbf{V}^{-1}| \leqslant \mathbf{I} + \gamma_B(N)\bar{\mathbf{B}}.$$

Hence,

$$\begin{aligned}|\mathbf{L}^{(N)-1}| &\leqslant |\mathbf{V}^{-1}|\,|\mathbf{U}^{-1}| \\ &\leqslant \alpha\mathbf{I} + \gamma_A(N)\bar{\mathbf{A}} + \alpha\gamma_B(N)\bar{\mathbf{B}} + \gamma_A(N)\gamma_B(N)\bar{\mathbf{B}}\bar{\mathbf{A}}.\end{aligned}$$

The result then follows from assumption (c) of Definition 3.3.

Corollary 3.3. If, in addition, the triangular matrices **A**, **B** satisfy the conditions of Corollary 3.2, then $|\mathbf{L}^{(N)-1}|$ is bounded uniformly in N.

Proof. In this case

$$\gamma_A(N) \leqslant \frac{\alpha^2 K_1}{1 - \alpha K_1 C_A}$$

$$\gamma_B(N) \leqslant \frac{K_2}{1 - K_2 C_B}$$

are bounded uniformly in N.

Let us now consider the $N \times N$ matrix $\mathbf{L}^{(N)}$ defined by (3.2.17). It is easy to show that **A**, **B** satisfy strong H1 with $C_A = C_B = 1/8$, and hypothesis (c) of Definition 3.3 with $\Omega = 1/8$, $\Omega_1 = 1/8$. Theorem 3.1 then yields the bounds $K_1 = 3/2$, $K_2 = 32/21$. Theorem 3.3 is applicable with $\alpha = 64/63$, and we find that $\gamma_A(\mathrm{N}) \leqslant 2048/1071 < 2$, and $\gamma_B(N) \leqslant 32/17$.

Hence, we find that, for all N,

$$|(L^{(N)-1})_{ij}| \leqslant \begin{cases} 68/63, & \text{for } i = j, \\ 5/(16ij), & \text{for } i \neq j. \end{cases}$$

3.4 Infinite matrices

The previous sections have considered the properties of the $N \times N$ matrix $\mathbf{L}^{(N)}$. We now consider the properties of the infinite matrix **L**, whose principal

$N \times N$ submatrix is $\mathbf{L}^{(N)}$. We assume, throughout this section, that for any N, $\mathbf{L}^{(N)}$ is normalised WAD, and that the constants involved satisfy the conditions of Corollary 3.3. These conditions ensure that, for every N, $\mathbf{L}^{(N)}$ is non-singular, and that $\mathbf{L}^{(N)-1}$ is uniformly bounded in N. In addition, we assume that the infinite matrix $\mathbf{L}$ and the image matrix $\bar{\mathbf{L}} = \mathbf{I} + \bar{\mathbf{A}} + \bar{\mathbf{B}}$ are bounded, and we use the notation

$$\sup_{i,j=1}^{\infty} |L_{ij}| = l,$$

$$\sup_{i,j=1}^{\infty} A_{ij} = a, \quad \sup_{i,j=1}^{\infty} \bar{A}_{ij} = \bar{a},$$

$$\sup_{i,j=1}^{\infty} B_{ij} = b, \quad \sup_{i,j=1}^{\infty} \bar{B}_{ij} = \bar{b}.$$

We now consider in detail the limiting behaviour of $\mathbf{L}^{(N)-1}$ as $N \to \infty$.

Theorem 3.4. *Under the conditions of this section:*
(a) for each $i, j = 1,2,\ldots,$ $\lim_{N\to\infty} (L^{(N)-1})_{ij} = (L^{-1})_{ij}$ exists;
(b) $(L^{-1})_{ij}$ is uniformly bounded in i and j;
(c) $\mathbf{L}\mathbf{L}^{-1} = \mathbf{L}^{-1}\mathbf{L} = \mathbf{I}$.

Proof.
(a) For any N, we may decompose $\mathbf{L}^{(N)}$ as

$$\mathbf{L}^{(N)} = \mathbf{U}\mathbf{V}.$$

Hence,

$$\mathbf{L}^{(N)-1} = \mathbf{V}^{-1}\mathbf{U}^{-1}$$

where

$$|\mathbf{U}^{-1}| \leqslant \alpha\mathbf{I} + \gamma_A\bar{\mathbf{A}},$$
$$|\mathbf{V}^{-1}| \leqslant \mathbf{I} + \gamma_B\bar{\mathbf{B}},$$

and γ_A, γ_B are independent of N.

Now, for any $M > N$ and $i, j \leqslant N$, we have

$$|(L^{(M)-1})_{ij} - (L^{(N)-1})_{ij}| = \left| \sum_{k=N+1}^{M} (V^{-1})_{ik} (U^{-1})_{kj} \right|$$
$$\leqslant \gamma_A\gamma_B \sum_{k=N+1}^{M} \bar{B}_{ik}\bar{A}_{kj}. \tag{3.4.1}$$

Now, $\sum_{k=\max\{i,j\}+1}^{M} \bar{B}_{ik}\bar{A}_{kj}$ is a series of positive terms, whose partial sum s_M is bounded above by hypothesis (c) of Definition 3.3. Hence the series converges as $M \to \infty$; it then follows that, for fixed i, j, $\{(L^{(N)-1})_{ij}\}$ is a Cauchy sequence in N, and $(L^{-1})_{ij}$ exists.

(b) Boundedness follows immediately from Theorem 3.3 and Corollary 3.3, since we assume that the infinite image matrix $\bar{\mathbf{L}} = (\mathbf{I} + \bar{\mathbf{A}} + \bar{\mathbf{B}})$ is bounded.

(c) For any N, and $i, j \leqslant N$, we have

$$(L^{-1})_{ij} = (L^{(N)-1})_{ij} + \sum_{k=N+1}^{\infty} (V^{-1})_{ik}(U^{-1})_{kj}.$$

Hence,

$$\begin{aligned}(LL^{-1})_{ij} &= \sum_{l=1}^{\infty} L_{il}(L^{-1})_{lj} \\ &= \sum_{l=1}^{N} L_{il}\left\{(L^{(N)-1})_{lj} + \sum_{k=N+1}^{\infty} (V^{-1})_{lk}(U^{-1})_{kj}\right\} \\ &\quad + \sum_{l=N+1}^{\infty} L_{il}(L^{-1})_{lj},\end{aligned}$$

and

$$\begin{aligned}|(LL^{-1})_{ij} - \delta_{ij}| &\leqslant \sum_{l=1}^{i-1} |L_{il}| \sum_{k=N+1}^{\infty} |(V^{-1})_{lk}|\,|(U^{-1})_{kj}| \\ &\quad + |L_{ii}| \sum_{k=N+1}^{\infty} |(V^{-1})_{ik}|\,|(U^{-1})_{kj}| \\ &\quad + \sum_{l=i+1}^{N} |L_{il}| \sum_{k=N+1}^{\infty} |(V^{-1})_{lk}|\,|(U^{-1})_{kj}| \\ &\quad + \sum_{l=N+1}^{\infty} |L_{il}|\,|(L^{-1})_{lj}| \\ &\leqslant \gamma_A\gamma_B\left\{\sum_{l=1}^{i-1} A_{il} \sum_{k=N+1}^{\infty} \bar{B}_{lk}\bar{A}_{kj}\right. \\ &\quad \left. + \sum_{k=N+1}^{\infty} \bar{B}_{ik}\bar{A}_{kj} + \sum_{l=i+1}^{N} B_{il} \sum_{k=N+1}^{\infty} \bar{B}_{lk}\bar{A}_{kj}\right\} \\ &\quad + \gamma_A(1 + \gamma_B\Omega) \sum_{l=N+1}^{\infty} \bar{B}_{il}\bar{A}_{lj}\end{aligned}$$

$$\leqslant [\gamma_A\gamma_B(1 + C_B) + \gamma_A(1 + \gamma_B\Omega)] \sum_{k=N+1}^{\infty} \bar{B}_{ik}\bar{A}_{kj}$$

$$+ \gamma_A\gamma_B\, ia \max_{l=1}^{i-1} \sum_{k=N+1}^{\infty} \bar{B}_{lk}\bar{A}_{kj}.$$

As in part (a), both these sums vanish as $N \to \infty$.

In a similar way, we can show that

$$|(L^{-1}L)_{ij} - \delta_{ij}| \to 0 \quad \text{as} \quad N \to \infty.$$

The second theorem of this section relates to the solution **b** of the infinite system (3.1.1).

Theorem 3.5. *Under the assumptions of this section, together with*

(a) $\|\mathbf{g}\|_1 = \sum_{i=1}^{\infty} |g_i|$ *exists,*

each element $a_i^{(N)}$ *of the solution of* (3.1.2) *is bounded uniformly in* N, *and for each* i

$$\lim_{N\to\infty} a_i^{(N)} = b_i.$$

If we assume additionally that
(*b*) $\|\bar{\mathbf{A}}\|_1$, $\|\bar{\mathbf{B}}\|_1$ *exist,*
where

$$\|\bar{\mathbf{A}}\|_1 = \sup_{j=1}^{\infty} \sum_{i=1}^{\infty} \bar{A}_{ij},$$

then $\|\mathbf{b}\|_1$ *exists, and* **b** *is the unique bounded solution of* (3.1.1).

Proof.
(i) *Boundedness of* $a_i^{(N)}$
From the bound on $|\mathbf{L}^{(N)-1}|$ given by Theorem 3.3 and Corollary 3.3,

$$|a_i^{(N)}| \leqslant \sum_{j=1}^{N} |(L^{(N)-1})_{ij}|\,|g_j|$$

$$\leqslant K_0'\,|g_i| + K_1' \sum_{j=1}^{i-1} \bar{A}_{ij}\,|g_j| + K_2' \sum_{j=i+1}^{N} \bar{B}_{ij}\,|g_j|$$

$$\leqslant \max\{K_0', K_1'\bar{a}, K_2'\bar{b}\}\, \|\mathbf{g}\|_1,$$

and this bound is independent of N.

(ii) Existence of b_i

For any $M > N$ and $i \leqslant N$,

$$a_i^{(M)} - a_i^{(N)} = \sum_{j=1}^{N} [(L^{(M)-1})_{ij} - (L^{(N)-1})_{ij}]g_j + \sum_{j=N+1}^{M} (L^{(M)-1})_{ij}g_j.$$

Hence, using (3.4.1),

$$\begin{aligned}|a_i^{(M)} - a_i^{(N)}| &\leqslant \sum_{j=1}^{N} \gamma_A\gamma_B \sum_{k=N+1}^{M} \bar{B}_{ik}\bar{A}_{kj}|g_j| + \sum_{j=N+1}^{M} K_2'\bar{B}_{ij}|g_j| \qquad (3.4.2)\\ &\leqslant \sum_{j=1}^{i-1} \gamma_A\gamma_B\Omega\bar{A}_{ij}|g_j| + \gamma_A\gamma_B\Omega\Omega_1|g_i|\\ &\quad + \sum_{j=i+1}^{N} \gamma_A\gamma_B\Omega\bar{B}_{ij}|g_j| + \sum_{j=N+1}^{M} K_2'\bar{B}_{ij}|g_j|.\end{aligned}$$

As in part (*a*) of Theorem 3.4, this series converges as $M \to \infty$, and hence

$$\lim_{N\to\infty} a_i^{(N)} = b_i \text{ exists.}$$

(iii) Existence of $\|\mathbf{b}\|_1$

The hypothesis and Theorem 3.3 immediately imply that

$$\begin{aligned}\|\mathbf{a}^{(N)}\|_1 &\leqslant \|\mathbf{L}^{(N)-1}\|_1 \|\mathbf{g}\|_1\\ &\leqslant \{K_0' + K_1'\|\bar{\mathbf{A}}\|_1 + K_2'\|\bar{\mathbf{B}}\|_1\}\|\mathbf{g}\|_1\end{aligned}$$

is bounded independent of N. Hence $\|\mathbf{b}\|_1$ exists.

(iv) **b** *is a solution of* (3.1.1)

We consider the i-th equation of (3.1.1), with $i \leqslant N$. Then

$$\begin{aligned}r_i &= \sum_{j=1}^{\infty} L_{ij}b_j - g_i\\ &= \sum_{j=1}^{\infty} L_{ij}b_j - \sum_{j=1}^{N} L_{ij}a_j^{(N)}.\end{aligned}$$

Hence,

$$\begin{aligned}|r_i| &\leqslant \sum_{j=1}^{i-1} A_{ij}|b_j - a_j^{(N)}| + |b_i - a_i^{(N)}|\\ &\quad + \sum_{j=i+1}^{N} B_{ij}|b_j - a_j^{(N)}| + \sum_{j=N+1}^{\infty} B_{ij}|b_j|.\end{aligned}$$

Under the given conditions, the first, second and fourth terms vanish as

$N \to \infty$. We now consider the remaining term, using (3.4.2),

$$\sum_{j=i+1}^{N} B_{ij}\,|b_j - a_j^{(N)}| \leqslant \sum_{j=i+1}^{N} B_{ij}\left\{\gamma_A\gamma_B \sum_{l=1}^{N}\sum_{k=N+1}^{\infty} \bar{B}_{jk}\bar{A}_{kl}\,|g_l| + K_2' \sum_{l=N+1}^{\infty} \bar{B}_{jl}\,|g_l|\right\}$$

$$\leqslant \sum_{l=1}^{N}\sum_{k=N+1}^{\infty} \gamma_A\gamma_B C_B \bar{B}_{ik}\bar{A}_{kl}\,|g_l| + K_2' \sum_{k=N+1}^{\infty} C_B\bar{B}_{ik}\,|g_k|$$

$$\leqslant \sum_{k=N+1}^{\infty} \{\gamma_A\gamma_B C_B \bar{a}\,\|\mathbf{g}\|_1\,\bar{B}_{ik} + K_2' C_B \bar{B}_{ik}|g_k|\}.$$

Hence, under the assumptions of the theorem,

$$\sum_{j=i+1}^{N} B_{ij}\,|b_j - a_j^{(N)}| \to 0 \quad \text{as} \quad N \to \infty.$$

It immediately follows that $\mathbf{b}$ is a solution of (3.1.1).

(v) $\mathbf{b}$ *is the unique solution of* (3.1.1)
This follows immediately from Theorem 3.4. Let $\mathbf{c}$ be any other bounded solution. Then

$$\mathbf{Lb} = \mathbf{Lc} = \mathbf{g},$$

and hence $\mathbf{L}^{-1}(\mathbf{Lb}) = \mathbf{L}^{-1}(\mathbf{Lc})$.

But, since $\mathbf{L}$, $\mathbf{L}^{-1}$, $\mathbf{b}$, $\mathbf{c}$ are all bounded, the matrix multiplications are associative, and therefore

$$\mathbf{b} = \mathbf{c}.$$

It is also possible to prove Theorem 3.5 under different assumptions (a) and (b): we may replace (a) by:

(i) $\|\mathbf{g}\|_\infty = \sup_{i=1}^{\infty}|g_i|$ exists and $\sigma_{Ri}(\bar{\mathbf{B}}) = \sum_{j=1}^{\infty}\bar{B}_{ij}$ exists for each $i = 1,2,\ldots$;

or (ii) $\|\mathbf{g}\|_2 = \{\sum_{i=1}^{\infty} g_i^2\}^{1/2}$ exists and $\sigma_i(\bar{\mathbf{B}}) = \{\sum_{j=1}^{\infty}\bar{B}_{ij}^2\}^{1/2}$ exists for each $i = 1,2,\ldots$.

Correspondingly, we may replace (b) by

(i) $\|\bar{\mathbf{A}}\|_\infty$, $\|\bar{\mathbf{B}}\|_\infty$ exist, where $\|\bar{\mathbf{A}}\|_\infty = \sup_{i=1}^{\infty}\sum_{j=1}^{\infty}\bar{A}_{ij}$;

or (ii) $\|\bar{\mathbf{A}}\|_2$, $\|\bar{\mathbf{B}}\|_2$ exist, where $\|\bar{\mathbf{A}}\|_2 = \{\sum_{i=1}^{\infty}\sum_{j=1}^{\infty}\bar{A}_{ij}^2\}^{1/2}$.

The proofs in these two alternative cases follow step by step that given above.

3.5 Behaviour of $\mathbf{a}^{(N)}$ and $\mathbf{b}$

The analysis of the previous sections considers in detail the structure of the WAD matrix $\mathbf{L}^{(N)}$ and the related infinite matrix $\mathbf{L}$, and of their inverses, and thus forms the generalisation of Section 2.2. We now consider the behaviour of the solutions $\mathbf{b}$ and $\mathbf{a}^{(N)}$ of (3.1.1) and (3.1.2) respectively, by extending the results of Theorems 2.2 and 2.3.

Definitions 3.1, 3.2 and 3.3 of a WAD matrix are rather obscure. Therefore in Definition 3.5 and Lemma 3.1 below, we give somewhat stronger, but more transparent sufficient conditions for a matrix to be WAD.

Definition 3.5. *Let the $N \times N$ positive matrix $\mathbf{A}$ have elements A_{ij} which are semi-monotone increasing in j, bounded above independent of N, and semi-monotone decreasing in i. Also let*

$$\|\mathbf{A}\|_\infty \leqslant C, \tag{3.5.1}$$

and

$$\|\mathbf{A}\|_1 \leqslant \Omega, \tag{3.5.2}$$

where C and Ω are independent of N. Then the matrix $\mathbf{A}$ has property MN (monotonicity and norm).

It is of course trivial to extend this definition to an infinite matrix $\mathbf{A}$.

Lemma 3.1. *Let the strictly lower triangular, $N \times N$ matrices $\mathbf{A}$ and $\mathbf{B}^t$ have property MN. Then they satisfy* (3.2.3) *and* (3.3.3) *with $\overline{\mathbf{A}} = \mathbf{A}$, $\overline{\mathbf{B}} = \mathbf{B}$, $C = \max\{\|\mathbf{A}\|_\infty, \|\mathbf{B}\|_1\}$, $\Omega = \max\{\|\mathbf{A}\|_1, \|\mathbf{B}\|_\infty\}$ and $C_1 = \Omega_1 = \max_{i=2}^N\{A_{i,i-1}, B_{i-1,i}\}$.*

Proof. For $i > j$,

$$\begin{aligned}(AB)_{ij} &= \sum_{k=1}^{j-1} A_{ik}B_{kj} \\ &= A_{ij}\sum_{k=1}^{j-1} \frac{A_{ik}}{A_{ij}} B_{kj} \\ &\leqslant A_{ij}\sum_{k=1}^{j-1} B_{kj} \\ &\leqslant A_{ij}\|\mathbf{B}\|_1 \end{aligned} \tag{3.5.3}$$

and

$$\begin{aligned}(BA)_{ij} &= \sum_{k=i+1}^{N} B_{ik}A_{kj} \\ &= A_{ij} \sum_{k=i+1}^{N} B_{ik} \frac{A_{kj}}{A_{ij}} \\ &\leqslant A_{ij} \sum_{k=i+1}^{N} B_{ik} \\ &\leqslant A_{ij} \, \|\mathbf{B}\|_{\infty}. \end{aligned} \tag{3.5.4}$$

The results for $i \leqslant j$ follow similarly.

Corollary 3.4. Let the matrix **L** satisfy (3.2.1), (3.2.2) with matrices **A**, $\mathbf{B}^t$ which have property MN. Then **L** is SAD.

Property MN is not strictly required for any of the theorems which follow. However, its use considerably simplifies some of the results, and we invoke it whenever the gain seems worthwhile. In these cases, it is not always necessary to invoke it in the strong form of Corollary 3.4 (property MN for matrices **A**, $\mathbf{B}^t$); it is often sufficient to invoke property MN for the image matrices $\overline{\mathbf{A}}$, $\overline{\mathbf{B}}^t$ of (3.3.3).

We assume throughout this section that the components of the infinite vector **g** of (3.1.1) satisfy

$$|g_i| \leqslant \hat{g}_i, \quad \text{for all} \quad i, \tag{3.5.5}$$

where $\hat{\mathbf{g}}$ is a positive infinite vector, whose components $\hat{g}_i$ are monotonically decreasing in i. We also assume that, for all N, the $N \times \mathrm{N}$ matrix $\mathbf{L}^{(N)}$ satisfies the conditions of Section 3.4.

We are now able to bound the components b_i of the solution of (3.1.1).

Theorem 3.6. *Under the conditions of this section,*

$$|b_i| \leqslant K_0'\hat{g}_i + K_1'\bar{\mathbf{a}}_i^t\hat{\mathbf{g}} + K_2'\bar{\mathbf{b}}_i^t\hat{\mathbf{g}}, \tag{3.5.6}$$

where $\bar{\mathbf{a}}_i^t$ *denotes the i-th row of* $\overline{\mathbf{A}}$, *and* $\bar{\mathbf{b}}_i^t$ *denotes the i-th row of* $\overline{\mathbf{B}}$.

Proof.

$$\mathbf{b} = \mathbf{L}^{-1}\mathbf{g}$$

Hence,

$$\begin{aligned}|b_i| &\leqslant \sum_{k=1}^{\infty} |(L^{-1})_{ik}|\,|g_k| \\ &\leqslant \sum_{k=1}^{\infty} (K'_0\delta_{ik} + K'_1\bar{A}_{ik} + K'_2\bar{B}_{ik})\,\hat{g}_k \\ &\leqslant K'_0\hat{g}_i + K'_1 \sum_{k=1}^{i-1} \bar{A}_{ik}\hat{g}_k + K'_2 \sum_{k=i+1}^{\infty} \bar{B}_{ik}\hat{g}_k.\end{aligned}$$

Corollary 3.5. If, additionally, the infinite matrix $\bar{\mathbf{B}}^t$ has property MN, then

$$|b_i| \leqslant (K'_0 + \Omega K'_2)\,\hat{g}_i + K'_1\bar{\mathbf{a}}_i^t\hat{\mathbf{g}}. \tag{3.5.7}$$

Proof. In (3.5.6), we may bound

$$\begin{aligned}\bar{\mathbf{b}}_i^t\hat{\mathbf{g}} &= \sum_{k=i+1}^{\infty} \bar{B}_{ik}\hat{g}_k \\ &\leqslant \hat{g}_i \sum_{k=i+1}^{\infty} \bar{B}_{ik} \\ &\leqslant \Omega\hat{g}_i.\end{aligned}$$

In particular cases, tight bounds on $|b_i|$ follow on inserting the relevant forms for $\bar{\mathbf{A}}$ and $\hat{\mathbf{g}}$, as will be shown in Chapter 4.

We can also extend Theorem 2.3 to bound the differences $|b_i - a_i^{(N)}|$.

Theorem 3.7. *Under the conditions of this section,*

$$|b_i - a_i^{(N)}| \leqslant \gamma_A\gamma_B \sum_{j=1}^{N} \hat{g}_j \sum_{k=N+1}^{\infty} \bar{B}_{ik}\bar{A}_{kj} + K'_2 \sum_{j=N+1}^{\infty} \bar{B}_{ij}\hat{g}_j. \tag{3.5.8}$$

Proof. The result follows immediately from (3.4.2) on allowing $M \to \infty$.

Corollary 3.6. If, in addition, $\bar{\mathbf{A}}$ satisfies property MN, then

$$|b_i - a_i^{(N)}| \leqslant C_{i,N+1}\,\{\gamma_A\gamma_B\bar{\mathbf{a}}_{N+1}^t\hat{\mathbf{g}} + K'_2\hat{g}_{N+1}\}, \tag{3.5.9}$$

where

$$C_{i,m} = \sum_{j=m}^{\infty} \bar{B}_{ij}.$$

Proof. In (3.5.8), we may bound

$$\sum_{j=1}^{N} \hat{g}_j \sum_{k=N+1}^{\infty} \bar{B}_{ik}\bar{A}_{kj} \leqslant C_{i,N+1} \sum_{j=1}^{N} \hat{g}_j\bar{A}_{N+1,j},$$

and

$$\sum_{j=N+1}^{\infty} \bar{B}_{ij}\hat{g}_j \leqslant C_{i,N+1}\hat{g}_{N+1}.$$

Again, in particular cases, tight bounds on $|b_i - a_i^{(N)}|$ follow on inserting the relevant forms for $\bar{\mathbf{A}}$, $\bar{\mathbf{B}}$ and $\hat{\mathbf{g}}$ (see Chapter 4).

3.6 Convergence of Galerkin calculations

If we assume that the operator $\mathscr{L}$ of (1 ᐣ.1) is positive definite, then it is possible to generalise the results of Theorems 2.4 and 2.5. These theorems can, however, be derived from the analysis of Chapter 7, and therefore, since the theorem proofs are somewhat lengthy, we proceed directly to a generalisation of the more powerful Theorem 2.7.

We first of all extend the definition of a "nice" system.

Definition 3.6. *Let $\mathscr{L}f = g$ be an inhomogeneous operator equation, and let $\mathbf{Lb} = \mathbf{g}$ be the corresponding infinite linear system of equations, with $\{h_i, i = 1,2,\ldots\}$ as expansion set. We then refer to this system as a weakly nice system (WNS) if, for the given expansion set, every submatrix $\mathbf{L}^{(N)}$ of $\mathbf{L}$ satisfies the conditions of Section 3.4, and the vector $\mathbf{g}$ satisfies (3.5.5). The system is a strongly nice system (SNS), if, in addition, $\mathbf{L}$ is SAD.*

For a weakly nice system it is possible to bound the error $e_N = f - f_N$ in the energy ($\mathscr{L}$-) (pseudo)norm, (2.4.7), by generalising Theorem 2.7. This involves inserting the bounds (3.5.6) and (3.5.8) into (2.4.6), and results in a very cumbersome expression. We therefore proceed directly to the special case when the matrices $\bar{\mathbf{A}}$ and $\bar{\mathbf{B}}^t$ satisfy property MN.

Theorem 3.8. *For a WNS for which the matrices $\bar{\mathbf{A}}$ and $\bar{\mathbf{B}}^t$ satisfy property MN, it follows that*

$$\|e_N\|_{\mathscr{L}}^2 \leqslant \{E_1D_1(N) + E_2D_2(N)\}\,(\mathbf{a}_{N+1}^{*t}\hat{\mathbf{g}})^2 + E_3 \sum_{i=N+1}^{\infty} (\mathbf{a}_i^{*t}\hat{\mathbf{g}})^2, \tag{3.6.1}$$

where $\mathbf{a}_i^{*t}$ *is the i-th row of* $\mathbf{A}^* = \overline{\mathbf{A}} + \mathbf{I}$,

$$D_1(N) = \sum_{i=1}^{N} C_{i,N+1},$$

$$D_2(N) = \sum_{i=1}^{N} (C_{i,N+1})^2,$$

and E_1, E_2, E_3 *are positive constants. Here* $C_{i,j}$ *is defined in Corollary* 3.6.

Proof. We first of all note that, in terms of the matrix $\mathbf{A}^*$, (3.5.7) can be rewritten as

$$|b_i| \leqslant \alpha_1 \mathbf{a}_i^{*t}\hat{\mathbf{g}}, \tag{3.6.2}$$

where

$$\alpha_1 = \max\{K_0' + \Omega K_2', K_1'\},$$

and (3.5.9) can be rewritten as

$$|b_i - a_i^{(N)}| \leqslant \alpha_2 C_{i,N+1}\mathbf{a}_{N+1}^{*t}\hat{\mathbf{g}}, \tag{3.6.3}$$

where

$$\alpha_2 = \max\{\gamma_A \gamma_B, K_2'\}.$$

We now insert these bounds into (2.4.6) to obtain

$$\begin{aligned}
\|e_N\|_{\mathscr{L}}^2 \leqslant & \sum_{i=1}^{N}\sum_{j=1}^{i} A_{ij}^* C_{i,N+1}C_{j,N+1}\alpha_2^2(\mathbf{a}_{N+1}^{*t}\hat{\mathbf{g}})^2 \\
& + \sum_{i=1}^{N-1}\sum_{j=i+1}^{N} \overline{B}_{ij}C_{i,N+1}C_{j,N+1}\alpha_2^2(\mathbf{a}_{N+1}^{*t}\hat{\mathbf{g}})^2 \\
& + \sum_{i=1}^{N}\sum_{j=N+1}^{\infty} \overline{B}_{ij}C_{i,N+1}\alpha_1\alpha_2(\mathbf{a}_{N+1}^{*t}\hat{\mathbf{g}})(\mathbf{a}_j^{*t}\hat{\mathbf{g}}) \\
& + \sum_{i=N+1}^{\infty}\sum_{j=1}^{N} \overline{A}_{ij}C_{j,N+1}\alpha_1\alpha_2(\mathbf{a}_{N+1}^{*t}\hat{\mathbf{g}})(\mathbf{a}_i^{*t}\hat{\mathbf{g}}) \\
& + \sum_{i=N+1}^{\infty}\sum_{j=N+1}^{i} A_{ij}^*\alpha_1^2(\mathbf{a}_i^{*t}\hat{\mathbf{g}})(\mathbf{a}_j^{*t}\hat{\mathbf{g}}) \\
& + \sum_{i=N+1}^{\infty}\sum_{j=i+1}^{\infty} \overline{B}_{ij}\alpha_1^2(\mathbf{a}_i^{*t}\hat{\mathbf{g}})(\mathbf{a}_j^{*t}\hat{\mathbf{g}}).
\end{aligned}$$

We label these terms (1)–(6), and consider them in turn:

(1) $\leqslant \alpha_2^2(\mathbf{a}_{N+1}^{*t}\hat{\mathbf{g}})^2 \sum_{i=1}^{N}\sum_{j=1}^{i} A_{ij}^* C_{i,N+1}C_{j,N+1}$

$\leqslant \alpha_2^2(\mathbf{a}_{N+1}^{*t}\hat{\mathbf{g}})^2 (1+C)\sum_{i=1}^{N}(C_{i,N+1})^2$, since $\|\overline{\mathbf{A}}\|_\infty \leqslant C$, and $C_{j,N+1}$ is semi-monotone increasing in j;

(2) $\leqslant \alpha_2^2(\mathbf{a}_{N+1}^{*t}\hat{\mathbf{g}})^2 \sum_{i=1}^{N-1}\sum_{j=i+1}^{N} \overline{B}_{ij}C_{i,N+1}C_{j,N+1}$

$\leqslant \alpha_2^2(\mathbf{a}_{N+1}^{*t}\hat{\mathbf{g}})^2 \sum_{j=2}^{N}\sum_{i=1}^{j-1} \overline{B}_{ij}C_{i,N+1}C_{j,N+1}$

$\leqslant \alpha_2^2(\mathbf{a}_{N+1}^{*t}\hat{\mathbf{g}})^2\, C\sum_{j=1}^{N}(C_{j,N+1})^2$, since $\|\overline{\mathbf{B}}^t\|_\infty \leqslant C$;

(3) $\leqslant \alpha_1\alpha_2(\mathbf{a}_{N+1}^{*t}\hat{\mathbf{g}}) \sum_{i=1}^{N}\sum_{j=N+1}^{\infty} \overline{B}_{ij}C_{i,N+1}\,(\mathbf{a}_j^{*t}\hat{\mathbf{g}})$

$\leqslant \alpha_1\alpha_2(\mathbf{a}_{N+1}^{*t}\hat{\mathbf{g}})^2 \sum_{i=1}^{N}(C_{i,N+1})^2$, since $(\mathbf{a}_j^{*t}\hat{\mathbf{g}})$ is semi-monotone decreasing in j;

(4) $\leqslant \alpha_1\alpha_2(\mathbf{a}_{N+1}^{*t}\hat{\mathbf{g}}) \sum_{i=N+1}^{\infty}\sum_{j=1}^{N} \overline{A}_{ij}C_{j,N+1}(\mathbf{a}_i^{*t}\hat{\mathbf{g}})$

$\leqslant \alpha_1\alpha_2(\mathbf{a}_{N+1}^{*t}\hat{\mathbf{g}})^2\,\Omega\sum_{j=1}^{N} C_{j,N+1}$, since $\|\overline{\mathbf{A}}\|_1 \leqslant \Omega$;

(5) $\leqslant \alpha_1^2 \sum_{j=N+1}^{\infty}\sum_{i=j}^{\infty} A_{ij}^*(\mathbf{a}_i^{*t}\hat{\mathbf{g}})\,(\mathbf{a}_j^{*t}\hat{\mathbf{g}})$

$\leqslant \alpha_1^2(1+\Omega)\sum_{j=N+1}^{\infty}(\mathbf{a}_j^{*t}\hat{\mathbf{g}})^2;$

(6) $\leqslant \alpha_1^2 \sum_{i=N+1}^{\infty}\sum_{j=i+1}^{\infty} \overline{B}_{ij}(\mathbf{a}_i^{*t}\hat{\mathbf{g}})\,(\mathbf{a}_j^{*t}\hat{\mathbf{g}})$

$\leqslant \alpha_1^2\Omega\sum_{i=N+1}^{\infty}(\mathbf{a}_i^{*t}\hat{\mathbf{g}})^2$, since $\|\overline{\mathbf{B}}^t\|_1 \leqslant \Omega$.

Hence, we obtain (3.6.1) with

$$E_1 = \alpha_1\alpha_2\Omega$$
$$E_2 = \alpha_2^2(1 + 2C) + \alpha_1\alpha_2$$
$$E_3 = \alpha_1^2(1 + 2\Omega).$$

Corollary 3.7. Further, under the conditions of Theorem 3.8,

$$\|e_N\|_{\mathscr{L}}^2 \leqslant E_4 N(\mathbf{a}_{N+1}^{*^t}\hat{\mathbf{g}})^2 + E_3 \sum_{i=N+1}^{\infty} (\mathbf{a}_i^{*^t}\hat{\mathbf{g}})^2,$$

where E_4 is a positive constant.

Proof.

$$D_1(N) = \sum_{i=1}^{N} C_{i,N+1}$$
$$= \sum_{i=1}^{N} \sum_{j=N+1}^{\infty} \bar{B}_{ij}$$
$$\leqslant N\Omega.$$

Similarly, $D_2(N) \leqslant N\Omega^2$.

We note that the bound on $\|e_N\|_{\mathscr{L}}^2$ depends on $\mathbf{A}^*$, but not on $\bar{\mathbf{B}}$, and hence depends only on the behaviour of the lower triangular half of $\mathbf{L}$.

In the final theorem of this chapter we generalise Theorem 2.8 to provide a bound on $\|e_N\|$, the natural norm of the error. A straightforward generalisation of this theorem, obtained by simply inserting the bounds (3.5.6) and (3.5.8) into (2.4.10), again gives a very inelegant result. We therefore again proceed directly to the special case when the matrices $\bar{\mathbf{A}}$ and $\bar{\mathbf{B}}^t$ satisfy property MN.

Theorem 3.9. *Under the conditions of Theorem* 3.8,

$$\|e_N\|^2 \leqslant \alpha_2^2(\mathbf{a}_{N+1}^{*^t}\hat{\mathbf{g}})^2 \sum_{i=1}^{N} \gamma_i^{-2}(C_{i,N+1})^2 + \alpha_1^2 \sum_{i=N+1}^{\infty} \gamma_i^{-2}(\mathbf{a}_i^{*^t}\hat{\mathbf{g}})^2,$$

where γ_i *is defined by* (2.4.9).

Proof. From (3.6.2), (3.6.3) and (2.4.10),

$$\|e_N\|^2 \leqslant \sum_{i=1}^{N} \gamma_i^{-2}\alpha_2^2(C_{i,N+1})^2 (\mathbf{a}_{N+1}^{*^t}\hat{\mathbf{g}})^2 + \sum_{i=N+1}^{\infty} \gamma_i^{-2}\alpha_1^2(\mathbf{a}_i^{*^t}\hat{\mathbf{g}})^2$$
$$\leqslant \alpha_2^2(\mathbf{a}_{N+1}^{*^t}\hat{\mathbf{g}})^2 \sum_{i=1}^{N} \gamma_i^{-2}(C_{i,N+1})^2 + \alpha_1^2 \sum_{i=N+1}^{\infty} \gamma_i^{-2}(\mathbf{a}_i^{*^t}\hat{\mathbf{g}})^2.$$

Corollary 3.8. If, in addition, there exists a constant α_3 such that

$$\sum_{i=1}^{\infty} \gamma_i^{-2} \leqslant \alpha_3,$$

then

$$\|e_N\|^2 \leqslant E_5(\mathbf{a}_{N+1}^{*t}\hat{\mathbf{g}})^2,$$

where E_5 is a positive constant.

Proof. Now,

$$\begin{aligned}
\mathbf{a}_{N+k+1}^{*t}\hat{\mathbf{g}} &= \sum_{j=1}^{N} \bar{A}_{N+k+1,j}\hat{g}_j + \sum_{j=N+1}^{N+k} \bar{A}_{N+k+1,j}\hat{g}_j + \hat{g}_{N+k+1} \\
&\leqslant \sum_{j=1}^{N} \bar{A}_{N+1,j}\hat{g}_j + \sum_{j=N+1}^{N+k} \bar{A}_{N+k+1,j}\hat{g}_{N+1} + \hat{g}_{N+1} \\
&\leqslant \bar{\mathbf{a}}_{N+1}^{t}\hat{\mathbf{g}} + (C+1)\,\hat{g}_{N+1} \\
&\leqslant (1+C)\,\mathbf{a}_{N+1}^{*t}\hat{\mathbf{g}}.
\end{aligned}$$

Hence,

$$\|e_N\|^2 \leqslant \alpha_2^2(\mathbf{a}_{N+1}^{*t}\hat{\mathbf{g}})^2 \sum_{i=1}^{N} \gamma_i^{-2}(C_{i,N+1})^2 + \alpha_1^2(\mathbf{a}_{N+1}^{*t}\hat{\mathbf{g}})^2 (1+C)^2 \sum_{i=N+1}^{\infty} \gamma_i^{-2}.$$

But

$$\max_{i=1}^{N} (C_{i,N+1})^2 = (C_{N,N+1})^2 = \left(\sum_{j=N+1}^{\infty} \bar{B}_{Nj}\right)^2 \leqslant \Omega^2.$$

Hence,

$$\|e_N\|^2 \leqslant (\mathbf{a}_{N+1}^{*t}\hat{\mathbf{g}})^2 \{\alpha_2^2\Omega^2 + \alpha_1^2(1+C)^2\}\alpha_3.$$

We again note that the result of Corollary 3.8 depends only on the lower triangular half of **L**.

A word of hope

The results of this chapter are perhaps somewhat formal. Their usefulness may become more apparent in the next chapter, when we consider specific classes of WAD matrices.

4

Particular Classes of Weakly Asymptotically Diagonal Systems

4.1 Introduction

In this chapter we demonstrate how the rather abstract results of Chapter 3 can be used to predict the convergence rates of expansion method solutions of operator equations. With this in mind, we introduce several particular classes of WAD systems and give numerical examples to show how such systems arise in practice.

Many operator equations which arise in practical situations are Hermitian. For simplicity we therefore assume in the first four sections of this chapter, as in Chapter 2, that the matrix $\mathbf{L}$ in the infinite system of equations

$$\mathbf{Lb} = \mathbf{g} \tag{4.1.1}$$

is symmetric, although the theory of Chapter 3 does not require such an assumption, as is shown by Section 4.5. We introduce the following particular types of AD matrix:

Definition 4.1. *The symmetric matrix* $\mathbf{L}$ *is AD of type A, B, C, E, if, for all* $i > j$,

$$\frac{|L_{ij}|}{\{|L_{ii}|\,|L_{jj}|\}^{1/2}} \leqslant A_{ij}, \tag{4.1.2}$$

and

$$A_{ij} = Ci^{-p}, \qquad \textit{type } A(p; C), \tag{4.1.3a}$$

$$A_{ij} = C(i-j)^{-r}i^{-p}, \qquad \textit{type } B(p, r; C), \tag{4.1.3b}$$

$$A_{ij} = Cj^{r}i^{-p}, \qquad \textit{type } C(p, r; C), \tag{4.1.3c}$$

$$A_{ij} = Ca^{i}b^{j}, \quad 0 < a, b < 1, \qquad \textit{type } E(a, b; C). \tag{4.1.3d}$$

We note, as in Chapter 2, that the bounds for the case $i < j$ follow from the symmetry.

Before applying the results of Chapter 3, we need to characterise the behaviour of the right-hand side vector $\mathbf{g}$; we therefore assume that the components g_i of $\mathbf{g}$ satisfy either

$$|g_i| \leqslant \mathscr{C} i^{-s}, \quad s > 1, \tag{4.1.4a}$$

or

$$|g_i| \leqslant \mathscr{C} d^i, \quad 0 < d < 1. \tag{4.1.4b}$$

In the next sections, we consider the convergence rates of the solutions of (4.1.1) and the related truncated system

$$\mathbf{L}^{(N)}\mathbf{a}^{(N)} = \mathbf{g}^{(N)}, \tag{4.1.5}$$

under the above assumptions.

We do not repeat the convergence analysis for type A systems, since the results obtained from the general theory of Chapter 3 are the same as those obtained from the direct analysis of Chapter 2. Instead, we proceed directly to the convergence analysis for type B systems.

4.2 Type B systems

We must first show that a matrix $\mathbf{L}$, which is AD of type B(p, r; C), is also WAD; the results of Chapter 3 are then applicable.

With notation as in Chapter 3, we proceed as follows. Since $\mathbf{L}$ is symmetric we take $\mathbf{B}^t = \mathbf{A}$, and we note that, provided $p \geqslant 0$, $r > 1$ or $p > 1$, $r \geqslant 0$, $\mathbf{A}$ has property MN, since

(a) $A_{ij} = C(i-j)^{-r} i^{-p}$ is semi-monotone increasing in j, bounded above by C, and semi-monotone decreasing in i;

(b)
$$\|\mathbf{A}\|_\infty = \max_{i=1}^{\infty} \sum_{j=1}^{i-1} C(i-j)^{-r} i^{-p},$$

$$\|\mathbf{A}\|_1 = \max_{j=1}^{\infty} \sum_{i=j+1}^{\infty} C(i-j)^{-r} i^{-p}$$

are both bounded, using the series bounds of the Appendix.

We also find, for $i > j$, that

$$(AA)_{ij} = \sum_{k=j+1}^{i-1} C^2(i-k)^{-r}i^{-p}(k-j)^{-r}k^{-p}$$

$$= C^2(i-j)^{-r}i^{-p} \sum_{k=j+1}^{i-1} \left(\frac{i-j}{(i-k)(k-j)}\right)^r k^{-p}$$

$$= C^2(i-j)^{-r}i^{-p} \sum_{k=j+1}^{i-1} \left(\frac{1}{i-k} + \frac{1}{k-j}\right)^r k^{-p}.$$

When $r \geqslant 0$, $p > 1$, we can again use the bounds of the Appendix to show that

$$(AA)_{ij} \leqslant C_A(i-j)^{-r}i^{-p}, \quad \text{where } C_A \leqslant \frac{C^2 2^{r+1-p}}{p-1}.$$

In a similar way, for $r > 1$, $p \geqslant 0$,

$$(AA)_{ij} \leqslant C'_A(i-j)^{-r}i^{-p}, \quad \text{where } C'_A \leqslant C^2\left\{2^{r+1} + 2\left(\frac{r+1}{r-1}\right)\right\}.$$

Hence the matrix $\mathbf{L}$ is SAD with property $M\tilde{N}$, and provided C of (4.1.3b) is sufficiently small, the results of Chapter 3 are applicable, since then the conditions of Section 3.5 are satisfied.

We find, from Corollary 3.5, that

$$|b_i| \leqslant \alpha_1|g_i| + \alpha_2 \sum_{k=1}^{i-1} A_{ik}|g_k|$$

$$\leqslant \alpha_1\mathscr{C}i^{-s} + \alpha_2 \sum_{k=1}^{i-1} C(i-k)^{-r}i^{-p}\mathscr{C}k^{-s}.$$

Bounding this summation, we obtain

$$|b_i| \leqslant \beta_1 i^{-q_1}, \quad p \geqslant 0, \quad r > 1, \tag{4.2.1a}$$

$$\leqslant \beta_2 i^{-q_2}, \quad p > 1, \quad r \geqslant 0, \tag{4.2.1b}$$

where

$$q_1 = \min\{s, p+r\},$$

and

$$q_2 = \min\{s, p\}.$$

We may also bound the differences $|b_i - a_i^{(N)}|$ using the result of Corollary 3.6,

$$|b_i - a_i^{(N)}| \leqslant C_{i,N+1}\left\{\alpha_3 \sum_{k=1}^{N} A_{N+1,k}|g_k| + \alpha_4|g_{N+1}|\right\},$$

where

$$C_{i,N+1} = \sum_{j=N+1}^{\infty} C(j-i)^{-r}j^{-p}.$$

Hence

$$|b_i - a_i^{(N)}| \leqslant \sum_{j=N+1}^{\infty} C(j-i)^{-r} j^{-p} \left\{ \alpha_3 \sum_{k=1}^{N} C(N+1-k)^{-r}(N+1)^{-p} \mathscr{C} k^{-s} + \alpha_4 \mathscr{C}(N+1)^{-s} \right\},$$

and we can again use the bounds of the Appendix to obtain the result,

$$|b_i - a_i^{(N)}| \leqslant \beta_3 N^{-(p+q_1)}(N+1-i)^{-r+1}, \quad p \geqslant 0, \quad r > 1, \quad (4.2.2a)$$

$$\leqslant \beta_4 N^{-(p+q_2)+1}(N+1-i)^{-r}, \quad p > 1, \quad r \geqslant 0, \quad (4.2.2b)$$

where q_1 and q_2 are defined above.

The bounds, (4.2.1), (4.2.2), allow the *a priori* prediction of vertical, horizontal and diagonal convergence rates of an algebraic system with a coefficient matrix which is AD of type B. Not surprisingly, on setting $r = 0$, we obtain the results of Section 2.3.

We may also consider the case when the given algebraic system arises from a variational or Galerkin solution of the operator equation (1.2.1) and apply the results of Section 3.6 to bound some norm of the error $e_N = f - f_N$. We then find from Corollary 3.7 that

$$\|e_N\|_{\mathscr{L}}^2 \leqslant \alpha_5 N \left\{ \sum_{k=1}^{N} A_{N+1,k}|g_k| + |g_{N+1}| \right\}^2 + \alpha_6 \sum_{i=N+1}^{\infty} \left\{ \sum_{k=1}^{i-1} A_{ik}|g_k| + |g_i| \right\}^2,$$

and we can again use the summation bounds of the Appendix to obtain

$$\|e_N\|_{\mathscr{L}}^2 \leqslant \beta_5 N^{-2q_1+1}, \quad p \geqslant 0, r > 1, \quad (4.2.3a)$$

$$\leqslant \beta_6 N^{-2q_2+1}, \quad p > 1, r \geqslant 0. \quad (4.2.3b)$$

Finally, Theorem 3.9 shows that the error in the natural norm $\|e_N\|^2$ also satisfies the bounds of (4.2.3a, b) provided that γ_i^{-2} is bounded above $\forall i$.

We now return to the second numerical example of Section 2.6. We see, from (1.5.9), that the matrix $\mathbf{L}$ is AD of type B(2, 2; $8\lambda/\pi$), and that in the normalisation employed in Chapter 3,

$$|g_i| \leqslant \mathscr{C} i^{-2}.$$

Hence for λ sufficiently small (4.2.1a), (4.2.2a) provide the bounds

$$|b_i| \leqslant \beta_1 i^{-2}$$

$$|b_i - a_i^{(N)}| \leqslant \beta_3 N^{-4}(N+1-i)^{-1}.$$

Then in the normalisation used in the solution of the problem in Section 1.5, we obtain the bounds

$$|\bar{b}_i| \leqslant \gamma_1 i^{-3} \quad (4.2.4)$$

$$|\bar{b}_i - \bar{a}_i^{(N)}| \leqslant \gamma_2 N^{-4} i^{-1}(N + 1 - i)^{-1}. \quad (4.2.5)$$

Hence, we estimate the vertical, horizontal and diagonal convergence rates to be 3, 5 and 5, respectively. In each case, the measured convergence rates of (1.5.7) agree well with those predicted above. Thus we see that taking full account of the structure of the matrix **L** allows us to find sharp convergence rate estimates.

4.3 Type C systems

As in Section 4.2 we can show that a matrix **L**, which is AD of type C($p, r; C$), is also SAD and has property MN, provided

$$p > r + 1 \geqslant 1. \quad (4.3.1)$$

Unfortunately, in a practical calculation, the type C matrices which occur seem usually to have parameters $p = r$, and our analysis is unable to cater for such a case. Nonetheless, we proceed to quote the results obtained when the parameters satisfy the restriction (4.3.1). These results are again obtained from those of Chapter 3, by inserting the appropriate summation bounds of the Appendix.

We find from Corollary 3.5,

$$|b_i| \leqslant \beta_7 i^{-q_3}, \quad (4.3.2)$$

where

$$q_3 = \min\{s, p\}. \quad (4.3.3)$$

From Corollary 3.6 we may bound the differences $|b_i - a_i^{(N)}|$ as,

$$|b_i - a_i^{(N)}| \leqslant \beta_8 i^r N^{-(p+q_3)+1}, \quad (4.3.4)$$

where q_3 is given by (4.3.3). These bounds provide vertical, horizontal and diagonal convergence rate estimates of q_3, $p + q_3 - 1$ and $p + q_3 - r - 1$. Contrast this result with Section 2.3, where we found rather surprisingly that the horizontal and diagonal convergence rate estimates were the same for a type A system.

When the algebraic system arises from a variational or Galerkin solution of (1.2.1), we obtain the following bounds on the norm of the error $e_N = f - f_N$. From Corollary 3.7,

$$\|e_N\|_{\mathscr{L}}^2 \leqslant \beta_9 N^{-2q_3+1}, \quad (4.3.5)$$

where q_3 is given by (4.3.3), while from Theorem 3.9,

$$\|e_N\|^2 \leqslant \beta_{10} N^{-2q_3+1}, \tag{4.3.6}$$

provided γ_i^{-2} is bounded above $\forall i$.

The results of Sections 4.2 and 4.3 can also be obtained by a direct analysis of AD matrices of types B and C (see Freeman, Delves and Reid, 1974).

4.4 Type E systems

The assumption that **L** is AD of type E($a, b; C'$) implies that the off-diagonal elements of **L** decay exponentially; we therefore assume that the right-hand side vector **g** displays a similar behaviour,

$$|g_i| \leqslant \mathscr{C} d^i, \quad 0 < d < 1, \tag{4.4.1}$$

and say that g_i exhibits an i-exponential convergence rate d.

It is easy to show that **L** is SAD provided $a \leqslant b$.

For $i \geqslant j$,

$$\begin{aligned}(AA^t)_{ij} &= \sum_{k=1}^{j-1} C'a^i b^k C' a^j b^k \\ &= C'^2 a^i b^j \left\{ \left(\frac{a}{b}\right)^j \frac{b^2}{(1-b^2)} - \frac{a^j b^j}{(1-b^2)} \right\} \\ &\leqslant C' a^i b^j \left\{ \frac{C' b^2}{(1-b^2)} \right\}.\end{aligned}$$

Using similar results for $i < j$, we find that (3.2.3), (3.2.4) are satisfied with

$$C = \frac{C'b^2}{1-b^2},$$

$$C_1 = C'a^2 b^2.$$

We also find that **A** satisfies hypothesis strong H1 (Definition 3.2) with

$$C_A = \frac{C'a^2b^2}{1-ab}, \quad \text{since, for } i > j+1,$$

$$\begin{aligned}(AA)_{ij} &= \sum_{k=j+1}^{i-1} C'a^i b^k C' a^k b^j \\ &= C'^2 a^i b^j \left\{ \frac{(ab)^{j+1} - (ab)^i}{1-ab} \right\} \\ &\leqslant C' a^i b^j \left\{ \frac{C'(ab)^{j+1}}{1-ab} \right\}.\end{aligned}$$

Finally, **A** satisfies (3.3.3) with $\Omega = C'ba^3/(1 - a^2)$ and $\Omega_1 = C'ab$, since, for $i \geqslant j$,

$$(A^t A)_{ij} = \sum_{k=i+1}^{\infty} C'a^k b^i C' a^k b^j$$
$$= C'^2 b^{i+j} \frac{a^{2i+2}}{1 - a^2}$$
$$= C'a^i b^j \left\{ C' \frac{b^i a^{i+2}}{1 - a^2} \right\}.$$

As in Section 4.2, as long as C' is sufficiently small, we can use the results of Chapter 3 to estimate the convergence rates.

We obtain from Theorem 3.6

$$|b_i| \leqslant K'_0 \mathscr{C} d^i + K'_1 \sum_{k=1}^{i-1} C'a^i b^k \mathscr{C} d^k + K'_2 \sum_{k=i+1}^{\infty} C'a^k b^i \mathscr{C} d^k$$
$$\leqslant \mathscr{C} K'_0 d^i + C'\mathscr{C} K'_1 a^i \frac{bd(1 - (bd)^{i-1})}{1 - bd} + C'\mathscr{C} K'_2 b^i \frac{(ad)^{i+1}}{1 - ad}$$
$$\leqslant E_1 m_1^i, \tag{4.4.2}$$

where

$$m_1 = \max\{a, b, d\},$$
$$E_1 = \mathscr{C} \max \left\{ K'_0, \frac{C'K'_1 bd}{1 - bd}, \frac{C'K'_2 (ad)^2}{1 - ad} \right\}.$$

Thus we see that the components b_i of the solution vector **b** exhibit i-exponential convergence at a rate bounded by the i- and j-exponential convergence rates of the elements L_{ij} of the matrix **L** and the i-exponential convergence rate of the components g_i of the right-hand side vector **g**.

In a similar way, we use Theorem 3.7 to bound the differences $|b_i - a_i^{(N)}|$. We note that A_{ij} is semi-monotone decreasing in i, and we may thus use Corollary 3.6 to obtain

$$|b_i - a_i^{(N)}| \leqslant C_{i, N+1} \left\{ \gamma_A \gamma_B \sum_{j=1}^{N} C'a^{N+1} b^j \mathscr{C} d^j + K'_2 \mathscr{C} d^{N+1} \right\},$$

where

$$C_{i, N+1} = \sum_{j=N+1}^{\infty} C'a^j b^i$$
$$= C'b^i \frac{a^{N+1}}{(1 - a)}.$$

Thus,

$$\begin{aligned}|b_i - a_i^{(N)}| &\leqslant \frac{C'b^i a^{N+1}}{(1-a)}\left\{\gamma_A\gamma_B C'\mathscr{C}a^{N+1}\frac{bd(1-(bd)^N)}{(1-bd)} + \mathscr{C}K_2'd^{N+1}\right\}\\ &\leqslant E_2 b^i a^{N+1} m_2^{N+1},\end{aligned} \tag{4.4.3}$$

where

$$m_2 = \max\{a, d\},$$

$$E_2 = \frac{\mathscr{C}C'}{(1-a)}\max\left\{\gamma_A\gamma_B C'\frac{bd}{(1-bd)}, K_2'\right\}.$$

We thus find that the horizontal convergence rate is described by the bound

$$|b_i - a_i^{(N)}| \leqslant E_3(am_2)^{N+1},$$

where $E_3 = E_2 b^i$, and i is fixed, and the diagonal convergence rate is described by the bound

$$|b_i - a_i^{(i)}| \leqslant E_2 b^i (am_2)^{i+1}.$$

We note that in general the diagonal convergence is faster than the horizontal convergence.

We also consider the case when a variational or Galerkin solution of (1.2.1) leads to the given algebraic equations. We then wish to bound some norm of the error $e_N = f - f_N$. Since $\mathbf{A}$ does not have property MN, we cannot use Theorem 3.8 to bound $\|e_N\|_{\mathscr{L}}$. We can, however, substitute the bounds (4.4.2), (4.4.3) into (2.4.6) to obtain

$$\begin{aligned}\|e_N\|_{\mathscr{L}}^2 \leqslant{}& \sum_{i=1}^{N} E_2^2 b^{2i}(am_2)^{2(N+1)} + 2\sum_{i=1}^{N}\sum_{j=i+1}^{N} E_2^2 b^i b^j (am_2)^{2(N+1)} C' a^j b^i\\ &+ 2\sum_{i=1}^{N}\sum_{j=N+1}^{\infty} E_1 E_2 b^i (am_2)^{N+1} m_1^j C' a^j b^i + \sum_{i=N+1}^{\infty} E_1^2 m_1^{2i}\\ &+ 2\sum_{i=N+1}^{\infty}\sum_{j=i+1}^{\infty} E_1^2 m_1^i m_1^j C' a^j b^i\\ \leqslant{}& E_3 m_1^{2(N+1)},\end{aligned} \tag{4.4.4}$$

where the result follows on bounding the sums and noting that $m_1 \geqslant m_2$.

In order to bound the error in the natural norm we insert the bounds (4.4.2), (4.4.3) into (2.4.10) to obtain

$$\begin{aligned}\|e_N\|^2 &\leqslant \sum_{i=1} \gamma_i^{-2} E_2^2 b^{2i}(am_2)^{2(N+1)} + \sum_{i=N+1}^{\infty} \gamma_i^{-2} E_1^2 m_1^{2i}\\ &\leqslant E_2^2 (am_2)^{2(N+1)} \sum_{i=1}^{N} \gamma_i^{-2} b^{2i} + E_1^2 \sum_{i=N+1}^{\infty} \gamma_i^{-2} m_1^{2i}.\end{aligned}$$

A bound for $\|e_N\|^2$ then follows on inserting the relevant form for γ_i^{-2}.

As an example of a type E system, we consider the integral equation

$$f(x) = g(x) + \lambda \int_0^a \frac{1}{1 - xt} f(t)\, dt, \tag{4.4.5}$$

where $0 < a < 1$ and $0 < x < a$. If we let

$$b_j = \int_0^a t^{j-1} f(t)\, dt, \tag{4.4.6}$$

then on expanding the kernel of (4.4.5) as a power series in xt, we see that (4.4.5) has the formal solution

$$f(x) = g(x) + \lambda \sum_{i=1}^{\infty} b_i x^{i-1}. \tag{4.4.7}$$

If we multiply (4.4.5) by x^{i-1}, $i = 1, 2, \ldots$, and integrate over $[0, a]$, we obtain the defining system of equations

$$\mathbf{Lb} = \mathbf{g}, \tag{4.4.8}$$

where

$$L_{ij} = \delta_{ij} - \lambda \frac{a^{i+j-1}}{i+j-1},$$

$$g_i = \int_0^a g(x) x^{i-1}\, dx.$$

An approximate solution to (4.4.5) is given by

$$f_N(x) = g(x) + \lambda \sum_{i=1}^{N} a_i^{(N)} x^{i-1},$$

where the coefficients $a_i^{(N)}$, $i = 1, 2, \ldots, N$, are given by the truncated $N \times N$ system

$$\mathbf{L}^{(N)} \mathbf{a}^{(N)} = \mathbf{g}^{(N)}.$$

If we let $\mathbf{b} = \mathbf{D}^{-1/2} \bar{\mathbf{b}}$, where $\mathbf{D}^{1/2} = \mathbf{diag}(L_{ii})^{1/2}$, then (4.4.8) is equivalent to the normalized system

$$\mathbf{D}^{-1/2} \mathbf{L} \mathbf{D}^{-1/2} \bar{\mathbf{b}} = \mathbf{D}^{-1/2} \mathbf{g},$$

$$\bar{\mathbf{L}} \bar{\mathbf{b}} = \bar{\mathbf{g}},$$

where

$$|\bar{L}_{ii}| = 1,$$

$$|\bar{L}_{ij}| \leqslant C' \frac{a^{i+j-1}}{i+j-1}$$

with

$$C' = |\lambda| \max_{i=1}^{\infty} |(L_{ii})^{-1}|$$

$$= \frac{|\lambda|}{1 - \lambda a}, \quad \lambda a < 1.$$

Hence $\bar{L}$ is AD of type E $(a, a; C'a^{-1})$. If we also assume that $g(x)$ is bounded on $[0, a]$, then there exists a constant M such that $|g(x)| < M$, $\forall x \in [0, a]$, and hence

$$|g_i| = \left| \int_0^a g(x)x^{i-1} \, dx \right| \leqslant \frac{M}{i} a^i,$$

and

$$|\bar{g}_i| = |D_{ii}^{-1/2}| \, |g_i|$$

$$\leqslant \frac{M}{i\sqrt{(1 - \lambda a)}} a^i \leqslant \mathscr{C} a^i.$$

The constants required to demonstrate that $\bar{L}$ is SAD are

$$C = \frac{C'a}{1 - a^2}, \; C_1 = C'a^3,$$

$$C_A = \frac{C'a^3}{1 - a^2},$$

$$\Omega = \frac{C'a^3}{1 - a^2}, \; \Omega_1 = C'a.$$

For sufficiently small λ, these constants clearly satisfy the conditions of Chapter 3; we therefore consider this case first.

Case 1: Small λ

We find the bounds

$$|\bar{b}_i| \leqslant E_4 a^i \tag{4.4.9}$$

and

$$|\bar{b}_i - \bar{a}_i^{(N)}| \leqslant E_5 a^{i+2(N+1)} \tag{4.4.10}$$

Table 4.1 lists numerical results obtained with $g(x) = x$, $a = 0{\cdot}5$ and $\lambda = 0{\cdot}1$. The observed vertical convergence rate is slightly faster than that predicted by (4.4.9), while the observed horizontal and diagonal convergence rates agree very well with those predicted by (4.4.10).

Case 2: Large λ

For any a and sufficiently large λ, condition (3.2.7) of Theorem 3.1 will fail

Table 4.1

$\mathbf{a}^{(N)}$

$N = 1$	2	3	4	5	6	7	8	60
0·13157895	0·13215131	0·13222295	0·13223369	0·13223547	0·13223579	0·13223585	0·13223586	0·13223587
	0·43499807 × 10^{-1}	0·43526210 × 10^{-1}	0·43530420 × 10^{-1}	0·43531149 × 10^{-1}	0·43531283 × 10^{-1}	0·43531308 × 10^{-1}	0·43531313 × 10^{-1}	0·43531314 × 10^{-1}
		0·16254097 × $10^{\ 1}$	0·16255840 × 10^{-1}	0·16256150 × 10^{-1}	0·16256208 × 10^{-1}	0·16256219 × 10^{-1}	0·16256221 × 10^{-1}	0·16256222 × 10^{-1}
			0·64887791 × 10^{-2}	0·64889143 × 10^{-2}	0·64889400 × 10^{-2}	0·64889451 × 10^{-2}	0·64889461 × 10^{-2}	0·64889464 × 10^{-2}
				0·27003398 × 10^{-2}	0·27003514 × 10^{-2}	0·27003536 × 10^{-2}	0·27003541 × 10^{-2}	0·27003542 × 10^{-2}
					0·11563323 × 10^{-2}	0·11563334 × 10^{-2}	0·11563336 × 10^{-2}	0·11563336 × 10^{-2}
						0·50559623 × 10^{-3}	0·50559633 × 10^{-3}	0·50559636 × 10^{-3}
							0·22460981 × 10^{-3}	0·22460982 × 10^{-3}

to hold, and the preceding convergence analysis is not directly applicable. The difficulty stems from the first few rows and columns of **L**, and we therefore partition (4.4.8) in the form

$$\begin{bmatrix} \mathbf{L}_{11} & \mathbf{L}_{12} \\ \mathbf{L}_{12}^t & \mathbf{L}_{22} \end{bmatrix} \begin{bmatrix} \mathbf{b}_1 \\ \mathbf{b}_2 \end{bmatrix} = \begin{bmatrix} \mathbf{g}_1 \\ \mathbf{g}_2 \end{bmatrix}, \tag{4.4.11}$$

where $\mathbf{L}_{11}$ is an $M \times M$ matrix and $\mathbf{b}_1$ and $\mathbf{g}_1$ are M vectors. Eliminating $\mathbf{b}_1$ from (4.4.11), we obtain the reduced system

$$\mathbf{L}'\mathbf{b}_2 = \mathbf{g}', \tag{4.4.12}$$

where

$$\mathbf{L}' = \mathbf{L}_{22} - \mathbf{L}_{12}^t \mathbf{L}_{11}^{-1} \mathbf{L}_{12},$$

$$\mathbf{g}' = \mathbf{g}_2 - \mathbf{L}_{12}^t \mathbf{L}_{11}^{-1} \mathbf{g}_1.$$

Now,

$$L'_{ij} = L_{i+M,\, j+M} - \sum_{k,\, l=1}^{M} L_{i+M,\, k} (L_{11}^{-1})_{kl} L_{l,\, j+M},$$

and it is simple to show that, for $i \neq j$,

$$|L'_{ij}| \leqslant \left\{ \frac{\lambda}{2M+1} + \lambda^2 Q_M \right\} a^{2M} a^{i+j-1},$$

where

$$Q_M = \max_{i,\, j} \sum_{k,\, l=1}^{M} |(L_{11}^{-1})_{kl}| \frac{a^{k+l-1}}{(i+k+M-1)(j+l+M-1)}.$$

Also,

$$|L'_{ii}| \geqslant d_M = \min_{i=1}^{\infty} \left| 1 - \left\{ \frac{\lambda}{2i+2M-1} + \lambda^2 Q_M \right\} a^{2M+2i-1} \right|$$

$$= 1 - \left(\frac{\lambda}{2M+1} + \lambda^2 Q_M \right) a^{2M+1},$$

provided that

$$\left(\frac{\lambda}{2M+1} + \lambda^2 Q_M \right) a^{2M+1} < 1.$$

Thus $\mathbf{L}'$ is AD of type $E(a, a; C_M)$, where

$$C_M = \frac{a^{2M-1}}{d_M} \left\{ \frac{\lambda}{2M+1} + \lambda^2 Q_M \right\}.$$

The constant Q_M can be evaluated numerically. It turns out to be only a slowly increasing function of M, except for values of λ in the neighbourhood of an eigenvalue of the integral operator. Hence, for "almost all" λ, we can pick M so that the constant C_M is sufficiently small for the conditions of Chapter 3 to be satisfied. Thus we can apply the earlier convergence theory to (4.4.12) and the related $(N-M) \times (N-M)$ system

$$\mathbf{L}^{(N)'}\mathbf{a}_2^{(N)} = \mathbf{g}^{(N)'}$$

We then obtain the bounds

$$|(\bar{b}_2)_i| \leqslant E_4' a^i,$$

and

$$|(\bar{b}_2)_i - (\bar{a}_2^{(N)})_i| \leqslant E_5' a^{i+2(N+1)}.$$

These bounds can be rewritten in the form

$$|\bar{b}_i| \leqslant E_4'' a^i, \tag{4.4.13}$$

$$|\bar{b}_i - \bar{a}_i^{(N)}| \leqslant E_5'' a^{i+2(N+1)}. \tag{4.4.14}$$

Comparing these bounds with (4.4.9), (4.4.10) we see that there is in fact no need to carry out the partitioning process explicitly, the convergence results remaining valid for (almost all) λ. Table 4.2 gives numerical results for the same system as before, but with $\lambda = 0{\cdot}5$, so that the analysis of case 1 is not applicable. Again, the observed horizontal and diagonal convergence rates agree well with those predicted by (4.4.14), while the observed vertical convergence rate is slightly faster than that predicted by (4.4.13).

4.5 Unsymmetric systems

We now remove the assumption that the matrix $\mathbf{L}$ is symmetric, having noted in Section 4.1 that the theory of Chapter 3 does not require such an assumption. We introduce the following types of unsymmetric AD matrices:

Definition 4.2. *The matrix* $\mathbf{L}$ *is LAD (lower AD) of type A, B, if, for all* $i > j$,

$$\frac{|L_{ij}|}{\{|L_{ii}|\,|L_{jj}|\}^{1/2}} \leqslant A_{ij}, \tag{4.5.1}$$

and

$$A_{ij} = Ci^{-p}, \qquad \textit{type } A(p;\, C), \tag{4.5.2a}$$

$$A_{ij} = C(i-j)^{-r} i^{-p}, \qquad \textit{type } B(p, r;\, C). \tag{4.5.2b}$$

TABLE 4.2

$\mathbf{a}^{(N)}$

$N = 1$	2	3	4	5	6	7	8	60
0·16666667	0·17112299	0·17168581	0·17176982	0·17178376	0·17178624	0·17178670	0·17178679	0·17178681
	$0·53475936 \times 10^{-1}$	$0·53668902 \times 10^{-1}$	$0·53699223 \times 10^{-1}$	$0·53704420 \times 10^{-1}$	$0·53705366 \times 10^{-1}$	$0·53705545 \times 10^{-1}$	$0·53705581 \times 10^{-1}$	$0·53705590 \times 10^{-1}$
		$0·19682584 \times 10^{-1}$	$0·19694752 \times 10^{-1}$	$0·19696890 \times 10^{-1}$	$0·19697287 \times 10^{-1}$	$0·19697363 \times 10^{-1}$	$0·19697378 \times 10^{-1}$	$0·19697382 \times 10^{-1}$
			$0·77897530 \times 10^{-2}$	$0·77906685 \times 10^{-2}$	$0·77908408 \times 10^{-2}$	$0·77908744 \times 10^{-2}$	$0·77908811 \times 10^{-2}$	$0·77908828 \times 10^{-2}$
				$0·32241620 \times 10^{-2}$	$0·32242384 \times 10^{-2}$	$0·32242534 \times 10^{-2}$	$0·32242565 \times 10^{-2}$	$0·32242573 \times 10^{-2}$
					$0·13755632 \times 10^{-2}$	$0·13755700 \times 10^{-2}$	$0·13755714 \times 10^{-2}$	$0·13755718 \times 10^{-2}$
						$0·59986216 \times 10^{-3}$	$0·59986280 \times 10^{-3}$	$0·59986297 \times 10^{-3}$
							$0·26595699 \times 10^{-3}$	$0·26595707 \times 10^{-3}$

Definition 4.3. *The matrix* $\mathbf{L}$ *is UAD (upper AD) of a given type if* $\mathbf{L}^t$ *is LAD of the same type.*

It is possible to construct matrices which are LAD of one type and UAD of another. For simplicity, however, we restrict our attention to matrices which are LAD and UAD of the same type, although with possibly different constants p, r, C which we label with subscripts L, U. In an obvious notation we then refer to matrices of the following types:

$$\text{type A}(p_L, p_U; C_L, C_U),$$

$$\text{type B}(p_L, p_U, r_L, r_U; C_L, C_U).$$

The results for type A systems can be obtained from the type B results by setting $r_L = r_U = 0$, and hence we simply consider the latter systems. Rather than repeating an analysis very similar to that of Section 4.2 we simply quote the results for the case when $\mathbf{L}$ is AD of type $\mathrm{B}(p_L, p_U, r_L, r_U; C_L, C_U)$ and $\mathbf{g}$ satisfies (4.1.4a). From Corollary 3.5, we find that

$$|b_i| \leqslant \beta_1 i^{-q_1}, \quad p_L, p_U \geqslant 0, \quad r_L, r_U > 1, \tag{4.5.3a}$$

$$\leqslant \beta_2 i^{-q_2}, \quad p_L, p_U > 1, \quad r_L, r_U \geqslant 0, \tag{4.5.3b}$$

where

$$q_1 = \min\{s, p_L + r_L\},$$

$$q_2 = \min\{s, p_L\}.$$

We also find from Corollary 3.6 that

$$|b_i - a_i^{(N)}| \leqslant \beta_3 N^{-(p_U + q_1)}(N + 1 - i)^{-r_U + 1}, \quad p_L, p_U \geqslant 0, \quad r_L, r_U > 1, \tag{4.5.4a}$$

$$\leqslant \beta_4 N^{-(p_U + q_2) + 1}(N + 1 - i)^{-r_U}, \quad p_L, p_U > 1, \quad r_L, r_U \geqslant 0, \tag{4.5.4b}$$

where q_1, q_2 are defined above.

If the algebraic system arises from a variational or Galerkin solution of the operator equation (1.2.1) then we may proceed as in Section 4.2 to bound the error $e_N = f - f_N$. We find that

$$\|e_N\|_{\mathscr{L}}^2 \leqslant \beta_5 N^{-2q_1 + 1}, \quad p_L, p_U \geqslant 0, r_L, r_U > 1, \tag{4.5.5a}$$

$$\leqslant \beta_6 N^{-2q_2 + 1}, \quad p_L, p_U > 1, \quad r_L, r_U \geqslant 0. \tag{4.5.5b}$$

We can show that $\|e_N\|^2$ also satisfies (4.5.5a, b) provided that γ_i^{-2} is bounded above $\forall i$.

It should be noted that the results of this section exhibit an interesting asymmetry between the lower and upper halves of the matrix. In particular, the bounds (4.5.3) on $|b_i|$, and (4.5.5) on $\|e_N\|_{\mathscr{L}}^2$ are independent of the upper half parameters p_U, r_U; while the bound on $\|e_N\|^2$ depends strongly on **g** and the lower half of **L**, but only weakly on the upper half of **L**. This asymmetry is particularly important in the applications of Chapter 13.

The AD of type C and E matrices of Definition 4.1, and the results of Sections 4.3 and 4.4 can also be easily extended to the unsymmetric case.

An example of an unsymmetric system of type B is a problem in the theory of water waves considered by Ursell (1974). It can be shown that the unknown coefficients, p_{2j}, $j = 1, 2, \ldots$, in a series expansion for the velocity potential satisfy the system of equations

$$c_{2i} = 2ip_{2i} + Ka \sum_{j=1}^{\infty} (2jp_{2j})b_{ij}, \tag{4.5.6}$$

where

$$|c_{2i}| < \alpha/i^2,$$

$$b_{ij} = \frac{32}{\pi}(-1)^{i+j} \frac{i^2(j-1)}{(4i^2-1)(2j-1)(2i+2j-1)(2i-2j+1)},$$

and $i = 1, 2, \ldots$.

Setting $y_i = c_{2i}$, $x_i = 2ip_{2i}$, (4.5.6) becomes

$$(\mathbf{I} + \lambda\mathbf{B})\mathbf{x} = \mathbf{y}, \tag{4.5.7}$$

where

$$\lambda = (32/\pi)\ Ka$$

$$B_{ij} = (-1)^{i+j} \frac{i^2(j-1)}{(4i^2-1)(2j-1)(2i+2j-1)(2i-2j+1)},$$

$$|y_i| \leqslant \alpha i^{-2}.$$

Now, for $i, j \geqslant 1$,

$$\frac{i^2}{4i^2-1} \leqslant \frac{1}{3} \quad \text{and} \quad \frac{j-1}{2j-1} < \frac{1}{2},$$

and for $j \geqslant i + 1$,

$$\left(j - i - \frac{1}{2}\right) \geqslant \frac{1}{2}(j-i).$$

Hence,

$$|B_{ij}| < \frac{1}{24}(i-j)^{-2}, \quad i > j,$$

$$< \frac{1}{12}(j-i)^{-2}, \quad i < j.$$

Also,

$$B_{ii} = 0, \qquad i = 1,$$
$$\geqslant \frac{1}{48i}, \quad i \geqslant 2,$$

and hence, for $\lambda > 0$, $\mathbf{A} = \mathbf{I} + \lambda\mathbf{B}$ satisfies

$$A_{ii} = 1, \qquad i = 1,$$
$$\geqslant 1 + \frac{\lambda}{48i} \geqslant 1, \quad i \geqslant 2,$$
$$|A_{ij}| = \lambda|B_{ij}| < \frac{\lambda}{24}(i-j)^{-2}, \quad i > j,$$
$$< \frac{\lambda}{12}(j-i)^{-2}, \quad i < j.$$

It therefore follows that $\mathbf{A}$ is AD of type B(0, 2, 0, 2; $\lambda/24$, $\lambda/12$).

Hence for $\lambda = (32/\pi)Ka$ sufficiently small we can apply the result (4.5.3a) to (4.5.7) to obtain

$$|x_i| \leqslant \beta_1 i^{-q_1},$$

where $q_1 = \min\{2, 0 + 2\} = 2$.
Therefore

$$|p_{2i}| = \frac{|x_i|}{2i} \leqslant \frac{\beta_1}{2} i^{-3},$$

which agrees with the result of Ursell (1974).

5

Stability Analysis of Weakly Asymptotically Diagonal Systems

5.1 Introduction

In this chapter we consider the stability properties of an expansion method solution of the linear operator equation (1.2.1). We recall from Section 1.2 that such a solution proceeds by approximating f by

$$f_N = \sum_{i=1}^{N} a_i^{(N)} h_i, \tag{5.1.1}$$

where the coefficients $a_i^{(N)}$ are given by the $N \times N$ system of equations

$$\mathbf{L}^{(N)}\mathbf{a}^{(N)} = \mathbf{g}^{(N)}, \tag{5.1.2}$$

where

$$L_{ij} = (h_i, \mathscr{L}h_j), \tag{5.1.3a}$$

$$g_i = (h_i, g). \tag{5.1.3b}$$

We assume for simplicity of presentation that $\mathscr{L}$ is a Hermitian operator, and therefore $\mathbf{L}^{(N)}$ is a symmetric matrix; most of the results extend to the case when $\mathbf{L}^{(N)}$ is not symmetric.

In Chapter 3, we show that the convergence properties of (5.1.2) and the related infinite system

$$\mathbf{Lb} = \mathbf{g}, \tag{5.1.4}$$

and the convergence properties of the approximate solution f_N to f, can be characterised by assuming that the matrix $\mathbf{L}$ is WAD. However, those discussions assume that the inner products (5.1.3a,b) are known exactly.

In a practical calculation, numerical errors will exist in these inner products, arising either from the finite machine arithmetic or from a numerical quadrature approximation to the inner products. We now look at the effects of these errors on the computed solutions of WAD systems; that is, we study the perturbed $N \times N$ system,

$$(\mathbf{L} + \delta\mathbf{L})(\mathbf{a} + \delta\mathbf{a}) = \mathbf{g} + \delta\mathbf{g}, \tag{5.1.5}$$

where, here and throughout this chapter, we omit the superscript N.

5.2 Bounds on $\mathbf{L}^{-1}$

We recall from Sections 3.2 and 3.3 that the matrix $\mathbf{L}$ of (5.1.2) is WAD if it can be factorised as

$$\mathbf{L} = \mathbf{\Lambda}\tilde{\mathbf{L}}\mathbf{\Lambda}, \tag{5.2.1}$$

and the matrix $\tilde{\mathbf{L}}$ is a normalised WAD matrix (see Definition 3.3 and equation (3.2.5)). We note that the symmetry of $\mathbf{L}$ allows us to take $\mathbf{B} = \mathbf{A}^t$ and $\overline{\mathbf{B}} = \overline{\mathbf{A}}^t$, and that when $\mathbf{\Lambda} = \mathbf{I}$, $\mathbf{L}$ is a normalised WAD matrix. Clearly, the matrix $\mathbf{\Lambda}$ depends on the normalisation of the set $\{h_i\}$, and in terms of the normalised matrix $\tilde{\mathbf{L}}$, (5.1.5) can be written as

$$(\tilde{\mathbf{L}} + \delta\tilde{\mathbf{L}})(\tilde{\mathbf{a}} + \delta\tilde{\mathbf{a}}) = (\tilde{\mathbf{g}} + \delta\tilde{\mathbf{g}}), \tag{5.2.2}$$

where

$$\tilde{\mathbf{a}} = \mathbf{\Lambda}\mathbf{a}, \qquad \delta\tilde{\mathbf{a}} = \mathbf{\Lambda}\delta\mathbf{a}, \tag{5.2.3a}$$

$$\tilde{\mathbf{g}} = \mathbf{\Lambda}^{-1}\mathbf{g}, \quad \delta\tilde{\mathbf{g}} = \mathbf{\Lambda}^{-1}\delta\mathbf{g}, \tag{5.2.3b}$$

$$\delta\tilde{\mathbf{L}} = \mathbf{\Lambda}^{-1}\delta\mathbf{L}\mathbf{\Lambda}^{-1}. \tag{5.2.3c}$$

In Chapter 3, we introduce the assumption

$$|\tilde{g}_i| \leqslant \hat{g}_i, \tag{5.2.4}$$

where $\hat{\mathbf{g}}$ is a positive vector whose components $\hat{g}_i$ are monotonically decreasing in i. We can then use the techniques of Theorem 3.6 on

$$\tilde{\mathbf{a}} = \tilde{\mathbf{L}}^{-1}\tilde{\mathbf{g}}$$

to show that

$$|\tilde{a}_i| \leqslant K'_0\hat{g}_i + K'_1\bar{\mathbf{a}}_i^t\hat{\mathbf{g}} + K'_2(\bar{\boldsymbol{\alpha}}_i)^t\,\hat{\mathbf{g}}, \tag{5.2.5}$$

where $\bar{\mathbf{a}}_i^t$ denotes the i-th row of $\overline{\mathbf{A}}$ and $\bar{\boldsymbol{\alpha}}_i$ denotes the i-th column of $\overline{\mathbf{A}}$, provided that the matrix $\tilde{\mathbf{L}}$ satisfies the conditions of Corollary 3.3. Throughout this chapter, we shall assume that the matrix $\tilde{\mathbf{L}}$ satisfies the conditions of this corollary, although we note that these conditions are sufficient, but not necessary, for the analysis which follows, as the example of Section 5.6 will illustrate.

From Corollary 3.3 we obtain the following lemma.

Lemma 5.1. *Let* $\tilde{\mathbf{L}}$ *be WAD and normalised to satisfy* (3.2.5), *and let* $\bar{\mathbf{A}}$ *have property MN. Then for some* K_1, K_2 *independent of N*,

$$\|\tilde{\mathbf{L}}^{-1}\|_\infty \leqslant K_1, \tag{5.2.6}$$

and the condition number

$$\|\tilde{\mathbf{L}}\|_\infty \|\tilde{\mathbf{L}}^{-1}\|_\infty \leqslant K_2. \tag{5.2.7}$$

Proof.

$$\begin{aligned}\|\tilde{\mathbf{L}}^{-1}\|_\infty &\leqslant K'_0 + K'_1\|\bar{\mathbf{A}}\|_\infty + K'_2\|\bar{\mathbf{A}}^t\|_\infty \\ &\leqslant K'_0 + K'_1C + K'_2\Omega = K_1.\end{aligned}$$

Now,

$$\begin{aligned}|\tilde{\mathbf{L}}| &\leqslant \mathbf{I} + \mathbf{A} + \mathbf{A}^t \\ &\leqslant \mathbf{I} + \bar{\mathbf{A}} + \bar{\mathbf{A}}^t.\end{aligned}$$

Therefore

$$\|\tilde{\mathbf{L}}\|_\infty \leqslant 1 + C + \Omega,$$

and

$$\|\tilde{\mathbf{L}}\|_\infty \|\tilde{\mathbf{L}}^{-1}\|_\infty \leqslant K_1(1 + C + \Omega) = K_2.$$

From Corollary 3.3 we also know that the individual elements of $\tilde{\mathbf{L}}^{-1}$ are bounded,

$$|\tilde{\mathbf{L}}^{-1}| \leqslant K'_0\mathbf{I} + K'_1\bar{\mathbf{A}} + K'_2\bar{\mathbf{A}}^t. \tag{5.2.8}$$

We note that the symmetry of $\mathbf{L}$ and $\tilde{\mathbf{L}}$ allows us to take $K'_2 = K'_1$ in (5.2.8) and (5.2.5).

5.3 Bounds on $\|\delta\mathbf{a}\|_\infty$

From (5.1.5) we find that, provided the denominator is positive,

$$\|\delta\mathbf{a}\| \leqslant \frac{\|\mathbf{L}^{-1}\|\,\{\|\delta\mathbf{g}\| + \|\delta\mathbf{L}\|\,\|\mathbf{a}\|\}}{1 - \|\mathbf{L}^{-1}\|\,\|\delta\mathbf{L}\|}, \tag{5.3.1}$$

where we have used the fact that $\mathbf{a}$ satisfies (5.1.2).

We can then use the bounds of (5.2.6) to prove the following theorem.

Theorem 5.1. *Let* $\mathbf{L}$ *be WAD and let* $\bar{\mathbf{A}}$ *have property MN. Also let* $\|\Lambda^{-1}\|_\infty \leqslant \gamma_N$ *and* $\tilde{\mathbf{g}}$ *satisfy* (5.2.4) *with* $\max_{i=1}^N \hat{g}_i = \hat{g}_p$, $p < N$. *Then*

$$\|\delta\mathbf{a}\|_\infty \leqslant \frac{\gamma_N^2 K_1(\|\delta\mathbf{g}\|_\infty + K_3\gamma_N \|\delta\mathbf{L}\|_\infty)}{1 - K_1\gamma_N^2 \|\delta\mathbf{L}\|_\infty}, \tag{5.3.2}$$

where K_3 *is a positive constant, independent of* N.

Proof. Now $\mathbf{L}^{-1} = \Lambda^{-1}\tilde{\mathbf{L}}^{-1}\Lambda^{-1}$ and hence, using (5.2.6),

$$\begin{aligned}\|\mathbf{L}^{-1}\|_\infty &\leqslant \|\Lambda^{-1}\|_\infty \|\tilde{\mathbf{L}}^{-1}\|_\infty \|\Lambda^{-1}\|_\infty \\ &\leqslant \gamma_N^2 K_1.\end{aligned}$$

Also $\mathbf{a} = \Lambda^{-1}\tilde{\mathbf{a}}$, and

$$\begin{aligned}\|\tilde{\mathbf{a}}\|_\infty &\leqslant \max_{i=1}^{N} \{K_0'\hat{g}_i + K_1'\bar{\mathbf{a}}_i^t\,\hat{\mathbf{g}} + K_1'(\bar{\boldsymbol{\alpha}}_i)^t\,\hat{\mathbf{g}}\} \\ &\leqslant \max_{i=1}^{N}\left\{K_0'\hat{g}_i + K_1' \sum_{k=1}^{i-1} \bar{A}_{ik}\hat{g}_k + K_1' \sum_{k=i+1}^{N} \bar{A}_{ki}\hat{g}_k\right\} \\ &\leqslant K_0'\hat{g}_p + K_1'C\hat{g}_p + K_1'\Omega\hat{g}_p = K_3.\end{aligned}$$

Hence, $\|\mathbf{a}\|_\infty \leqslant \gamma_N K_3$, and we thus find,

$$\|\delta\mathbf{a}\|_\infty \leqslant \frac{\gamma_N^2 K_1(\|\delta\mathbf{g}\|_\infty + K_3\gamma_N \|\delta\mathbf{L}\|_\infty)}{1 - \gamma_N^2 K_1 \|\delta\mathbf{L}\|_\infty}.$$

Comments

Theorem 5.1 yields a direct estimate of the effect of round-off errors on the solution vector $\mathbf{a}$, and hence characterises the numerical *stability* of the calculation. It is convenient to be able to split calculations into two classes: those which are *stable*, and those which are not; to do this we require a formal definition of stability. The standard treatment of the stability of variational methods for Hermitian operators is that of Mikhlin (1971), who introduces the following definition of stability:

Definition 5.1. (*Mikhlin*, 1971) *The variational process represented by* (5.1.2) *is* stable *if there exist constants* α, β, γ *independent of* N *such that for* $\|\delta\mathbf{L}\| \leqslant \gamma$ *and arbitrary* $\delta\mathbf{g}$, (5.1.5) *is solvable (for each finite* N*) and*

$$\|\delta\mathbf{a}\| \leqslant \alpha\|\delta\mathbf{L}\| + \beta\|\delta\mathbf{g}\|. \tag{5.3.3}$$

Introducing the following weaker form of a WNS:

Definition 5.2. *A WADS (weakly asymptotically diagonal system) is a WNS (Definition 3.6), which is not restricted to satisfy the normalisation condition (3.2.5).*

We find the following corollaries.

Corollary 5.1. A WADS is stable (in the sense of Mikhlin) if for some $\hat{\gamma}$ and all N, $\gamma_N \leqslant \hat{\gamma}$.

Corollary 5.2. A WNS is stable, since $\mathbf{\Lambda} = \mathbf{I}$ and $\gamma_N = 1$.

We now introduce further definitions from Mikhlin (1971). We assume that the matrix **L** of (5.1.2) is positive semi-definite and has ordered eigenvalues

$$0 \leqslant \lambda_1^{(N)} \leqslant \lambda_2^{(N)} \leqslant \ldots \leqslant \lambda_N^{(N)}.$$

Definition 5.3. *The set* $\{h_i\}$ *is* strongly minimal *w.r.t.* $\mathscr{L}$ *if for some* λ_0,

$$\lim_{N\to\infty} \lambda_1^{(N)} \geqslant \lambda_0 > 0.$$

We then refer to the matrix $\mathbf{L} = (h_i, \mathscr{L}h_j)$ as strongly minimal, and note that this implies that **L** is strictly positive definite for all N.

Definition 5.4. *The set* $\{h_i\}$ *is* almost orthonormal *w.r.t.* $\mathscr{L}$ *if it is strongly minimal w.r.t.* $\mathscr{L}$, *and for some* Λ_0,

$$\lim_{N\to\infty} \lambda_N^{(N)} \leqslant \Lambda_0.$$

We then refer to the matrix $\mathbf{L} = (h_i, \mathscr{L}h_j)$ as almost orthonormal.

In Chapter 8, we use the above definitions with reference to a non-Hermitian operator $\mathscr{L}$. In that case we assume that the unsymmetric matrix **L** has ordered eigenvalues

$$0 \leqslant |\lambda_1^{(N)}| \leqslant |\lambda_2^{(N)}| \leqslant \ldots \leqslant |\lambda_N^{(N)}|.$$

We can then extend Definitions 5.3 and 5.4.

Definition 5.5. *The set* $\{h_j\}$ *is* strongly minimal *w.r.t. the non-Hermitian operator* $\mathscr{L}$ *if for some* λ_0,

$$\lim_{N\to\infty} |\lambda_1^{(N)}| \geqslant \lambda_0 > 0.$$

We note that this implies that the unsymmetric matrix **L** is non-singular for all N.

Definition 5.6. *The set* $\{h_i\}$ *is* almost orthonormal *w.r.t. the non-Hermitian operator* $\mathscr{L}$ *if it is strongly minimal w.r.t.* $\mathscr{L}$, *and for some* Λ_0,

$$\lim_{N\to\infty} |\lambda_N^{(N)}| \leqslant \Lambda_0.$$

We then say that the matrix $\mathbf{L}$ is almost orthonormal.

With these definitions, we obtain the following two lemmas:

Lemma 5.2. *If* $\{h_i\}$ *is strongly minimal w.r.t.* $\mathscr{L}$, *then the positive definite matrix* $\mathbf{L}$ *of* (5.1.2) *satisfies*

$$\|\mathbf{L}^{-1}\| \leqslant \lambda_0^{-1}, \text{ for all } N.$$

Lemma 5.3. *If* $\{h_i\}$ *is almost orthonormal w.r.t.* $\mathscr{L}$, *then the positive definite matrix* $\mathbf{L}$ *of* (5.1.2) *satisfies the following inequalities for all* N,

$$\|\mathbf{L}\| \leqslant \Lambda_0$$
$$\rho(\mathbf{L}) \leqslant \Lambda_0/\lambda_0,$$

where $\rho(\mathbf{L})$ *is the spectral radius of* $\mathbf{L}$.

Mikhlin (1971) shows that if $\mathbf{L}$ is positive definite, then a necessary and sufficient condition for the process (5.1.2) to be stable is that the expansion set $\{h_i\}$ is strongly minimal w.r.t. $\mathscr{L}$.

We thus find that if $\mathbf{L}$ is positive definite and $\gamma_N \leqslant \hat{\gamma}$, then the expansion set of a WADS is strongly minimal w.r.t. the operator $\mathscr{L}$.

These results are interesting but not representative of the numerical behaviour of a typical calculation. The diagonal matrix $\mathbf{\Lambda}$ and hence the eigenvalues $\lambda_i^{(N)}$, $i = 1, \ldots, N$, and the stability of the system on this criterion, depend on the normalisation of the expansion set. In practice so do the error matrices $\delta\mathbf{g}$ and $\delta\mathbf{L}$, but the overall behaviour of the calculation, and in particular its observed stability, is independent of the normalisation. We illustrate this assertion by a numerical example in Section 5.6. This suggests that the following definition of stability is more appropriate numerically.

Definition 5.7. *The variational process represented by* (5.1.2) *is* relatively stable *if there exists a diagonal matrix* $\mathbf{\Gamma}$ *such that*

$$(\mathbf{\Gamma}\mathbf{L}\mathbf{\Gamma})\mathbf{a}' = \mathbf{\Gamma}\mathbf{g}$$

is stable.

We now introduce a further definition from Mikhlin (1971).

Definition 5.8. *Let* $h_1, h_2, \ldots,$ *be elements of a Hilbert space* $\mathbf{H}$, *and let* $\mathbf{H}_k$ *be the subspace spanned by* $h_1, h_2, \ldots, h_{k-1}, h_{k+1}, \ldots$. *The set* $\{h_i\}$ *is* minimal *if, for all k, the element* h_k *does not lie in* $\mathbf{H}_k$.

This definition leads to the following lemma.

Lemma 5.4. *For* $\mathbf{L}$ *positive definite, the process represented by* (5.1.2) *is relatively stable if and only if the expansion set* $\{h_i\}$ *is minimal.*

Proof. Theorem 2.2 of Mikhlin guarantees the sufficiency of the condition.
If in some normalisation, the set $\{h_i\}$ is strongly minimal w.r.t. $\mathscr{L}$, then it is minimal by Theorem 2.1 of Mikhlin (1971). Since minimality is independent of the normalisation used, necessity follows immediately.

Definition 5.7 and Lemma 5.4 then imply that, if $\mathbf{L}$ is positive definite, a WADS is relatively stable, and the expansion set $\{h_i\}$ is minimal, in any normalisation.

5.4 Bounds on the individual elements of $\delta\mathbf{a}$

We can extend the discussion of the previous section to obtain bounds on each element of $\delta\mathbf{a}$. The distribution of errors in $\delta\mathbf{a}$ depends on the structure, as well as the size, of $\delta\mathbf{L}$ and $\delta\mathbf{g}$, and this is largely under the user's control. He may try to achieve accuracy in some elements at the expense of others, or he may try to achieve some sort of uniform accuracy in all the elements. We identify two extreme "user strategies":

(a) *Fixed relative errors:*

$$|\delta g_i| \leqslant \varepsilon_g |g_i|; \quad |\delta L_{ij}| \leqslant \varepsilon_L |L_{ij}|. \tag{5.4.1}$$

(b) *Fixed absolute errors:*

$$|\delta g_i| \leqslant \varepsilon_g; \quad |\delta L_{ij}| \leqslant \varepsilon_L. \tag{5.4.2}$$

In each case, the difficulty of achieving these bounds depends on the normalisation of the expansion set $\{h_i\}$. For definiteness we therefore assume that the set $\{h_i\}$ is normalised such that the matrix $\mathbf{L}$ satisfies

$$|L_{ii}| = 1, \quad i = 1,2,\ldots,N. \tag{5.4.3}$$

For a WNS, for which $\overline{\mathbf{A}}$ has property MN, these two strategies then lead to the following bounds valid for all N.

$$\begin{aligned} \|\delta\mathbf{L}\|_\infty &\leqslant (1 + C + \Omega)\,\varepsilon_L, && \text{relative errors,} \\ &\leqslant N\varepsilon_L, && \text{absolute errors,} \end{aligned}$$

and hence, via (5.3.2), to overall error bounds of the form

$$\|\delta\mathbf{a}\|_\infty \leqslant \frac{A_1\varepsilon_g + A_2\varepsilon_L}{1 - A_3\varepsilon_L}, \quad \text{relative errors,} \tag{5.4.4}$$

$$\|\delta\mathbf{a}\|_\infty \leqslant \frac{A'_1\varepsilon_g + NA'_2\varepsilon_L}{1 - NA'_3\varepsilon_L}, \quad \text{absolute errors,} \tag{5.4.5}$$

where the constants A_i, A'_i, $i = 1,2,3$, depend on the parameters of the WNS, but not on N.

For a fixed ε_L, ε_g, (5.4.5) shows, not surprisingly, that an "absolute errors" calculation may well explode as N is increased. Maintaining a fixed relative error accuracy ensures that the calculation will retain its accuracy for all N. However, such a strategy could be expensive to achieve since, typically, the elements L_{ij}, g_i become more difficult to evaluate accurately with increasing i and j.

We now consider bounds on the individual elements of $\delta\mathbf{a}$. Since $\mathbf{L}$ is normalised to satisfy (5.4.3), we assume that $\delta\mathbf{L}$ satisfies

$$\delta L_{ii} = 0, \quad i = 1,2,\ldots,N. \tag{5.4.6}$$

Theorem 5.2. *Let* (5.1.2) *be a SNS and* $\mathbf{A}$ *have property MN. Then with fixed relative errors in* $\mathbf{L}$ *and* $\mathbf{g}$, *and for* $\varepsilon_L < \varepsilon_{\max}$, *there exist positive constants* D_1, D_2, *independent of* N, *such that*

$$|\delta a_i| \leqslant (\hat{g}_i + \mathbf{a}_i^t\hat{\mathbf{g}} + (\boldsymbol{\alpha}_i)^t\,\hat{\mathbf{g}})\,(D_1\varepsilon_g + D_2\varepsilon_L) \tag{5.4.7}$$

where D_1, D_2, $\varepsilon_{\max}$ *are defined in the proof.*

Comment

The proof of this theorem is considerably simplified if we assume that the constants C_1 and Ω_1 of (3.2.4) and (3.3.3) satisfy

$$C_1, \Omega_1 \leqslant 1.$$

We shall therefore assume that such a restriction is satisfied, although we note that this is *not* a necessary condition of the theorem.

Proof. From (5.1.5),

$$\mathbf{La} + \mathbf{L}\delta\mathbf{a} + \delta\mathbf{L}(\mathbf{a} + \delta\mathbf{a}) = \mathbf{g} + \delta\mathbf{g},$$

but from (5.1.2)

$$\mathbf{La} = \mathbf{g}.$$

Hence,

$$\delta\mathbf{a} = \mathbf{L}^{-1}(\delta\mathbf{g} - \delta\mathbf{L}(\mathbf{a} + \delta\mathbf{a})). \tag{5.4.8}$$

This can be written as

$$\mathbf{Z} = \mathbf{F}(\mathbf{Z}), \tag{5.4.9}$$

where $Z_i = \delta a_i$.

We seek a region $\mathscr{R}$ in which lies the solution of (5.4.8). If we obtain a region $\mathscr{R}$ such that $\mathbf{F}: \mathscr{R} \to \mathscr{R}$ and over which $\mathbf{F}$ has a Lipschitz constant which is less than unity, then an elementary theorem of contraction mappings implies that the solution of (5.4.8) lies in $\mathscr{R}$.

Consider the following region $\mathscr{R}$,

$$|Z_i| \leqslant (\hat{g}_i + \mathbf{a}_i^t\hat{\mathbf{g}} + (\boldsymbol{\alpha}_i)^t\hat{\mathbf{g}})(D_1\varepsilon_g + D_2\varepsilon_L).$$

Using (5.2.5), with $K_2' = K_1'$, this leads to

$$|a_i| + |\delta a_i| \leqslant \beta_1\hat{g}_i + \beta_2(\mathbf{a}_i^t\hat{\mathbf{g}} + (\boldsymbol{\alpha}_i)^t\hat{\mathbf{g}}),$$

where

$$\beta_1 = K_0' + D_1\varepsilon_g + D_2\varepsilon_L,$$

$$\beta_2 = K_1' + D_1\varepsilon_g + D_2\varepsilon_L.$$

Now, from (5.4.8), (5.4.9),

$$|F_i| \leqslant \sum_{k=1}^{N} |(L^{-1})_{ik}|(|\delta g_k| + M_k),$$

where

$$M_k = \sum_{l=1}^{N} |\delta L_{kl}|(|a_l| + |\delta a_l|)$$

$$\leqslant \sum_{l=1}^{k-1} \varepsilon_L C A_{kl}\{\beta_1\hat{g}_l + \beta_2(\mathbf{a}_l^t\hat{\mathbf{g}} + (\boldsymbol{\alpha}_l)^t\hat{\mathbf{g}})\}$$

$$+ \sum_{l=k+1}^{N} \varepsilon_L C A_{lk}\{\beta_1\hat{g}_l + \beta_2(\mathbf{a}_l^t\hat{\mathbf{g}} + (\boldsymbol{\alpha}_l)^t\hat{\mathbf{g}})\}.$$

After reversing the order of the summations, and using the bounds of Definitions 3.1, 3.2 and 3.3, we obtain,

$$M_k \leqslant \beta_3\hat{g}_k + \beta_4(\mathbf{a}_k^t\hat{\mathbf{g}} + (\boldsymbol{\alpha}_k)^t\hat{\mathbf{g}}),$$

where

$$\beta_3 = \varepsilon_L C\beta_2(\Omega + C),$$

$$\beta_4 = \varepsilon_L C(\beta_1 + \beta_2(C + C_A + \Omega)).$$

Hence,

$$|F_i| \leqslant \sum_{k=1}^{i-1} K'_1 A_{ik}\{(\varepsilon_g + \beta_3)\hat{g}_k + \beta_4(\mathbf{a}_k^t\hat{\mathbf{g}} + (\boldsymbol{\alpha}_k)^t\hat{\mathbf{g}})\}$$

$$+ K'_0\{(\varepsilon_g + \beta_3)\hat{g}_i + \beta_4(\mathbf{a}_i^t\hat{\mathbf{g}} + (\boldsymbol{\alpha}_i)^t\hat{\mathbf{g}})\}$$

$$+ \sum_{k=i+1}^{N} K'_1 A_{ki}\{(\varepsilon_g + \beta_3)\hat{g}_k + \beta_4(\mathbf{a}_k^t\hat{\mathbf{g}} + (\boldsymbol{\alpha}_k)^t\hat{\mathbf{g}})\}.$$

We again reverse the order of the summations, and use the bounds of Definitions 3.1, 3.2 and 3.3, to obtain

$$|F_i| \leqslant \{\beta_4 K'_1 C + K'_0(\varepsilon_g + \beta_3) + \beta_4 K'_1\Omega\}\hat{g}_i$$
$$+ \{K'_1(\varepsilon_g + \beta_3) + \beta_4 K'_1 C_A + \beta_4 K'_1 C + \beta_4 K'_0 + \beta_4 K'_1\Omega\}(\mathbf{a}_i^t\hat{\mathbf{g}} + (\boldsymbol{\alpha}_i)^t\hat{\mathbf{g}}).$$

A sufficient condition for $\mathbf{F}: \mathscr{R} \to \mathscr{R}$ is

$$|F_i| \leqslant (\hat{g}_i + \mathbf{a}_i^t\hat{\mathbf{g}} + (\boldsymbol{\alpha}_i)^t\hat{\mathbf{g}})(D_1\varepsilon_g + D_2\varepsilon_L),$$

which is obviously satisfied if

$$D_1 = \max\{K'_0, K'_1\} \tag{5.4.10}$$

$$D_2\varepsilon_L = \max\{\beta_4 K'_1(C + \Omega) + \beta_3 K'_0, \beta_4 K'_1(C + C_A + \Omega) + \beta_4 K'_0 + \beta_3 K'_1\}$$
$$= \beta_4 K'_1(C + C_A + \Omega) + \beta_4 K'_0 + \beta_3 K'_1,$$

since $\beta_3 \leqslant \beta_4$. Hence

$$D_2\varepsilon_L = C\varepsilon_L(K'_0 + K'_1(C + C_A + \Omega) + (D_1\varepsilon_g + D_2\varepsilon_L)(1 + C + C_A + \Omega))$$
$$\times (K'_0 + K'_1(C + C_A + \Omega)) + C\varepsilon_L(K'_1 + D_1\varepsilon_g + D_2\varepsilon_L)(C + \Omega)K'_1,$$

and

$$D_2 = \frac{C\{(K'_0 + K'_1(C + C_A + \Omega) + D_1\varepsilon_g(1 + C + C_A + \Omega)) \times (K'_0 + K'_1(C + C_A + \Omega)) + (K'_1 + D_1\varepsilon_g)(C + \Omega)K'_1\}}{1 - C\varepsilon_L\{(1 + C + C_A + \Omega)(K'_0 + K'_1(C + C_A + \Omega)) + (C + \Omega)K'_1\}} \tag{5.4.11}$$

provided

$$\varepsilon_L < \varepsilon_1 = [C\{(1 + C + C_A + \Omega)(K'_0 + K'_1(C + C_A + \Omega)) + (C + \Omega)K'_1\}]^{-1} \tag{5.4.12}$$

A Lipschitz constant for $\mathbf{F}$ is given by

$$M = \sup_{\mathscr{R}}\left\|\frac{\partial F_\nu}{\partial Z_\mu}\right\| = \sup_{\mathscr{R}}\max_{i=1}^{N}\sum_{m=1}^{N}\left|\frac{\partial F_i}{\partial Z_m}\right|.$$

But

$$F_i = \sum_{k=1}^{N} (L^{-1})_{ik} \left(\delta g_k - \sum_{l=1}^{N} \delta L_{kl} (a_l + \delta a_l) \right),$$

and therefore,

$$\frac{\partial F_i}{\partial Z_m} = - \sum_{k=1}^{N} (L^{-1})_{ik} \, \delta L_{km}.$$

Hence,

$$\begin{aligned}
\sum_{m=1}^{N} \left| \frac{\partial F_i}{\partial Z_m} \right| &\leqslant \sum_{m=1}^{N} \sum_{k=1}^{N} |(L^{-1})_{ik}| \, |\delta L_{km}| \\
&\leqslant \sum_{k=1}^{i-1} \sum_{m=1}^{k-1} K_1' A_{ik} C \varepsilon_L A_{km} + \sum_{k=1}^{i-1} \sum_{m=k+1}^{N} K_1' A_{ik} C \varepsilon_L A_{mk} \\
&\quad + K_0' \left\{ \sum_{m=1}^{i-1} C \varepsilon_L A_{im} + \sum_{m=i+1}^{N} C \varepsilon_L A_{mi} \right\} \\
&\quad + \sum_{k=i+1}^{N} \sum_{m=1}^{k-1} K_1' A_{ki} C \varepsilon_L A_{km} + \sum_{k=i+1}^{N} \sum_{m=k+1}^{N} K_1' A_{ki} C \varepsilon_L A_{mk}.
\end{aligned}$$

We again reverse the order of the summations, and use the bounds of Definitions 3.1, 3.2 and 3.3, to obtain

$$\begin{aligned}
\sum_{m=1}^{N} \left| \frac{\partial F_i}{\partial Z_m} \right| &\leqslant \sum_{m=1}^{i-2} C K_1' \varepsilon_L C_A A_{im} + \sum_{m=2}^{i-1} C K_1' \varepsilon_L C A_{im} \\
&\quad + C K_1' \varepsilon_L C + \sum_{m=i+1}^{N} C K_1' \varepsilon_L C A_{mi} + \sum_{m=1}^{i-1} C K_0' \varepsilon_L A_{im} \\
&\quad + \sum_{m=i+1}^{N} C K_0' \varepsilon_L A_{mi} + \sum_{m=1}^{i-1} C K_1' \varepsilon_L \Omega A_{im} + C K_1' \varepsilon_L \Omega \\
&\quad + \sum_{m=i+1}^{N-1} C K_1' \varepsilon_L \Omega A_{mi} + \sum_{m=i+2}^{N} C K_1' \varepsilon_L C_A A_{mi}.
\end{aligned}$$

But $\mathbf{A}$ has property MN, and hence

$$\|\mathbf{A}\|_\infty = \max_{i=1}^{N} \sum_{j=1}^{i-1} A_{ij} \leqslant C,$$

$$\|\mathbf{A}\|_1 = \max_{j=1}^{N} \sum_{i=j+1}^{N} A_{ij} \leqslant \Omega.$$

Hence,

$$M = \sup_{\mathscr{R}} \max_{i=1}^{N} \sum_{m=1}^{N} \left| \frac{\partial F_i}{\partial Z_m} \right|$$
$$\leqslant C\varepsilon_L(C + \Omega)\{K_1'(1 + C + C_A + \Omega) + K_0'\}.$$

Hence a sufficient condition for $M < 1$ is

$$\varepsilon_L < \varepsilon_2 = [C(C + \Omega)\{K_1'(1 + C + C_A + \Omega) + K_0'\}]^{-1}. \quad (5.4.13)$$

Both (5.4.12) and (5.4.13) are satisfied if $\varepsilon_L < \varepsilon_{\max} = \min\{\varepsilon_1, \varepsilon_2\} = \varepsilon_1$.
This completes the proof of the theorem.

When there are fixed absolute errors in **L** and **g**, we can prove the following theorem.

Theorem 5.3. *Let* (5.1.2) *be a WNS,* $\bar{\mathbf{A}}$ *have property MN, and* $\hat{\mathbf{g}}$ *satisfy* $\max_{i=1}^{N} \hat{g}_i = \hat{g}_p$, $p < N$. *Then with fixed absolute errors in* **L** *and* **g**, *and for* $\varepsilon_L < \varepsilon'_{\max}$, *there exist positive constants* D_1', D_2', *such that*

$$|\delta a_i| \leqslant D_1'\varepsilon_g + D_2'\varepsilon_L. \quad (5.4.14)$$

Comment
In this case D_1' is independent of N, but D_2' increases monotonically with N. Additionally $\varepsilon'_{\max}$, which restricts the magnitude of absolute errors for which the theorem is valid, decreases monotonically with N.
The proof of Theorem 5.3 is similar to, but somewhat simpler than, the proof of Theorem 5.2. Full details of the proof are given in Freeman (1974).

5.5 The stability of f_N

The results of the previous section describe the errors induced in the expansion coefficients $a_i^{(N)}$ under the normalisation condition $\Lambda = \mathbf{I}$. The total error induced in the approximate solution f_N is

$$E_N = \sum_{i=1}^{N} \delta a_i h_i. \quad (5.5.1)$$

Boundedness of δa_i does not necessarily imply boundedness of $\|E_N\|$, and Mikhlin (1971) therefore introduces the following definition of stability, appropriate to a positive definite operator $\mathscr{L}$.

Definition 5.9. *Let* $\mathscr{L}$ *be positive definite, then the approximate solution*

f_N *of* (5.1.1) *is* stable *if there exist constants* α', β', γ' *independent of N such that for* $\|\boldsymbol{\delta}\mathbf{L}\| < \gamma'$ *and all* $\boldsymbol{\delta}\mathbf{g}$,

$$\|E_N\|_{\mathscr{L}} \equiv (E_N, \mathscr{L}E_N)^{1/2} \leqslant \alpha'\|\boldsymbol{\delta}\mathbf{L}\| + \beta'\|\boldsymbol{\delta}\mathbf{g}\|. \tag{5.5.2}$$

We note that stability in this sense is again dependent on the normalisation used, since $\|\boldsymbol{\delta}\mathbf{L}\|$ and $\|\boldsymbol{\delta}\mathbf{g}\|$ depend on this normalisation.

Theorem 5.4. *Let* $\mathscr{L}$ *be positive definite,* (5.1.2) *be a WNS, and* $\overline{\mathbf{A}}$ *have property MN. Then, provided* $\|\hat{\mathbf{g}}\|_1 \leqslant D_3$ *independent of N, the approximate solution* f_N *is stable.*

Proof.

$$\begin{aligned}(E_N, \mathscr{L}E_N) &= \left(\sum_{i=1}^{N} \delta a_i h_i, \mathscr{L}\left(\sum_{j=1}^{N} \delta a_j h_j\right)\right) \\ &= \sum_{i=1}^{N}\sum_{j=1}^{N} \delta a_i \delta a_j L_{ij} \\ &\leqslant \max_{k=1}^{N} \max_{l=1}^{N} \sum_{i=1}^{N}\sum_{j=1}^{N} |\delta a_i|\,|\delta a_k|\,|L_{lj}| \\ &\leqslant \|\boldsymbol{\delta}\mathbf{a}\|_1\,\|\boldsymbol{\delta}\mathbf{a}\|_\infty\,\|\mathbf{L}\|_\infty.\end{aligned}$$

Now,

$$\|\boldsymbol{\delta}\mathbf{a}\|_\infty \leqslant \|\boldsymbol{\delta}\mathbf{a}\|_1,$$

and by Lemma 5.1,

$$\|\mathbf{L}\|_\infty \leqslant 1 + C + \Omega = K_3.$$

Hence,

$$\begin{aligned}\|E_N\|_{\mathscr{L}} &\leqslant K_3^{1/2}\|\boldsymbol{\delta}\mathbf{a}\|_1 \\ &\leqslant \frac{K_3^{1/2}\|\mathbf{L}^{-1}\|_1\{\|\boldsymbol{\delta}\mathbf{g}\|_1 + \|\boldsymbol{\delta}\mathbf{L}\|_1\,\|\mathbf{a}\|_1\}}{1 - \|\mathbf{L}^{-1}\|_1\,\|\boldsymbol{\delta}\mathbf{L}\|_1}\end{aligned}$$

provided the denominator is positive, using the bound of (5.3.1).

By Lemma 5.1,

$$\|\mathbf{L}^{-1}\|_1 = \|\mathbf{L}^{-1}\|_\infty \leqslant K_1, \text{ since } \mathbf{L}^{-1} \text{ is symmetric.}$$

Now, under the conditions of the theorem, $\mathbf{a}$ satisfies (5.2.5), and hence

$$\begin{aligned}\|\mathbf{a}\|_1 &\leqslant \sum_{i=1}^{N} \{K_0'\hat{g}_i + K_1'\bar{\mathbf{a}}_i^t\,\hat{\mathbf{g}} + K_1'\,(\boldsymbol{\alpha}_i)^t\,\hat{\mathbf{g}}\} \\ &\leqslant K_0'\sum_{i=1}^{N}\hat{g}_i + K_1'\sum_{i=1}^{N}\sum_{j=1}^{i-1}\bar{A}_{ij}\hat{g}_j + K_1'\sum_{i=1}^{N}\sum_{j=i+1}^{N}\bar{A}_{ji}\hat{g}_j.\end{aligned}$$

On changing the order of the summations, and using the fact that $\bar{\mathbf{A}}$ has property MN, we obtain

$$\|\mathbf{a}\|_1 \leqslant \{K_0' + K_1'(C + \Omega)\} \|\hat{\mathbf{g}}\|_1$$
$$\leqslant D_4 = \{K_0' + K_1'(C + \Omega)\} D_3.$$

Hence,

$$\|E_N\|_{\mathscr{L}} \leqslant \frac{K_3^{1/2} K_1\{\|\delta\mathbf{g}\|_1 + D_4\|\delta\mathbf{L}\|_1\}}{1 - K_1 \|\delta\mathbf{L}\|_1},$$

provided that the denominator is positive.

The stability of the approximate solution immediately follows from Definition 5.9.

We may also wish to consider (for example, if $\mathscr{L}$ is not positive definite) the error in the natural norm,

$$\|E_N\|^2 = (E_N, E_N).$$

In this case,

$$(E_N, E_N) = \left(\sum_{i=1}^{N} \delta a_i h_i, \sum_{j=1}^{N} \delta a_j h_j\right)$$
$$= \sum_{i=1}^{N} \sum_{j=1}^{N} \delta a_i \delta a_j (h_i, h_j). \tag{5.5.3}$$

We again assume that $\mathbf{L}$ is normalised, and introduce the renormalised expansion set $\{\bar{h}_i\}$:

$$\bar{h}_i = \gamma_i h_i; (\bar{h}_i, \bar{h}_i) = 1; (\bar{h}_i, \mathscr{L}\bar{h}_j) = \pm\gamma_i^2. \tag{5.5.4}$$

If the set $\{h_i\}$ is orthogonal, then

$$(h_i, h_j) = (\gamma_i^{-1}\bar{h}_i, \gamma_j^{-1}\bar{h}_j)$$
$$= \gamma_i^{-1}\gamma_j^{-1}\delta_{ij},$$

and (5.5.3) becomes

$$(E_N, E_N) = \sum_{i=1}^{N} \gamma_i^{-2}\delta a_i^2. \tag{5.5.5}$$

Using this expression we can consider the stability of the approximate solution in the natural norm.

Theorem 5.5. *Let* $\mathbf{L}$, $\mathbf{g}$ *satisfy the conditions of Theorem* 5.2, *and let* $\{h_i\}$ *be an orthogonal set. Then with fixed relative errors in* $\mathbf{L}$ *and* $\mathbf{g}$, *and for* $\varepsilon_L < \varepsilon_{\max}$, *the approximate solution f_N is uniformly stable (E_N is bounded independent of N) in the natural norm, provided that $\max_{i=1}^{N} |\gamma_i^{-1}| = |\gamma_p^{-1}|$, $p < N$, and $\|\hat{\mathbf{g}}\|_1 \leqslant D_3$, independent of N.*

Proof. From (5.5.5),

$$\|E_N\| = (E_N, E_N)^{1/2} \leqslant \sum_{i=1}^{N} |\gamma_i^{-1}|\,|\delta a_i|.$$

Using the bound on $|\delta a_i|$ given by Theorem 5.2, we obtain

$$\|E_N\| \leqslant \sum_{i=1}^{N} |\gamma_i^{-1}| \left\{ \hat{g}_i + \sum_{j=1}^{i-1} A_{ij}\hat{g}_j + \sum_{j=i+1}^{N} A_{ji}\hat{g}_j \right\} (D_1\varepsilon_g + D_2\varepsilon_L)$$

$$\leqslant |\gamma_p^{-1}| \{\|\hat{\mathbf{g}}\|_1 + \Omega\|\hat{\mathbf{g}}\|_1 + C\|\hat{\mathbf{g}}\|_1\} (D_1\varepsilon_g + D_2\varepsilon_L),$$

which is independent of N and hence the theorem follows.

Similarly uniform stability in the energy norm can be considered.

Theorem 5.6. *Let $\mathscr{L}$ be positive definite, and let* **L** *and* **g** *satisfy the conditions of Theorem 5.4. Then with fixed relative errors in* **L** *and* **g**, *the approximate solution is uniformly stable in the energy norm, provided $\varepsilon_L < (K_1 (C + \Omega))^{-1}$, where K_1 is given by Lemma 5.1.*

Proof. From Theorem 5.4,

$$\|E_N\|_{\mathscr{L}} \leqslant \frac{K_3^{1/2} K_1 \{\|\delta\mathbf{g}\|_1 + D_4\|\delta\mathbf{L}\|_1\}}{1 - K_1\|\delta\mathbf{L}\|_1},$$

provided the denominator is positive.

Now, $\|\delta\mathbf{L}\|_1 = \|\delta\mathbf{L}\|_\infty \leqslant \varepsilon_L(C + \Omega)$, where we assume that $\delta\mathbf{L}$ satisfies (5.4.6).

Also, $\|\delta\mathbf{g}\|_1 \leqslant \varepsilon_g\|\hat{\mathbf{g}}\|_1 \leqslant \varepsilon_g D_3$. Hence,

$$\|E_N\|_{\mathscr{L}} \leqslant \frac{K_3^{1/2} K_1 \{D_3\varepsilon_g + D_4\,\varepsilon_L(C + \Omega)\}}{1 - K_1\,\varepsilon_L(C + \Omega)},$$

provided the denominator is positive. Hence $\|E_N\|_{\mathscr{L}}$ is bounded independent of N.

5.6 Discussion. Numerical examples

From the results of the preceding sections of this chapter, we note two important points:

(a) The great inherent stability of a WNS. This stability is epitomised by Lemma 5.1, which shows that the condition number of the coefficient matrix is bounded independent of N.
(b) The suggestion that *relative stability* is more important numerically than *stability* in the sense of Mikhlin.

We consider this second point in detail with reference to a numerical example. We consider the boundary value problem of Section 1.5, with $\lambda = 1$, and attempt a Galerkin solution with the expansion set,

$$h_i(x) = C_i \sin ix, \quad i = 1,2,\ldots,$$

for a number of different choices of the normalising constants C_i:
(1) $C_i = (2/\pi)^{1/2}$ (this corresponds to the orthonormal expansion set of Section 1.5),
(2–6) $C_i = i^{-l+1}$, $l = 2,3,\ldots,6$.
It is straightforward to compute $\mathbf{L}$ and $\mathbf{g}$ exactly (as in Section 1.5) and to show that $\mathbf{L}$ is AD of type $B(2,2;8/\pi)$ and that with normalisation (1), $\mathbf{g}$ satisfies

$$|g_i| \leqslant \mathscr{C} i^{-1}, \quad \text{where } \mathscr{C} = \begin{cases} (6\pi^2 - 52)(2/\pi)^{1/2}, & i \text{ odd} \\ 6\pi^2(2/\pi)^{1/2}, & i \text{ even.} \end{cases}$$

We show in Section 4.2 that the matrix $\mathbf{L}$ is SAD and $\mathbf{A}$ has property MN, and it therefore follows from Theorem 5.1 that the calculation is stable in the normalisations (1) and (2), but only relatively stable in the normalisations (3) to (6). Nonetheless, we expect all six normalisations to appear stable numerically. We test this by solving the boundary value problem in all six normalisations using a single precision Fortran program with no special precautions against round-off errors.

Figures 5.1 and 5.2 show some typical results from these calculations. In Fig. 5.1 we consider the observed "horizontal" rate of convergence by plotting the computed values of $|b_i - a_i^{(N)}|$ for $i = 3$ and various values of N, while in Fig. 5.2 we consider the observed "diagonal" rate of convergence by plotting the computed values of $|b_N - a_N^{(N)}|$ for various values of N. Using the results of Section 4.2, we predict these rates of convergence to be

$$|b_i - a_i^{(N)}| \leqslant D_5 N^{-h}, \quad i \text{ fixed},$$

where $h = 5$ for each normalisation (1) to (6), and

$$|b_N - a_N^{(N)}| \leqslant D_6 N^{-d},$$

where $d = 6 - l$, $l = 1,2,\ldots,6$, for normalisation (l).

We see that the measured slopes are in general agreement with these predictions, and indeed the agreement is further improved by increasing N.

More importantly, there are no signs of instability in any of the calculations, which suggests that the concept of relative stability is a useful one. That this observed stability is *not* expected from the discussion of Mikhlin (1971) is emphasized by the behaviour of the exact expansion coefficients b_i. From the exact analytical solution,

$$f(x) = -x^3 + (1+\pi)x^2 - \pi x,$$

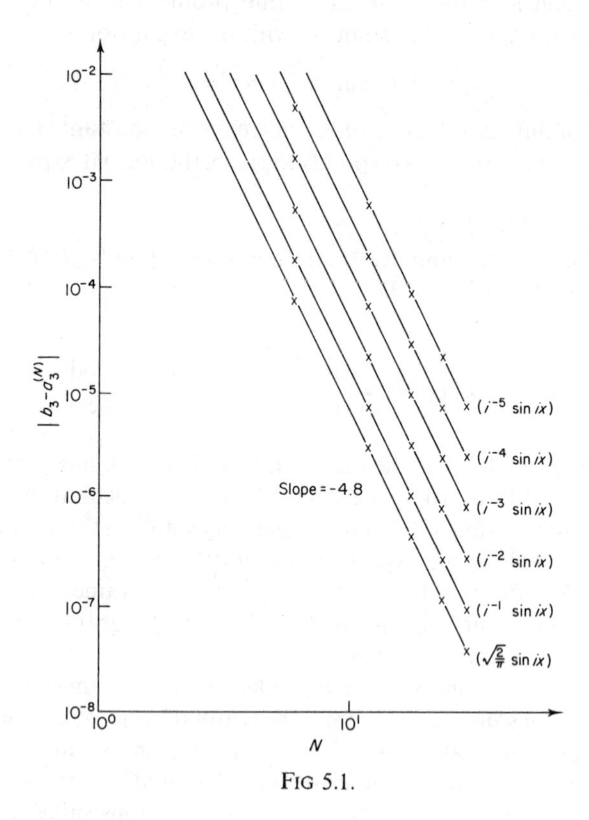

FIG 5.1.

we can show that these coefficients behave as

$$b_i = \mathcal{O}(i^{-r}),$$

with $r = 4 - l$ for normalisation (l), $l = 1,2,\ldots,6$. For a stable system, Mikhlin shows that $b_i \to 0$, while for normalisations (5) and (6), we have $b_i = \mathcal{O}(i)$ and $\mathcal{O}(i^2)$, respectively, both of which are unbounded.

It should be noted that the above system is not strictly a WNS, since the constants involved do not satisfy (3.2.7). This leads to speculation that a restriction such as (3.2.7) is *not* necessary to guarantee the stability (or relative stability) of the variational calculation. However, the next example shows that instability can arise when (3.2.7) is not satisfied, and we therefore conclude that some restriction on the constants C, C_1 is necessary for the (relative) stability of the calculation.

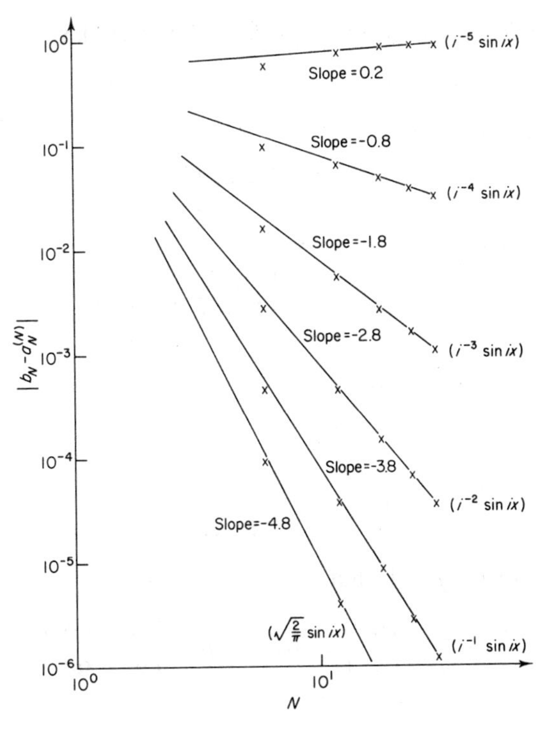

FIG 5.2.

Consider the Galerkin solution of the boundary value problem

$$\mathscr{L} f(x) = -\frac{d^2 f(x)}{dx^2} = 1, \quad x \in [0, \pi],$$

with homogeneous boundary conditions

$$f(0) = f(\pi) = 0.$$

This example was considered by Anderssen and Omodei (1974), and they introduce the expansion set

$$h_1(x) = \frac{x^2(x-\pi)}{\alpha}, \text{ where } \alpha = (2\pi^5/15)^{1/2}$$

$$h_i(x) = \left(\frac{2}{\pi}\right)^{1/2} \frac{1}{(i-1)} \sin(i-1)\,x, \quad i = 2,3,\ldots.$$

An approximate solution is given by (5.1.1) and (5.1.2), where

$$L_{ii} = 1,$$

$$L_{i+1,j+1} = 0, \quad i \neq j,$$

$$L_{1,i+1} = L_{i+1,1} = \begin{cases} -\dfrac{2\sqrt{(2\pi)}}{i^2\alpha}, & i \text{ odd}, \\[2ex] \dfrac{6\sqrt{(2\pi)}}{i^2\alpha}, & i \text{ even}. \end{cases}$$

$$g_1 = \frac{-\pi^4}{12\alpha}$$

$$g_{i+1} = \begin{cases} 2\sqrt{\dfrac{2}{\pi}}\, i^{-2}, & i \text{ odd}, \\[2ex] 0, & i \text{ even}. \end{cases}$$

We thus find, for $i > j \geqslant 1$, that

$$|L_{ij}| \leqslant \frac{\beta\sqrt{(2\pi)}}{\alpha(i-1)^2}, \text{ where } \beta = \begin{cases} 2, & i \text{ even}, \\ 6, & i \text{ odd}, \end{cases}$$

$$\leqslant \frac{\beta\sqrt{(2\pi)}}{\alpha}\left(\frac{i}{i-1}\right)^2 \frac{1}{i^2}.$$

But

$$\left(\frac{i}{i-1}\right)^2 \leqslant 4, \quad i \text{ even}, i \geqslant 2,$$

and

$$\left(\frac{i}{i-1}\right)^2 \leqslant \tfrac{9}{4}, \quad i \text{ odd}, i \geqslant 2.$$

Hence the matrix $\mathbf{L}$ is AD of type A(2; C), where

$$C = \frac{27\sqrt{(2\pi)}}{2\alpha} = \frac{27\sqrt{15}}{2\pi^2} \approx 5{\cdot}3.$$

TABLE 5.1
Exact and approximate coefficients $a_i^{(N)}$ and $\bar{a}_i^{(N)}$ for the example of Anderssen and Omodei (1974)

	$N = 5$	$N = 10$	$N = 15$	$N = 20$	$N = 25$	$N = 30$	$N = 35$	$N = 40$
$i=1$	$-3{\cdot}517259 \times 10^{-1}$ $-3{\cdot}551313 \times 10^{-1}$	$-1{\cdot}529826 \times 10^{-1}$ $-1{\cdot}741036 \times 10^{-1}$	$-2{\cdot}441850 \times 10^{-1}$ $-3{\cdot}357901 \times 10^{-1}$	$-1{\cdot}769356 \times 10^{-1}$ $-3{\cdot}694753 \times 10^{-1}$	$-2{\cdot}268062 \times 10^{-1}$ $-6{\cdot}526245 \times 10^{-1}$	$-1{\cdot}854837 \times 10^{-1}$ $-8{\cdot}648082 \times 10^{-1}$	$-2{\cdot}197839 \times 10^{-1}$ $-1{\cdot}390970$	$-1{\cdot}898548 \times 10^{-1}$ $-1{\cdot}836431$
$i=5$	$5{\cdot}175848 \times 10^{-2}$ $5{\cdot}225960 \times 10^{-2}$	$2{\cdot}251226 \times 10^{-2}$ $2{\cdot}562034 \times 10^{-2}$	$3{\cdot}593322 \times 10^{-2}$ $4{\cdot}941343 \times 10^{-2}$	$2{\cdot}603708 \times 10^{-2}$ $5{\cdot}437040 \times 10^{-2}$	$3{\cdot}337582 \times 10^{-2}$ $9{\cdot}603744 \times 10^{-2}$	$2{\cdot}729499 \times 10^{-2}$ $1{\cdot}272615 \times 10^{-1}$	$3{\cdot}234247 \times 10^{-2}$ $2{\cdot}046891 \times 10^{-1}$	$2{\cdot}793822 \times 10^{-2}$ $2{\cdot}702413 \times 10^{-1}$
$i=10$		$1{\cdot}821856 \times 10^{-2}$ $1{\cdot}801306 \times 10^{-2}$	$1{\cdot}733488 \times 10^{-2}$ $1{\cdot}644644 \times 10^{-2}$	$1{\cdot}798648 \times 10^{-2}$ $1{\cdot}612006 \times 10^{-2}$	$1{\cdot}750327 \times 10^{-2}$ $1{\cdot}337655 \times 10^{-2}$	$1{\cdot}790365 \times 10^{-2}$ $1{\cdot}132064 \times 10^{-2}$	$1{\cdot}757131 \times 10^{-2}$ $6{\cdot}222527 \times 10^{-3}$	$1{\cdot}786130 \times 10^{-2}$ $1{\cdot}906335 \times 10^{-3}$
$i=15$			$2{\cdot}933324 \times 10^{-3}$ $4{\cdot}033750 \times 10^{-3}$	$2{\cdot}125476 \times 10^{-3}$ $4{\cdot}438400 \times 10^{-3}$	$2{\cdot}724557 \times 10^{-3}$ $7{\cdot}839791 \times 10^{-3}$	$2{\cdot}228163 \times 10^{-3}$ $1{\cdot}038869 \times 10^{-2}$	$2{\cdot}640201 \times 10^{-3}$ $1{\cdot}670932 \times 10^{-2}$	$2{\cdot}280671 \times 10^{-3}$ $2{\cdot}206051 \times 10^{-2}$
$i=20$				$4{\cdot}035747 \times 10^{-3}$ $3{\cdot}616744 \times 10^{-3}$	$3{\cdot}927326 \times 10^{-3}$ $3{\cdot}001164 \times 10^{-3}$	$4{\cdot}017163 \times 10^{-3}$ $2{\cdot}539867 \times 10^{-3}$	$3{\cdot}942593 \times 10^{-3}$ $1{\cdot}395969 \times 10^{-3}$	$4{\cdot}007660 \times 10^{-3}$ $4{\cdot}275156 \times 10^{-4}$
$i=25$					$9{\cdot}271062 \times 10^{-4}$ $2{\cdot}667707 \times 10^{-3}$	$7{\cdot}581942 \times 10^{-4}$ $3{\cdot}535041 \times 10^{-3}$	$8{\cdot}984018 \times 10^{-4}$ $5{\cdot}685809 \times 10^{-3}$	$7{\cdot}760618 \times 10^{-4}$ $7{\cdot}506702 \times 10^{-3}$
$i=30$						$1{\cdot}724371 \times 10^{-3}$ $1{\cdot}089951 \times 10^{-3}$	$1{\cdot}692361 \times 10^{-3}$ $5{\cdot}989318 \times 10^{-4}$	$1{\cdot}720292 \times 10^{-3}$ $1{\cdot}832225 \times 10^{-4}$
$i=35$							$4{\cdot}476466 \times 10^{-4}$ $2{\cdot}833067 \times 10^{-3}$	$3{\cdot}866882 \times 10^{-4}$ $3{\cdot}740364 \times 10^{-3}$
$i=40$								$9{\cdot}511935 \times 10^{-4}$ $1{\cdot}014084 \times 10^{-4}$

We see that C fails to satisfy the restrictions (2.2.3) and hence the system of equations (5.1.2) is not a "nice system of type A", and in fact is not a WNS. However, we might nonetheless expect that the calculation would be stable, even though the constant C is rather large. Anderssen and Omodei (1974) give explicit formulae for the "exact" coefficients $a_i^{(N)}$ and calculate these coefficients to seven significant digits of accuracy (Table 5.1). Anderssen and Omodei also form the "calculated" coefficients $\bar{a}_i^{(N)}$ from the linear system (5.1.2), having first rounded the right-hand side vector $\mathbf{g}^{(N)}$ to four significant figures. These results are displayed under the corresponding "exact" coefficients in Table 5.1.

It is apparent from the results contained in Table 5.1 that the calculation of the coefficients $\bar{a}_i^{(N)}$ from (5.1.2) is unstable, and we therefore conclude that some restriction on the constant C in this particular case, and in the general case on C, C_1 of (3.2.3), (3.2.4) is necessary.

6

Eigenvalue Problems

6.1 Introduction

The preceding and succeeding chapters of this book are devoted to an analysis of expansion methods for the solution of the inhomogeneous operator equation

$$\mathscr{L}f = g. \tag{6.1.1}$$

However, in this chapter we consider the calculation of eigenvalues and eigenfunctions of the linear homogeneous operator equation

$$(\mathscr{L} - \lambda\mathscr{M})f = 0. \tag{6.1.2}$$

An expansion method solution of (6.1.2) leads to the approximations

$$\lambda^{(N)} \text{ and } f^{(N)} = \sum_{i=1}^{N} a_i^{(N)} h_i \tag{6.1.3}$$

to a true eigenvalue and corresponding eigenfunction

$$\lambda \text{ and } f = \sum_{i=1}^{\infty} b_i h_i \tag{6.1.4}$$

of (6.1.2). The unknowns in (6.1.3) are given by the $N \times N$ algebraic eigenvalue problem

$$(\mathbf{L}^{(N)} - \lambda^{(N)}\mathbf{M}^{(N)})\,\mathbf{a}^{(N)} = \mathbf{0}, \tag{6.1.5}$$

where

$$L_{ij} = (h_i, \mathscr{L}h_j), \quad \text{and} \quad M_{ij} = (h_i, \mathscr{M}h_j). \tag{6.1.6}$$

Similarly, the unknowns in (6.1.4) are given by the corresponding infinite system

$$(\mathbf{L} - \lambda\mathbf{M})\,\mathbf{b} = \mathbf{0}. \tag{6.1.7}$$

We assume throughout this chapter that the operators $\mathscr{L}$ and $\mathscr{M}$ are Hermitian, and that $\mathscr{M}$ is positive definite. Then $\mathbf{L}$ and $\mathbf{M}$ are symmetric matrices and $\mathbf{M}$ is positive definite; hence $\mathbf{M}^{(N)}$ is positive definite for each N, and N real eigenvalues and corresponding real and linearly independent eigenvectors of (6.1.5) exist for all N.

We are then interested in the N-convergence rates of $\lambda - \lambda^{(N)}$ and $b_i - a_i^{(N)}$, since this allows us to consider the performance of the underlying expansion method solution of (6.1.2).

The analysis of the eigenvalue problem (6.1.2) is intrinsically more difficult than that of the inhomogeneous problem (6.1.1); for this reason the assumptions made are stronger than those which are needed in other chapters, and the theorem proofs are technically more difficult, although still straightforward. We assume that the matrices $\mathbf{L}$ and $\mathbf{M}$ are AD of type A, with parameters $(p; C)$, $(p_M; C_M)$ respectively (Definition 2.1), and that the matrix $\mathbf{M}$ is normalised so that

$$M_{ii} = 1, \quad \forall i. \tag{6.1.8}$$

We also assume that the diagonal elements of $\mathbf{L}$ satisfy

$$|L_{ii}| \leqslant C_1 i^{2t}, \quad \forall i,$$

which, together with (2.1.3), leads directly to the following bounds on the off-diagonal elements of $\mathbf{L}$,

$$|L_{ij}| \leqslant CC_1 i^{-p+t} j^t, \quad i > j. \tag{6.1.9}$$

Bounds for $i < j$ follow from the symmetry of $\mathbf{L}$. In practical examples, the exponent t may be positive (when $\mathscr{L}$ is a differential operator) or negative (when $\mathscr{L}$ is an integral operator); we introduce the notation

$$[t]_+ = \begin{cases} t, & t > 0, \\ 0, & t \leqslant 0. \end{cases} \tag{6.1.10}$$

The results of this chapter may be extended to the case when $\mathbf{L}$ and $\mathbf{M}$ are both AD of type B (Definition 4.1), and we quote the extended results as companions to the theorems of this chapter; details of the proofs of these companion theorems are given in Freeman (1974). Further extensions to the case when $\mathbf{L}$, $\mathbf{M}$ are SAD are presumably possible, but have not so far been reported.

6.2 Eigenvalue convergence

In this section we consider the behaviour of the spectrum $\{\lambda_i\}$ of (6.1.7), and the N-convergence rate of $\lambda_i - \lambda_i^{(N)}$ for fixed i.

Provided C_M satisfies the conditions of Theorem 2.1, and $p_M > 1$, then by Theorem 2.1, **M** may be factorised as

$$\mathbf{TMT}^t = \mathbf{I}, \tag{6.2.1}$$

where, using the notation of Lemma 2.1, the lower triangular matrix **T** may be written as $\mathbf{T} = \mathbf{I} + \mathbf{F}$ and **F** satisfies

$$|F_{ij}| \leqslant K'_M i^{-p_M}, \quad i \geqslant j. \tag{6.2.2}$$

Equations (6.1.5), (6.1.7) may then be rewritten as

$$[\bar{\mathbf{L}}^{(N)} - \lambda^{(N)}\mathbf{I}^{(N)}]\,\bar{\mathbf{a}}^{(N)} = \mathbf{0}, \tag{6.2.3}$$

where

$$\mathbf{a}^{(N)} = \mathbf{T}_N^t \bar{\mathbf{a}}^{(N)}, \quad \text{and} \quad \bar{\mathbf{L}}^{(N)} = \mathbf{T}_N \mathbf{L}^{(N)} \mathbf{T}_N^t, \tag{6.2.4}$$

and

$$[\bar{\mathbf{L}} - \lambda\mathbf{I}]\,\bar{\mathbf{b}} = \mathbf{0}, \tag{6.2.5}$$

where

$$\mathbf{b} = \mathbf{T}^t\bar{\mathbf{b}}, \quad \text{and} \quad \bar{\mathbf{L}} = \mathbf{TLT}^t, \tag{6.2.6}$$

and $\mathbf{T}_N$ is the leading $N \times N$ minor of **T**.

The elements of the transformed matrix $\bar{\mathbf{L}}$ are characterised by the following lemma.

Lemma 6.1. *Let the matrix* **L** *be AD of type* $A(p; C)$ *such that*

$$C_2 i^{2t} \leqslant |L_{ii}| \leqslant C_1 i^{2t};$$

also let **M** *be AD of type* $A(p_M; C_M)$ *and be normalised by* (6.1.8). *Then, provided* C_M *satisfies the conditions of Theorem* 2.1 *and* $p_M > 1$,

(1)
$$A_2 i^{2t} \leqslant |\bar{L}_{ii}| \leqslant A_1 i^{2t}, \tag{6.2.7}$$

where A_1, A_2 *are positive constants independent of i, provided*

$$p_M > [-t]_+ + 1,$$

and

$$C_1/C_2 < [K'_M\{2(C+1) + K'_M(2C+1)\}]^{-1}, \tag{6.2.8}$$

and

(2) *for* $i > j$,

$$|\bar{L}_{ij}| \leqslant A_3 i^{-p+t} j^{-v_1} + A_4 i^{-p_M} j^{-v_2}, \tag{6.2.9}$$

where

$$v_1 = \min\{-t, p_M - 1\}$$
$$v_2 = \min\{p - t - [t]_+ - 1, -2t, p_M - 1\},$$

and the positive constants A_3, A_4 are independent of i and j.

Proof.

(1) The (i,i)-th component of (6.2.6) gives

$$\begin{aligned}\bar{L}_{ii} &= \sum_{k=1}^{i}\sum_{l=1}^{i} T_{ik}L_{kl}T_{il}\\ &= \sum_{k=1}^{i}\sum_{l=1}^{i} (\delta_{ik} + F_{ik})\, L_{kl}(\delta_{il} + F_{il})\\ &= L_{ii} + \sum_{l=1}^{i} L_{il}F_{il} + \sum_{k=1}^{i} F_{ik}L_{ki} + \sum_{k=1}^{i}\sum_{l=1}^{i} F_{ik}L_{kl}F_{il}. \end{aligned} \tag{6.2.10}$$

Inserting the appropriate bounds, we find

$$\begin{aligned}\sum_{l=1}^{i} |L_{il}||F_{il}| &= \sum_{k=1}^{i} |F_{ik}||L_{ki}|\\ &\leqslant \sum_{k=1}^{i-1} K'_M i^{-p_M} CC_1 i^{-p+t}k^t + K'_M i^{-p_M} C_1 i^{2t}\\ &\leqslant CC_1K'_M i^{-p-p_M+t+[t]_++1} + C_1K'_M i^{-p_M+2t}.\end{aligned}$$

Also,

$$\begin{aligned}&\sum_{k=1}^{i}\sum_{l=1}^{i} |F_{ik}||L_{kl}||F_{il}|\\ &\leqslant \sum_{k=1}^{i}\sum_{l=1}^{k-1} K'_M i^{-p_M} CC_1 k^{-p+t} l^t K'_M i^{-p_M} + \sum_{k=1}^{i} K'_M i^{-p_M} C_1 k^{2t} K'_M i^{-p_M}\\ &\quad + \sum_{k=1}^{i}\sum_{l=k+1}^{i} K'_M i^{-p_M} CC_1 l^{-p+t} k^t K'_M i^{-p_M}\\ &\leqslant C_1 K'^2_M \Bigg[C i^{-2p_M+[-p+t+[t]_++1]_++1} + i^{-2p_M+[2t]_++1}\\ &\quad + C\begin{Bmatrix} i^{-2p_M+[-p+2t]_++2}, & p \geqslant t\\ i^{-2p_M-p+2t+2}, & p < t\end{Bmatrix}\Bigg].\end{aligned}$$

Hence,

$$\sum_{l=1}^{i} |L_{il}||F_{il}| + \sum_{k=1}^{i} |F_{ik}||L_{ki}| + \sum_{k=1}^{i}\sum_{l=1}^{i} |F_{ik}||L_{kl}||F_{il}| \leqslant D_1 i^{2t},$$

where

$$D_1 = C_1 K'_M[2(C+1) + K'_M(2C+1)],$$

provided $p_M > [-t]_+ + 1$. Hence, from (6.2.10),

$$A_2 i^{2t} \leqslant |\bar{L}_{ii}| \leqslant A_1 i^{2t},$$

where

$$A_1 = C_1 + D_1,$$

$$A_2 = C_2 - D_1,$$

and

$$A_2 > 0 \text{ provided } C_1/C_2 < [K'_M\{2(C+1) + K'_M(2C+1)\}]^{-1}.$$

(2) The (i, j)-th component of (6.2.6) gives

$$\begin{aligned}\bar{L}_{ij} &= \sum_{k=1}^{i}\sum_{l=1}^{j} T_{ik} L_{kl} T_{jl} \\ &= \sum_{k=1}^{i}\sum_{l=1}^{j} (\delta_{ik} + F_{ik}) L_{kl} (\delta_{jl} + F_{jl}) \\ &= L_{ij} + \sum_{l=1}^{j} L_{il} F_{jl} + \sum_{k=1}^{i} F_{ik} L_{kj} + \sum_{k=1}^{i}\sum_{l=1}^{j} F_{ik} L_{kl} F_{jl}.\end{aligned}$$

If, for $i > j$, we insert the bounds (6.1.9) and (6.2.2), and use the summation bounds of the Appendix, then we obtain

$$|\bar{L}_{ij}| \leqslant A_3 i^{-p+t} j^{-v_1} + A_4 i^{-p_M} j^{-v_2}$$

where

$$A_3 = CC_1(1 + K'_M)\left[1 + K'_M\left\{\begin{matrix} 1, & t \geqslant p \\ \left(\dfrac{t-p+2}{t-p+1}\right), & p-1 < t < p \\ 0, & t < p-1 \end{matrix}\right\}\right]$$

$$A_4 = C_1 K'_M(1 + K'_M)\left(1 + C\left[1 + \left\{\begin{matrix} 0, & t > p-1 \\ \left(\dfrac{p-t}{p-t-1}\right), & t < p-1 \end{matrix}\right\}\right]\right)$$

$$\begin{aligned} v_1 &= \min\{-t, p_M - [t]_+ - 1\} \\ &= \min\{-t, p_M - 1\} \\ v_2 &= \min\{p - t - [t]_+ - 1, -2t, p_M - [-p + t + [t]_+ + 1]_+ - 1, \\ &\qquad\qquad p_M - [2t]_+ - 1\} \\ &= \min\{p - t - [t]_+ - 1, -2t, p_M - 1\}. \end{aligned}$$

Comments

(a) The assumption $C_2 i^{2t} \leqslant |L_{ii}|$ is only required in part (1) of the lemma.
(b) The restriction (6.2.8) is implicitly a restriction on the size of C and C_M.
(c) The case $t = p - 1$ is easily catered for in the proof of part (2) of the lemma (Freeman, 1974).

Using this lemma, we proceed to a theorem which characterises the eigenvalues.

Theorem 6.1. *Let the matrices* **L** *and* **M** *satisfy the conditions of Lemma* 6.1. *Then to every eigenvalue* λ_i *of* $\{\mathbf{L}, \mathbf{M}\}$ *there corresponds a diagonal element* $\bar{L}_{ii}$ *of* $\bar{\mathbf{L}}$ *such that for some positive constant* A_5,

$$|\lambda_i - \bar{L}_{ii}| \leqslant A_5(i-1)^{-v_3}, \quad i \geqslant 2,$$

where

$$v_3 = \min\{p - t - [t]_+ - 1, \quad p_M - [2t]_+ - 1\},$$

provided

$$v_3 > 0.$$

Proof. Since **T** is lower triangular, $\bar{\mathbf{L}}^{(N)}$ is the leading $N \times N$ submatrix of $\bar{\mathbf{L}}$. Hence we may partition $\bar{\mathbf{L}}$ in the form

$$\bar{\mathbf{L}} = \bar{\mathbf{L}}^{(N)'} + \Delta\bar{\mathbf{L}},$$

where

$$\bar{\mathbf{L}}^{(N)'} = \left[\begin{array}{c|c} \bar{\mathbf{L}}^{(N)} & \mathbf{0} \\ \hline \mathbf{0} & \mathbf{diag}(\bar{\mathbf{L}}) \end{array}\right]$$

and

$$\mathbf{diag}(\bar{\mathbf{L}}) = \begin{bmatrix} \bar{L}_{N+1,N+1} & 0 \cdots\cdots \\ 0 & \bar{L}_{N+2,N+2} \\ \vdots & & \ddots \end{bmatrix}.$$

The eigenvalues of $\overline{\mathbf{L}}^{(N)'}$ are clearly

$$\lambda_i^{(N)'} = \begin{cases} \lambda_i^{(N)}, & i \leqslant N, \\ \overline{L}_{ii}, & i > N. \end{cases}$$

Hence, using the well-known result that the corresponding eigenvalues of the symmetric matrices $\mathbf{L}$ and $\mathbf{L} + \boldsymbol{\delta}\mathbf{L}$ differ by no more than $\|\boldsymbol{\delta}\mathbf{L}\|$ (Wilkinson, 1965, Section 2.44), we obtain

$$|\lambda_i - \lambda_i^{(N)'}| \leqslant \|\Delta\overline{\mathbf{L}}\|, \tag{6.2.11}$$

where λ_i is the i-th ordered eigenvalue of $\{\mathbf{L}, \mathbf{M}\}$. Setting $i = N + 1$ in (6.2.11), and taking the ∞-norm gives

$$\begin{aligned} |\lambda_{N+1} - \overline{L}_{N+1,N+1}| &\leqslant \|\Delta\overline{\mathbf{L}}\|_\infty \\ &\leqslant \max\left\{\max_{i=1}^{N} \sum_{j=N+1}^{\infty} |\overline{L}_{ij}|, \max_{i=N+1}^{\infty} \sum_{\substack{j=1 \\ j\neq i}}^{\infty} |\overline{L}_{ij}|\right\}. \end{aligned}$$

Using the bounds on $|\overline{L}_{ij}|$ of Lemma 6.1 leads to

$$\begin{aligned} |\lambda_{N+1} - \overline{L}_{N+1,N+1}| \leqslant \max\Bigg\{ &\max_{i=1}^{N} \sum_{j=N+1}^{\infty} (A_3 j^{-p+t} i^{-v_1} + A_4 j^{-p_M} i^{-v_2}), \\ &\max_{i=N+1}^{\infty} \left[\sum_{j=1}^{i-1} (A_3 i^{-p+t} j^{-v_1} + A_4 i^{-p_M} j^{-v_2}) \right. \\ &\left. + \sum_{j=i+1}^{\infty} (A_3 j^{-p+t} i^{-v_1} + A_4 j^{-p_M} i^{-v_2})\right]\Bigg\} \\ \leqslant \max\Bigg\{ &\left[A_3 \frac{N^{-p+t+[-v_1]_+ +1}}{p-t-1} + A_4 \frac{N^{-p_M+[-v_2]_+ +1}}{p_M - 1}\right], \\ &\left[A_3 N^{-p+t+[-v_1]_+ +1} + A_4 N^{-p_M+[-v_2]_+ +1} \right. \\ &\left. + A_3 \frac{N^{-p-v_1+t+1}}{p-t-1} + A_4 \frac{N^{-p_M-v_2+1}}{p_M - 1}\right]\Bigg\}, \end{aligned}$$

provided $p > t + [-v_1]_+ + 1$, $p_M > [-v_2]_+ + 1$.

But $[-v_1]_+ = [t]_+$ and $[-v_2]_+ = [2t]_+$, hence

$$|\lambda_{N+1} - \overline{L}_{N+1,N+1}| \leqslant A_5 N^{-v_3},$$

where

$$A_5 = A_3\left(\frac{p-t}{p-t-1}\right) + A_4\left(\frac{p_M}{p_M - 1}\right),$$

and

$$v_3 = \min\{p - t - [t]_+ - 1, \quad p_M - [2t]_+ - 1\}.$$

This result is true for all $N \geqslant 1$, and the theorem follows immediately.

In a similar way we can consider the convergence of an approximate eigenvalue $\lambda_i^{(N)}$ to λ_i and derive a simple bound on $|\lambda_i - \lambda_i^{(N)}|$.

Theorem 6.2. *Let the matrices* **L** *and* **M** *satisfy the conditions of Lemma* 6.1. *Then to every eigenvalue* $\lambda_i^{(N)}$ *of* $\{\mathbf{L}^{(N)}, \mathbf{M}^{(N)}\}$ *there corresponds an eigenvalue* λ_i *of* $\{\mathbf{L}, \mathbf{M}\}$ *such that*

$$|\lambda_i - \lambda_i^{(N)}| \leqslant A_5 N^{-v_3}, \quad i \leqslant N,$$

where A_5 *and* v_3 *are given in Theorem* 6.1, *provided that* $v_3 > 0$.

Proof. The result follows immediately from (6.2.11) with $i \leqslant N$ on inserting the bound on $\|\Delta\bar{\mathbf{L}}\|_\infty$ given by Theorem 6.1.

We now quote the corresponding results for the case when **L** and **M** are AD of type B.

Theorem 6.1′. *Let the matrix* **L** *be AD of type* $B(p, r; C)$ *such that*

$$|L_{ii}| \leqslant C_1 i^{2t}.$$

We identify the parameter ranges:

type B1: $p > 1, r \geqslant 0$,

type B2: $p \geqslant 0, r > 1$.

Also, let **M** *be AD of type* $B(p_M, r_M; C_M)$ *and be normalised by* (6.1.8). *Then, provided* C_M *is sufficiently small, and* $p_M > 1$, $r_M \geqslant 0$ (*type B*1) *or* $p_M \geqslant 0$, $r_M > 1$ (*type B*2), *to every eigenvalue* λ_i *of* $\{\mathbf{L}, \mathbf{M}\}$ *there corresponds a diagonal element* $\bar{L}_{ii}$ *of* $\bar{\mathbf{L}}$ *such that for some positive constant* A_6,

$$|\lambda_i - \bar{L}_{ii}| \leqslant A_6 (i-1)^{-v_4}, \quad i \geqslant 2,$$

where

$$v_4 = \begin{cases} \min\{p - t - [t]_+ - 1, p_M - [2t]_+ - 1\}, & \text{type } B1, \\ \min\{p - t - [t]_+, p_M - [2t]_+\}, & \text{type } B2, \end{cases}$$

provided

$$\begin{cases} v_4 > 0, & \text{type } B1, \\ v_4 \geqslant 0, & \text{type } B2. \end{cases}$$

Comment

The restriction that C_M should be sufficiently small ensures that Theorem 3.1 can be invoked to factorise $\mathbf{M}$ as

$$\mathbf{T}\mathbf{M}\mathbf{T}^t = \mathbf{I}.$$

Precise details of this restriction are given in Freeman (1974).

Theorem 6.2′. *Let the matrices* $\mathbf{L}$ *and* $\mathbf{M}$ *satisfy the conditions of Theorem* 6.1′. *Then to every eigenvalue* $\lambda_i^{(N)}$ *of* $\{\mathbf{L}^{(N)}, \mathbf{M}^{(N)}\}$ *there corresponds an eigenvalue* λ_i *of* $\{\mathbf{L}, \mathbf{M}\}$ *such that*

$$|\lambda_i - \lambda_i^{(N)}| \leqslant A_6 N^{-v_4}, \quad i \leqslant N,$$

where A_6 *and* v_4 *are given in Theorem* 6.1′, *provided that* $v_4 > 0$, *type B*1 *or* $v_4 \geqslant 0$, *type B*2.

Comments

(a) Results for the special case $\mathbf{M} = \mathbf{I}$ are obtained by letting $p_M \to \infty$ in Theorems 6.1 and 6.2, and by letting $p_M, r_M \to \infty$ in Theorems 6.1′ and 6.2′.
(b) The results of Theorem 6.2 illustrate one inevitable limitation of theorems of this type. The eigenvalues $\lambda_i^{(N)}$ of $\{\mathbf{L}^{(N)}, \mathbf{M}^{(N)}\}$ are invariant under any non-singular linear transformation of $\mathbf{M}$, in particular under a Gram-Schmidt orthogonalisation; that is, if $\mathbf{A}$ is any non-singular $N \times N$ matrix, then the system

$$[\mathbf{A}\mathbf{L}^{(N)}\mathbf{A}^{-1} - \lambda \mathbf{A}\mathbf{M}^{(N)}\mathbf{A}^{-1}]\,\mathbf{y} = \mathbf{0},$$
$$\mathbf{y} = \mathbf{A}\mathbf{a}^{(N)},$$

has the same eigenvalue spectrum as (6.1.5). In particular, we may choose $\mathbf{A}$ to be triangular and such that for every N,

$$\mathbf{A}\mathbf{M}^{(N)}\mathbf{A}^{-1} = \mathbf{I}.$$

However, the results of Theorem 6.2 are *not* invariant under such a transformation. For matrices $\mathbf{M}^{(N)}$ which are *not* close to a unit matrix, the results predicted by Theorem 6.2 may therefore be *pessimistic*; often, better error bounds will result from considering the corresponding orthogonalised expansion directly. Even in this case we have no proof that Theorem 6.2 yields the best possible bound; indeed, Theorem 6.5 suggests that it does not. This is to be expected since the bounds of Theorem 6.2 apply uniformly to every eigenvalue of the system, while in practice we might hope to obtain more rapid convergence for some dominant eigenvalue. These comments are equally applicable to the results of Theorem 6.2′.

6.3 Eigenvector convergence

We now consider the eigenvector **b** of (6.1.7), and in the next theorem bound the i-convergence rate of the expansion coefficients b_i of the eigenfunction f of (6.1.4). As stated, the theorem singles out the element L_{11} of **L** and a corresponding eigenvalue and eigenvector λ_1, $\mathbf{b}_1$ of (6.1.7). However, the rows and columns of **L** could be permuted to interchange the elements L_{11} and L_{ii}, say; the resulting matrices **L** and **M** would remain AD of type A, although with possibly larger parameters C, C_M, and it would thus be possible to characterise the eigenvector $\mathbf{b}_i$ of (6.1.7). In this way the theorem could be used to characterise the first few eigenvectors of (6.1.7).

Theorem 6.3. *Let the eigenvector* $\mathbf{b}_1$ *be normalised so that* $\mathbf{b}_1^t\mathbf{M}\mathbf{b}_1 = 1$, *and let* **L** *and* **M** *satisfy the conditions of Lemma* 6.1; *also let* $\bar{\mathbf{L}}$, *defined by* (6.2.6), *have diagonal elements satisfying* $|\bar{L}_{11} - \bar{L}_{ii}| > 0$, $i \geqslant 2$. *Then there exists a positive constant* A_7 *such that*

$$|(b_1)_i| \leqslant A_7 i^{-v_5},$$

where

$$v_5 = \min\{p + |t|, p_M + [2t]_+\},$$

provided $p > 1$, $p_M > [-t]_+ + 1$, *and* C, C_M *are sufficiently small,* C_1, C_2 *are sufficiently large, and* C_1/C_2 *is sufficiently close to unity.*

We refer to Freeman (1974) for a proof of this theorem; the proof, which is straightforward but very lengthy, is based on the Newton-Kantorovich Theorem (see Tapia, 1971, for example). The proof illustrates more clearly the need for the restrictions on the constants C, C_M, C_1, C_2 and considers these restrictions in more detail.

Theorem 6.3′. *Let the eigenvector* $\mathbf{b}_1$ *be normalised by* $\mathbf{b}_1^t\mathbf{M}\mathbf{b}_1 = 1$, *and let* **L** *and* **M** *satisfy the conditions of Theorem* 6.1′, *with* $C_2 i^{2t} \leqslant |L_{ii}|$; *also let* $\bar{\mathbf{L}}$, *defined by* (6.2.6), *have diagonal elements satisfying* $|\bar{L}_{11} - \bar{L}_{ii}| > 0$, $i \geqslant 2$. *Then there exists a positive constant* A_8 *such that*

$$|(b_1)_i| \leqslant A_8 i^{-v_6},$$

where

$$v_6 = \begin{cases} \min\{p + |t|, p_M + [2t]_+\}, & \text{type B1}, \\ \min\{p + r_1 + |t|, p_M + r_1 + [2t]_+\}, & \text{type B2}, \end{cases}$$

$$r_1 = \min\{r, r_M\},$$

provided

$$p_M > [-t]_+ + 1, \quad \textit{type B1},$$

$$p_M \geqslant [-t]_+, \qquad \textit{type B2},$$

and C, C_M, C_1, C_2 satisfy restrictions of the type described in Theorem 6.3.

Comments

(a) The result embodied in Theorem 6.3 agrees with that which would be obtained from Theorem 2.2 by setting the right-hand side exponent to infinity and making the normalisations consistent.

(b) Although the theorem is stated for the infinite system (6.1.7), the proof and hence the result is valid uniformly for all N (and $i \leqslant N$) in (6.1.5).

6.4 Prediction coefficients for the eigenvectors

Theorem 6.3 provides an *a priori* bound on the i-convergence rate of the coefficients b_i. We now show that we may also predict this convergence rate numerically, using the results of a small pilot computation. The theorem which follows introduces a set of easily computable "prediction coefficients" d_i and shows that under suitable conditions $d_i \underset{i}{\rightarrow} b_i$. Analogous coefficients for the inhomogeneous problem are introduced in Section 8.5.

We suppose that we have available approximations $\lambda^* = \lambda + \delta$ and d_i, $i = 1, 2, \ldots, N$, to λ and b_i, $i = 1, 2, \ldots, N$; we then define the set $\{d_i, i > N\}$ as the solutions of the triangular equations

$$\sum_{j=1}^{i} (L_{ij} - \lambda^* M_{ij})\, d_j = 0, \quad i > N. \tag{6.4.1}$$

Theorem 6.4. *Let the matrices* $\mathbf{L}$ *and* $\mathbf{M}$ *satisfy the conditions of Lemma 6.1, with $|L_{ii} - \lambda| > 0$, $\forall i$, and let*

$$|b_i| \leqslant \kappa i^{-q},$$

with $q > [t]_+ + 1$. Then for some positive constant A_9, and all $i > N$,

$$|d_i - b_i| \leqslant A_9 i^{-v_7},$$

where

$$v_7 = \min\{p + [-t]_+, p_M + [t]_+, q + [t]_+\} + [t]_+,$$

provided $p > 1$, and $|\delta|$, C, C_M are sufficiently small (this restriction is considered in detail in the proof of the theorem).

Proof. The off-diagonal elements of **L** are bounded as in (6.1.9), and hence for $i > j$,

$$|L_{ij} - \lambda M_{ij}| \leqslant CC_1 i^{-p+t} j^{t} + |\lambda| C_M i^{-p_M}.$$

Since $|L_{ii} - \lambda| > 0$, $\forall i$, it follows that there exist positive constants B_1, B_2 such that

$$B_2 i^{[2t]_+} \leqslant |L_{ii} - \lambda M_{ii}| \leqslant B_1 i^{[2t]_+}.$$

We now introduce the renormalised expansion set $\{\bar{h}_i\}$ such that

$$\bar{L}_{ii} - \lambda \bar{M}_{ii} = J_i = \pm 1, \quad \forall i.$$

Let $\bar{h}_i = \alpha_i h_i$, then $|L_{ii} - \lambda M_{ii}| = \alpha_i^{-2}$, and hence

$$B_1^{-1} i^{-[2t]_+} \leqslant \alpha_i^2 \leqslant B_2^{-1} i^{-[2t]_+}.$$

Hence, for $i > j$,

$$\begin{aligned} |\bar{L}_{ij} - \lambda \bar{M}_{ij}| &\leqslant |L_{ij} - \lambda M_{ij}|\, |\alpha_i|\, |\alpha_j|, \\ &\leqslant B_3 i^{-p-[-t]_+} j^{-[-t]_+} + B_4 i^{-p_M-[t]_+} j^{-[t]_+}, \end{aligned}$$

where $B_3 = CC_1 B_2^{-1}$ and $B_4 = C_M |\lambda| B_2^{-1}$. Also

$$\begin{aligned} |\bar{b}_i| &= \frac{|b_i|}{|\alpha_i|} \\ &\leqslant B_5 i^{-q+[t]_+}, \end{aligned}$$

where $B_5 = \kappa B_1^{1/2}$. Also, for $i > j$,

$$\begin{aligned} |\bar{M}_{ij}| &= |M_{ij}|\, |\alpha_i|\, |\alpha_j| \\ &\leqslant B_6 i^{-p_M-[t]_+} j^{-[t]_+}, \end{aligned}$$

where $B_6 = C_M B_2^{-1}$, and

$$\begin{aligned} |\bar{M}_{ii}| &= |M_{ii}|\, |\alpha_i|^2 \\ &\leqslant B_2^{-1} i^{-[2t]_+}. \end{aligned}$$

With $\{\bar{h}_i\}$ as expansion set, (6.4.1) becomes

$$\sum_{j=1}^{i} (\bar{L}_{ij} - \lambda^* \bar{M}_{ij})\, \bar{d}_j = 0.$$

Therefore

$$\begin{aligned} \bar{d}_i &= \left(- \sum_{j=1}^{i-1} (\bar{L}_{ij} - \lambda^* \bar{M}_{ij})\, \bar{d}_j \right) \Big/ (\bar{L}_{ii} - \lambda^* \bar{M}_{ii}) \\ &= \left(- \sum_{j=1}^{i-1} (\bar{L}_{ij} - \lambda^* \bar{M}_{ij})\, \bar{d}_j \right) \Big/ (J_i - \delta \bar{M}_{ii}), \end{aligned} \tag{6.4.2}$$

since $\bar{L}_{ii} - \lambda \bar{M}_{ii} = J_i$, and $\lambda^* = \lambda + \delta$.

With $\{\bar{h}_i\}$ as expansion set, the i-th equation of (6.1.7) becomes

$$J_i\bar{b}_i = -\sum_{j=1}^{i-1} (\bar{L}_{ij} - \lambda\bar{M}_{ij})\,\bar{b}_j - \sum_{j=i+1}^{\infty} (\bar{L}_{ij} - \lambda\bar{M}_{ij})\,\bar{b}_j.$$

Hence,

$$J_i\bar{b}_i - \delta\bar{M}_{ii}\bar{b}_i = -\sum_{j=1}^{i-1} (\bar{L}_{ij} - \lambda\bar{M}_{ij})\,\bar{b}_j - \sum_{j=i+1}^{\infty} (\bar{L}_{ij} - \lambda\bar{M}_{ij})\,\bar{b}_j - \delta\bar{M}_{ii}\bar{b}_i$$

and

$$\bar{b}_i = \left[-\sum_{j=1}^{i-1} (\bar{L}_{ij} - \lambda^*\bar{M}_{ij})\,\bar{b}_j - \sum_{j=i+1}^{\infty} (\bar{L}_{ij} - \lambda^*\bar{M}_{ij})\,\bar{b}_j - \sum_{j=1}^{\infty} \delta\bar{M}_{ij}\bar{b}_j\right] \Big/ (J_i - \delta\bar{M}_{ii}) \qquad (6.4.3)$$

If we define $\bar{c}_i = \bar{d}_i - \bar{b}_i$ and substract (6.4.3) from (6.4.2), we obtain

$$|\bar{c}_i| \leqslant |(J_i - \delta\bar{M}_{ii})^{-1}| \left\{ \sum_{j=1}^{i-1} |\bar{L}_{ij} - \lambda^*\bar{M}_{ij}|\,|\bar{c}_j| + \sum_{j=i+1}^{\infty} |\bar{L}_{ij} - \lambda^*\bar{M}_{ij}|\,|\bar{b}_j| + \sum_{j=1}^{\infty} |\delta|\,|\bar{M}_{ij}|\,|\bar{b}_j| \right\}. \qquad (6.4.4)$$

Now, $|\bar{M}_{ii}| \leqslant B_2^{-1} i^{-[2t]_+}$, and hence

$$|(J_i - \delta\bar{M}_{ii})^{-1}| \leqslant |(1 - J_i\delta B_2^{-1} i^{-[2t]_+})^{-1}| \leqslant 1 + \beta i^{-[2t]_+},$$

where

$$\beta = \frac{|\delta|\,B_2^{-1}}{(1 - |\delta|\,B_2^{-1} i^{-[2t]_+})},$$

provided $|\delta|\,B_2^{-1} i^{-[2t]_+} < 1$. Also, for $i > j$,

$$|\bar{L}_{ij} - \lambda^*\bar{M}_{ij}| \leqslant |\bar{L}_{ij} - \lambda\bar{M}_{ij}| + |\delta|\,|\bar{M}_{ij}| \leqslant B_3 i^{-p-[-t]_+} j^{-[-t]_+} + B_7 i^{-p_M-[t]_+} j^{-[t]_+},$$

where $B_7 = B_4 + |\delta|\,B_6$.

Suppose $|\bar{c}_j| \leqslant B_8 j^{-w}, j = 1,2,\ldots, i-1$, then on inserting the appropriate bounds in (6.4.4) and using the summation bounds of the Appendix we obtain

$$|\bar{c}_i| \leqslant (1 + \beta i^{-[2t]_+}) \left[B_3 B_8 i^{-p-[-t]_+} \left(\frac{w + [-t]_+}{w + [-t]_+ - 1} \right) \right.$$

$$+ B_7 B_8 i^{-p_M - [t]_+} \left(\frac{w + [t]_+}{w + [t]_+ - 1} \right)$$

$$+ B_3 B_5 \frac{i^{-p-q-2[-t]_+ + [t]_+ + 1}}{(p + q + [-t]_+ - [t]_+ - 1)} + B_7 B_5 \frac{i^{-p_M - q - [t]_+ + 1}}{(p_M + q - 1)}$$

$$+ |\delta| B_6 B_5 i^{-p_M - [t]_+} \left(\frac{q}{q - 1} \right) + |\delta| B_2^{-1} B_5 i^{-q-[t]_+}$$

$$\left. + |\delta| B_6 B_5 \frac{i^{-p_M - q - [t]_+ + 1}}{(p_M + q - 1)} \right],$$

provided $w + [-t]_+ > 1, w + [t]_+ > 1, p + q > t + 1, p_M + q > 1, q > 1$. Hence,

$$|\bar{c}_i| \leqslant B_8 i^{-w},$$

provided

(1) $w = \min\{p + [-t]_+, p_M + [t]_+, q + [t]_+\}$.

(2) $$B_8 \geqslant (1 + \beta) \left[B_8 \left\{ B_3 \left(\frac{w + [-t]_+}{w + [-t]_+ - 1} \right) + B_7 \left(\frac{w + [t]_+}{w + [t]_+ - 1} \right) \right\} \right.$$

$$+ B_5 \left\{ \frac{B_3}{(p + q - t - 1)} + \frac{B_7}{(p_M + q - 1)} \right\}$$

$$\left. + |\delta| B_5 \left\{ B_6 \left(\frac{q}{q - 1} \right) + \frac{B_6}{(p_M + q - 1)} + B_2^{-1} \right\} \right].$$

Note that B_8 can always be chosen to satisfy this inequality provided

$$(1 + \beta) \left[B_3 \left(\frac{w + [-t]_+}{w + [-t]_+ - 1} \right) + B_7 \left(\frac{w + [t]_+}{w + [t]_+ - 1} \right) \right] < 1,$$

which in turn is immediately satisfied provided $|\delta|$, C, C_M are sufficiently small.

(3) $q > [t]_+ + 1$.

Now, the inductive process may begin at any integer $i = N$, say, since the inequality $|\bar{c}_i| \leqslant B_8 i^{-w}$ can be satisfied for all $i \leqslant N$ by choosing B_8 appropriately.

Hence

$$|\bar{c}_i| = |\bar{d}_i - \bar{b}_i| \leqslant B_8 i^{-w},$$

and therefore

$$|c_i| = |d_i - b_i| = |\alpha_i|\,|\bar{d}_i - \bar{b}_i| \leqslant A_9 i^{-v_7}$$

where

$$v_7 = w + [t]_+ \quad \text{and} \quad A_9 = B_8 B_2^{-1/2}.$$

A similar result holds if **L** and **M** are AD of type B.

Theorem 6.4′. *Let the matrices* **L** *and* **M** *satisfy the conditions of Theorem 6.1′, with* $C_2 i^{2t} \leqslant |L_{ii}|$, *and with* $|L_{ii} - \lambda| > 0$, $\forall i$, *and let*

$$|b_i| \leqslant \kappa i^{-q},$$

with $q > [t]_+ + 1$. *Then for some positive constant* A_{10}, *and all* $i > N$

$$|d_i - b_i| \leqslant A_{10} i^{-v_8},$$

where

$$v_8 = \begin{cases} \min\{p + [-t]_+, p_M + [t]_+, q + [t]_+\} + [t]_+, & \textit{type B1}, \\ \min\{p + r + [-t]_+, p_M + r_M + [t]_+, \\ \qquad p + q + [-t]_+ - t, q + [t]_+\} + [t]_+, & \textit{type B2}, \end{cases}$$

provided $|\delta|$, C, C_M *are sufficiently small.*

6.5 An alternative bound on the convergence rate of the dominant eigenvalue

Theorem 6.1 shows directly that if $t < 0$, or if $t > 0$ and L_{ii} is single signed for i sufficiently large, then the system (6.1.7) has a smallest or largest eigenvalue. Without essential loss of generality we take this dominant eigenvalue to be λ_1, and assume that it is the smallest eigenvalue. Using the Rayleigh quotient we then have

$$\lambda_1 = \min_{\mathbf{b}} \frac{\mathbf{b}^t \mathbf{L} \mathbf{b}}{\mathbf{b}^t \mathbf{M} \mathbf{b}}, \tag{6.5.1}$$

and

$$\lambda_1^{(N)} = \min_{\mathbf{a}^{(N)}} \frac{\mathbf{a}^{(N)t}\mathbf{L}^{(N)}\mathbf{a}^{(N)}}{\mathbf{a}^{(N)t}\mathbf{M}^{(N)}\mathbf{a}^{(N)}} \tag{6.5.2}$$

$$= \lambda_1 + \frac{\boldsymbol{\varepsilon}_N^t(\mathbf{L} - \lambda_1\mathbf{M})\,\boldsymbol{\varepsilon}_N}{\bar{\mathbf{a}}^{(N)t}\mathbf{M}\bar{\mathbf{a}}^{(N)}}, \tag{6.5.3}$$

where

$$\boldsymbol{\varepsilon}_N = \mathbf{b} - \bar{\mathbf{a}}^{(N)},$$

$$\bar{\mathbf{a}}^{(N)} = \begin{bmatrix} \mathbf{a}^{(N)} \\ 0 \\ 0 \\ \vdots \end{bmatrix},$$

and $\mathbf{b}$, $\mathbf{a}^{(N)}$ are the minimising vectors in (6.5.1), (6.5.2).

Theorem 6.5. *Let* (6.1.7) *have a dominant eigenvalue* λ_1 *and let the corresponding eigenvector* $\mathbf{b}$ *be normalised so that* $\mathbf{b}^t\mathbf{M}\mathbf{b} = 1$; *also let* $\mathbf{L}$ *and* $\mathbf{M}$ *be defined as in Lemma* 6.1. *Further, let*

$$|b_i| \leqslant \kappa i^{-q}, \quad q > [t]_+ + 1.$$

Then, for some positive constant A_{11}, *and all* N,

$$|\lambda_1 - \lambda_1^{(N)}| \leqslant A_{11} N^{-v_9}$$

where

$$v_9 = 2q - [2t]_+ - 1,$$

provided $p \geqslant 1$.

Proof. From (6.5.3), for any $\bar{\mathbf{a}}^{(N)}$,

$$|\lambda_1 - \lambda_1^{(N)}| \leqslant \frac{|\boldsymbol{\varepsilon}_N^t(\mathbf{L} - \lambda_1\mathbf{M})\,\boldsymbol{\varepsilon}_N|}{|\bar{\mathbf{a}}^{(N)t}\mathbf{M}\bar{\mathbf{a}}^{(N)}|}.$$

We choose $\bar{a}_i^{(N)} = b_i$, $i = 1, 2, \ldots, N$, and it then follows that

$$|\boldsymbol{\varepsilon}_N^t(\mathbf{L} - \lambda_1\mathbf{M})\,\boldsymbol{\varepsilon}_N| \leqslant \sum_{i=N+1}^{\infty} \sum_{j=N+1}^{\infty} |b_i|(|L_{ij}| + |\lambda_1|\,|M_{ij}|)\,|b_j|,$$

and $|\bar{\mathbf{a}}^{(N)t}\mathbf{M}\bar{\mathbf{a}}^{(N)}| > B_1$, where B_1 is a positive constant, since $\mathbf{b}^t\mathbf{M}\mathbf{b} = 1$. Hence,

$$|\lambda_1 - \lambda_1^{(N)}| \leqslant \sum_{i=N+1}^{\infty} \sum_{j=N+1}^{\infty} |b_i|(|L_{ij}| + |\lambda_1||M_{ij}|)|b_j|/B_1,$$

and, using the bound (6.1.9), we obtain

$$|\lambda_1 - \lambda_1^{(N)}| \leqslant \sum_{i=N+1}^{\infty} \left\{ \sum_{j=N+1}^{i-1} \kappa^2 i^{-q} j^{-q} (CC_1 i^{-p+t} j^t + |\lambda_1| C_M i^{-p_M}) \right.$$
$$+ \kappa^2 i^{-2q}(C_1 i^{2t} + |\lambda_1|)$$
$$\left. + \sum_{j=i+1}^{\infty} \kappa^2 i^{-q} j^{-q} (CC_1 j^{-p+t} i^t + |\lambda_1| C_M j^{-p_M} \right\} \Big/ B_1.$$

Using the summation bounds of the Appendix gives

$$|\lambda_1 - \lambda_1^{(N)}| \leqslant \kappa^2 \left\{ \frac{CC_1 N^{-(p+2q)+2t+2}}{(q-t-1)(p+q-t-1)} + \frac{|\lambda_1| C_M N^{-(p_M+2q)+2}}{(q-1)(p_M+q-1)} \right.$$
$$+ \frac{C_1 N^{-2q+2t+1}}{(2q-2t-1)} + \frac{|\lambda_1| N^{-2q+1}}{(2q-1)}$$
$$+ \frac{CC_1 N^{-(p+2q)+2t+2}}{(p+q-t-1)(p+2q-2t-2)}$$
$$\left. + \frac{|\lambda_1| C_M N^{-(p_M+2q)+2}}{(p_M+q-1)(p_M+2q-2)} \right\} \Big/ B_1,$$

provided $q > [t]_+ + 1$. Hence,

$$|\lambda_1 - \lambda_1^{(N)}| \leqslant A_{11} N^{-v_9},$$

where

$$v_9 = 2q - [2t]_+ - 1,$$
$$A_{11} = \kappa^2 \left\{ \frac{CC_1}{(p+q-t-1)} \left[\frac{1}{(q-t-1)} + \frac{1}{(p+2q-2t-2)} \right] \right.$$
$$+ \frac{|\lambda_1| C_M}{(p_M+q-1)} \left[\frac{1}{(q-1)} + \frac{1}{(p_M+2q-2)} \right] + \frac{C_1}{(2q-2t-1)}$$
$$\left. + \frac{|\lambda_1|}{(2q-1)} \right\}.$$

Corollary 6.1. If the conditions of Theorem 6.3 hold, then

$$|\lambda_1 - \lambda_1^{(N)}| \leqslant A_{12} N^{-v_{10}},$$

where A_{12} is a positive constant independent of N, and

$$v_{10} = 2\min\{p + [-t]_+, p_M + [t]_+\} - 1.$$

We note that this result is much stronger than that given by Theorem 6.2. We may again extend this result to the case when **L** and **M** are AD matrices of type B.

Theorem 6.5′. *If the conditions of Theorem* 6.3′ *hold, then*

$$|\lambda_1 - \lambda_1^{(N)}| \leqslant A_{13} N^{-v_{11}},$$

where A_{13} *is a positive constant independent of* N, *and*

$$v_{11} = \begin{cases} 2\min\{p + [-t]_+, p_M + [t]_+\} - 1, & \textit{type B1}, \\ 2\min\{p + r_1 + [-t]_+, p_M + r_1 + [t]_+\} - 1, & \textit{type B2}, \end{cases}$$

$$r_1 = \min\{r, r_M\}.$$

6.6 Stability against round-off errors

We conclude the theoretical part of this chapter with a brief consideration of the stability of the eigenvalue calculation. In this section, we restrict our attention to the case when **L** and **M** are AD of type A, and refer to Delves and Freeman (1975) for the corresponding type B results.

In the presence of round-off errors we solve the perturbed equations

$$[(\mathbf{L} + \delta\mathbf{L}) - (\lambda + \delta\lambda)(\mathbf{M} + \delta\mathbf{M})](\mathbf{a} + \delta\mathbf{a}) = \mathbf{0}, \tag{6.6.1}$$

where throughout this section we omit the superscript N.

Introducing the triangular matrix **T** of (6.2.1), it is straightforward to derive the bound (Delves, 1968),

$$|\delta\lambda| \leqslant \frac{\|\mathbf{T}\|\,\|\mathbf{T}^t\|(\|\delta\mathbf{L}\| + |\lambda|\,\|\delta\mathbf{M}\|)}{1 - \|\mathbf{T}\|\,\|\mathbf{T}^t\|\,\|\delta\mathbf{M}\|}, \tag{6.6.2}$$

provided the denominator is positive. It follows from (6.2.2) that $\|\mathbf{T}\|_\infty$, $\|\mathbf{T}^t\|_\infty$ are bounded uniformly in N, provided $p_M > 1$. Hence, for some constant K, independent of N, and for any eigenvalue λ,

$$|\delta\lambda| \leqslant \frac{K(\|\delta\mathbf{L}\|_\infty + |\lambda|\|\delta\mathbf{M}\|_\infty)}{1 - K\|\delta\mathbf{M}\|_\infty}, \tag{6.6.3}$$

provided $K\|\delta\mathbf{M}\|_\infty < 1$. The process of calculating λ is then clearly stable with respect to increasing N.

More formally, if $\mathscr{L}$ is positive definite, we can introduce the definition of stability due to Mikhlin (1971).

Definition 6.1. *If $\mathscr{L}$, $\mathscr{M}$ are positive definite, then the process* (6.1.5) *of determining the k-th eigenvalue is stable if*
(a) $\|\mathbf{L}^{-1}\| \leqslant B_1$;
(b) $|\lambda_k| \leqslant B_2$;
(c) for every n-vector $\boldsymbol{\alpha}$, $\boldsymbol{\alpha}^t\mathbf{M}\boldsymbol{\alpha}/\boldsymbol{\alpha}^t\mathbf{L}\boldsymbol{\alpha} \leqslant B_3$,
where B_1, B_2, B_3 *are independent of* N *(but* B_2 *may depend on* k*).*

Theorem 6.6. *Let $\mathscr{L}$ be positive definite, and let* **L**, **M** *be defined as in Lemma* 6.1. *Then the process* (6.1.5) *of determining the k-th eigenvalue is stable provided that* $t \geqslant 0$, (CC_1/C_2) *satisfies the conditions of Theorem* 2.1, *and* $p, p_M > 2t + 1$.

Proof. We first introduce the renormalised matrix $\tilde{\mathbf{L}}$:

$$\mathbf{L} = \mathbf{\Lambda}\tilde{\mathbf{L}}\mathbf{\Lambda},$$

where $\mathbf{\Lambda}$ is a diagonal matrix such that $|\tilde{L}_{ii}| = 1$.
Then

$$C_2^{1/2}i^t \leqslant |\mathbf{\Lambda}_{ii}| \leqslant C_1^{1/2}i^t,$$

and

$$|\tilde{L}_{ij}| \leqslant \frac{CC_1}{C_2} i^{-p}, \quad i > j.$$

Hence, by Theorem 2.1, there exists a lower triangular matrix $\mathbf{V} = \mathbf{I} + \mathbf{G}$ such that

$$\mathbf{V}\tilde{\mathbf{L}}\mathbf{V}^t = \mathbf{I},$$

and

$$|G_{ij}| \leqslant K'i^{-p}, \quad i \geqslant j.$$

Thus,

$$\begin{aligned} \mathbf{V}\tilde{\mathbf{L}}\mathbf{V}^t &= \mathbf{V}\mathbf{\Lambda}^{-1}\mathbf{L}\mathbf{\Lambda}^{-1}\mathbf{V}^t = \mathbf{I}, \\ \mathbf{L}^{-1} &= (\mathbf{V}\mathbf{\Lambda}^{-1})^t\mathbf{V}\mathbf{\Lambda}^{-1}, \\ \|\mathbf{L}^{-1}\|_\infty &\leqslant \|\mathbf{\Lambda}^{-1}\|_\infty \|\mathbf{V}\|_1 \|\mathbf{V}\|_\infty \|\mathbf{\Lambda}^{-1}\|_\infty. \end{aligned}$$

Also, for any $\boldsymbol{\alpha}$, we set $\boldsymbol{\alpha} = \mathbf{\Lambda}^{-1}\mathbf{V}^t\boldsymbol{\beta}$; then

$$\frac{\boldsymbol{\alpha}^t\mathbf{M}\boldsymbol{\alpha}}{\boldsymbol{\alpha}^t\mathbf{L}\boldsymbol{\alpha}} = \frac{\boldsymbol{\beta}^t\mathbf{V}\mathbf{\Lambda}^{-1}\mathbf{M}\mathbf{\Lambda}^{-1}\mathbf{V}^t\boldsymbol{\beta}}{\boldsymbol{\beta}^t\boldsymbol{\beta}} \leqslant \|\mathbf{V}\mathbf{\Lambda}^{-1}\mathbf{M}\mathbf{\Lambda}^{-1}\mathbf{V}^t\|_\infty.$$

Hence we find that conditions (a) and (c) of Definition 6.1 are satisfied provided that $\|\mathbf{\Lambda}^{-1}\|_\infty$, $\|\mathbf{V}\|_1$, $\|\mathbf{V}\|_\infty$ and $\|\mathbf{M}\|_\infty$ are bounded independent of N. It is easy to show, using the summation bounds of the Appendix, that these norms are indeed bounded independent of N, provided that $t \geqslant 0$, $p, p_M > 1$.

Further, condition (b) follows from Theorems 6.1 and 6.2, provided that $p > t + [t]_+ + 1$, $p_M > [2t]_+ + 1$.

6.7 Numerical example

We conclude this chapter by considering the homogeneous eigenproblem

$$\mathscr{L}f(x) - \lambda f(x) = 0, \tag{6.7.1}$$

where $\mathscr{L}$ is the differential operator

$$\mathscr{L} = -\left(\frac{\mathrm{d}^2}{\mathrm{d}x^2} + \sin x\right),$$

subject to the homogeneous boundary conditions

$$f(0) = f(\pi) = 0.$$

The eigenfunctions of this problem are either symmetric or antisymmetric about $x = \pi/2$. We seek here antisymmetric eigenfunctions using the procedure described in Section 6.1, with the expansion set

$$h_i(x) = \sqrt{(2/\pi)}\sin(2ix), \quad i = 1, 2, \ldots.$$

We then find

$$L_{ij} = (h_i, \mathscr{L}h_j) = \begin{cases} 4i^2 - \dfrac{2}{\pi} - \dfrac{2}{\pi(16i^2 - 1)}, & \text{for } i = j, \\[2ex] \dfrac{2}{\pi}\left[\dfrac{1}{(4(i-j)^2 - 1)} - \dfrac{1}{(4(i+j)^2 - 1)}\right], & \text{for } i \neq j, \end{cases}$$

and hence it follows that $\mathbf{L}$ is a symmetric AD matrix of type B(2, 2; $32/33\pi$). Also

$$3i^2 \leqslant |L_{ii}| \leqslant 4i^2.$$

In addition, for this orthogonal expansion set, $\mathbf{M} = \mathbf{I}$.

The theorems for type B systems given in this chapter are thus directly applicable to this example, with $p = r = 2$; $p_M = r_M = \infty$; $t = 1$.

TABLE 6.1.

N	$\lambda_1^{(N)}$	$\lambda_4^{(N)}$	i	$(b_l)i$	$\hat{\lambda}_i$	L_{ii}
1	0·1511736368		1	$9{\cdot}997857151 \times 10^{-1}$	0·1476400446	0·1511736368
2	0·1476579148		2	$2{\cdot}068368865 \times 10^{-2}$	8·346113098	8·345191091
3	0·1476412854		3	$8{\cdot}275332019 \times 10^{-4}$	24·35769485	24·35694973
4	0·1476402205	48·36195067	4	$1{\cdot}484295408 \times 10^{-4}$	48·36054767	48·36011551
5	0·1476400826	48·36057186	5	$4{\cdot}142176167 \times 10^{-5}$	80·36168313	80·36140927
6	0·1476400553	48·36055049	6	$1{\cdot}503520775 \times 10^{-5}$	120·3622496	120·3620622
7	0·1476400483	48·36054828	7	$6{\cdot}484036986 \times 10^{-6}$	168·3625730	168·3624371
8	0·1476400460	48·36054785	8	$3{\cdot}158879288 \times 10^{-6}$	224·3627750	224·3626721
9	0·1476400452	48·36054773	9	$1{\cdot}685368373 \times 10^{-6}$	288·3629095	288·3628290
10	0·1476400449	48·36054770	10	$9{\cdot}647971394 \times 10^{-7}$	360·3630037	360·3629391
11	0·1476400447	48·36054768	11	$5{\cdot}842146506 \times 10^{-7}$	440·3630722	440·3630191
12	0·1476400446	48·36054767	12	$3{\cdot}703636034 \times 10^{-7}$	528·3631236	528·3630792

Table 6.1 lists the calculated values of $\lambda_1^{(N)}$ and $\lambda_4^{(N)}$, $N = 1(1)12$; and $(b_1)_i$, λ_i and L_{ii}, $i = 1(1)12$ (we take $(a_1^{(30)})_i$ and $\lambda_i^{(30)}$ as approximations to $(b_1)_i$ and λ_i). We can now compare the predicted convergence rates of the theorems with the observed convergence rates of the numerical results of Table 6.1.

Eigenvalue spectrum
Theorem 6.1′ predicts

$$|\lambda_i - L_{ii}| \leqslant A_6(i-1)^{-v_4}, \tag{6.7.2}$$

where $v_4 = 0$. Hence

$$\lambda_i \sim Ci^2. \tag{6.7.3}$$

The results agree with these predictions, but give in fact $v_4 = 2$ in (6.7.2). Hence Theorem 6.1′ is pessimistic in this case.

Eigenvalue convergence
Theorem 6.5′ leads to the prediction

$$|\lambda_1 - \lambda_1^{(N)}| \leqslant A_{13}N^{-v_{11}}, \tag{6.7.4}$$

where $v_{11} = 7$.

The results of Table 6.1 yield a convergence rate parameter $v_{11} = 6.6$ in good agreement with that predicted.

For interest we note that a fit to $|\lambda_4 - \lambda_4^{(N)}|$ yields a parameter $v_{11} = 8{\cdot}2$ for that eigenvalue. We conjecture that Theorems 6.5 and 6.5′ should be extendable to cover eigenvalues other than the dominant eigenvalue.

Eigenvector convergence
Theorem 6.3′ gives the prediction

$$|(b_1)_i| \leqslant A_8 i^{-v_6},$$

where $v_6 = 5$.

The measured value of v_6 from the results of Table 6.1 is 5.05.

7

Block Weakly Asymptotically Diagonal Systems

7.1 Introduction

The examples given in Chapters 2 and 4 show that the theory of WAD matrices is well suited to the analysis of the convergence rates of expansion methods for single operator equations in one dimension.

Consider, however, the solution of a set of p coupled operator equations in one dimension,

$$\begin{aligned} \mathscr{L}_{11}f_1 + \mathscr{L}_{12}f_2 + \ldots + \mathscr{L}_{1p}f_p &= g_1 \\ \vdots \qquad\qquad\qquad\quad \vdots \quad &\quad \vdots \\ \mathscr{L}_{p1}f_1 + \mathscr{L}_{p2}f_2 + \ldots + \mathscr{L}_{pp}f_p &= g_p, \end{aligned} \tag{7.1.1}$$

where $g_1, g_2, \ldots, g_p$ are known and $f_1, f_2, \ldots, f_p$ are to be determined. It is then natural to write expansions of the form

$$\begin{aligned} f_1 &= \sum_{i=1}^{\infty} b_{1i}h_{1i}(x) \\ f_2 &= \sum_{i=1}^{\infty} b_{2i}h_{2i}(x) \\ &\vdots \\ f_p &= \sum_{i=1}^{\infty} b_{pi}h_{pi}(x) \end{aligned} \tag{7.1.2}$$

as the solutions of (7.1.1). In vector form, (7.1.1), (7.1.2) may be written as

$$\mathscr{L}\mathbf{f} = \mathbf{g}, \tag{7.1.3}$$

and

$$\mathbf{f} = \sum_{i=1}^{\infty} \mathbf{H}_i\mathbf{b}_i, \tag{7.1.4}$$

where $\mathscr{L}$ is a matrix operator of order $p \times p$, $\mathbf{H}_i$ is a $p \times p$ diagonal matrix with $(H_i)_{jj} = h_{ji}(x)$ and $\mathbf{b}_i$ is a p-vector.

Consider also the solution of the operator equation

$$\mathscr{L}f(\mathbf{x}) = g(\mathbf{x}) \tag{7.1.5}$$

in more than one independent variable. Suppose, for example, that (7.1.5) represents a partial differential equation in two dimensions x, y. In this case it is natural to consider the expansion

$$f(x, y) = \sum_{i=1}^{\infty} b_i h_i(x, y), \tag{7.1.6}$$

where the complete set $\{h_i(x, y)\}$ may typically be constructed as a product

$$h_i(x, y) = g_j(x) g_k(y). \tag{7.1.7}$$

With an expansion set of this form, the structure of the resulting matrix problem depends on the ordering of the integer pairs $(j, k) \rightarrow i$. One such ordering is that given by the modified speedometer ordering:

i	j	k	$Q = j + k - 1$
1	1	1	1
2	2	1	2
3	1	2	2
4	3	1	3
5	2	2	3
6	1	3	3
⋮	⋮	⋮	⋮

Here, the final column lists the integer $Q = j + k - 1$, and it is useful, when solving multi-dimensional problems, to group together all terms with a given value of Q, since they are likely to be of comparable importance. It is then preferable to use Q, rather than i, as a counting index for the expansion, and then (7.1.6) becomes

$$f(x, y) = \sum_{Q=1}^{\infty} \mathbf{h}_Q^T \mathbf{b}_Q, \tag{7.1.8}$$

where the vectors $\mathbf{h}_Q$, $\mathbf{b}_Q$ now have variable dimension ν_Q, say. In the above two-dimensional example $\nu_Q = Q$, but in problems of higher dimension ν_Q increases more rapidly.

A variational or Galerkin method solution of (7.1.3) or (7.1.5), with expansion (7.1.4) or (7.1.8) respectively, leads to a system of algebraic equations

$$\mathbf{Lb} = \mathbf{g}, \tag{7.1.9}$$

with the truncated form

$$\mathbf{L}^{(N)}\mathbf{a}^{(N)} = \mathbf{g}^{(N)}, \tag{7.1.10}$$

where $\mathbf{L}$, $\mathbf{L}^{(N)}$ are conforming block matrices of the form

$$\mathbf{L}^{(N)} \equiv \begin{bmatrix} \mathbf{L}_{11} & \mathbf{L}_{12} & \mathbf{L}_{13}\cdots \\ \mathbf{L}_{21} & \mathbf{L}_{22} & \mathbf{L}_{23}\cdots \\ \mathbf{L}_{31} & \mathbf{L}_{32} & \mathbf{L}_{33}\cdots \\ \vdots & \vdots & \vdots \end{bmatrix}, \tag{7.1.11}$$

and the diagonal blocks $\mathbf{L}_{ii}$ are square matrices of orders v_i, $i = 1, 2, \ldots$.

We could use the results of Chapter 3 to analyse the convergence properties of (7.1.9), (7.1.10), in the case when $\mathbf{L}$ is of the form (7.1.11). However, such an analysis would take no account of the block structure of $\mathbf{L}$. We therefore introduce the concept of a block weakly asymptotically diagonal (BWAD) matrix (see Definition 7.1 below) and show that the analysis of Chapter 3 extends in a straightforward way to such matrices.

Throughout this chapter we refer to $N \times N$ block matrices; we define the matrix $\mathbf{L}^{(N)}$ to be an $N \times N$ block matrix if its (i, j) element is the $v_i \times v_j$ matrix $\mathbf{L}_{ij}$, for $i, j = 1, 2, \ldots, N$. Hence, we see that the $N \times N$ block matrix $\mathbf{L}^{(N)}$ is an $M \times M$ matrix, where $M = \sum_{i=1}^{N} v_i$. We use the notation $(\mathbf{L}^t)_{ij}$ to denote the (i, j) block of the transpose of $\mathbf{L}$, $\mathbf{L}_{ii}^{-1}$ to denote the inverse of the matrix $\mathbf{L}_{ii}$, and $(\mathbf{L}^{-1})_{ij}$ to denote the (i, j) block of the inverse of $\mathbf{L}$.

For simplicity, and without loss of generality, we shall assume that the block matrix $\mathbf{L}$ considered is normalised by condition (7.1.12) below.

Definition 7.1. *Let the $N \times N$ block matrix $\mathbf{L}^{(N)}$ with blocks $\mathbf{L}_{ij}$ be such that in some norm, $\|\cdot\|$,*

$$\|\mathbf{L}_{ii}\| = 1, \tag{7.1.12}$$

$$\|\mathbf{L}_{ij}\| \leqslant A_{ij}, \quad \text{for all } i > j, \tag{7.1.13}$$

and

$$\|\mathbf{L}_{ij}\| \leqslant B_{ij}, \quad \text{for all } i < j. \tag{7.1.14}$$

The matrix $\mathbf{L}^{(N)}$ is said to be (normalised) BWAD if the zero diagonal, lower triangular matrices $\mathbf{A}$ and $\mathbf{B}^t$ satisfy (in the notation of Chapter 3),

(a)
$$\mathbf{AB} \leqslant C(\mathbf{A} + \mathbf{I} + \mathbf{B}), \tag{7.1.15}$$

where C is a positive constant;

(b) *there exist zero diagonal, lower triangular (image) matrices* $\bar{\mathbf{A}}$ *and* $\bar{\mathbf{B}}^t$ *such that*

(i) $$\mathbf{A} \leqslant \bar{\mathbf{A}}, \mathbf{B} \leqslant \bar{\mathbf{B}}, \tag{7.1.16}$$

(ii) $$\mathbf{A}\bar{\mathbf{A}} \leqslant C_A\bar{\mathbf{A}} \text{ and } \mathbf{B}\bar{\mathbf{B}} \leqslant C_B\bar{\mathbf{B}}, \tag{7.1.17}$$

$$\text{or } \bar{\mathbf{A}}\mathbf{A} \leqslant C_A\bar{\mathbf{A}} \text{ and } \bar{\mathbf{B}}\mathbf{B} \leqslant C_B\bar{\mathbf{B}}, \tag{7.1.18}$$

(iii) $$\bar{\mathbf{A}}\bar{\mathbf{B}} \leqslant \bar{C}(\bar{\mathbf{A}} + \mathbf{I} + \bar{\mathbf{B}}), \tag{7.1.19}$$

$$\bar{\mathbf{B}}\bar{\mathbf{A}} \leqslant \Omega(\bar{\mathbf{A}} + \mathbf{I} + \bar{\mathbf{B}}). \tag{7.1.20}$$

It should be noted that Definition 7.1 is a very natural extension of Definition 3.3, except that, for simplicity, the constants C_1 of (3.2.4) and Ω_1 of (3.3.3) have been absorbed into the constants C, Ω respectively, and that some of the earlier theorems of this section do not require all of conditions (7.1.13)–(7.1.20) to be satisfied. The normalisation condition, (7.1.12), can be assured by a block diagonal transformation of the matrix $\mathbf{L}^{(N)}$.

Definition 3.4 also has the obvious extension:

Definition 7.2. *The matrix* $\mathbf{L}^{(N)}$ *is block strongly asymptotically diagonal (BSAD) if it is BWAD with* $\bar{\mathbf{A}} = \mathbf{A}$ *and* $\bar{\mathbf{B}} = \mathbf{B}$.

Neither of these definitions assumes that $\mathbf{L}^{(N)}$ is symmetric.

7.2 Block triangular decomposition of BWAD matrices

We now generalise Theorem 3.1 to consider the decomposition of the $N \times N$ block matrix $\mathbf{L}^{(N)}$, (7.1.11), as

$$\mathbf{L}^{(N)} = \mathbf{U}\mathbf{V}, \tag{7.2.1}$$

where $\mathbf{U}$ and $\mathbf{V}^t$ are $N \times N$ block lower triangular matrices. We show, in Theorem 7.1, that if C, (7.1.15), and the norms of the inverses of the diagonal blocks of $\mathbf{L}^{(N)}$ are suitably restricted, then the blocks of $\mathbf{U}$ and $\mathbf{V}$ satisfy relationships similar to those satisfied by the blocks of $\mathbf{L}^{(N)}$.

Theorem 7.1. *Let the* $N \times N$ *block matrix* $\mathbf{L}^{(N)}$ *be (normalised) BWAD, and let the diagonal blocks of* $\mathbf{L}^{(N)}$ *satisfy*

$$\|\mathbf{L}_{ii}^{-1}\| \leqslant M, \quad i = 1, 2, \ldots, N, \tag{7.2.2}$$

where M is a positive constant. Then $\mathbf{L}^{(N)}$ may be factorised in the form (7.2.1), *with*

$$\mathbf{V}_{ii} = \mathbf{I}, \quad i = 1, 2, \ldots, N,$$

and the matrices $\mathbf{U}$ *and* $\mathbf{V}$ *satisfy*

$$\|\mathbf{U}_{ij}\| \leqslant K_1 A_{ij}, \quad i > j, \tag{7.2.3}$$

$$\|\mathbf{U}_{ii}\| \leqslant 1 + CK_1K_2, \tag{7.2.4}$$

$$\|\mathbf{U}_{ii}^{-1}\| \leqslant \frac{M}{1-\delta}, \tag{7.2.5}$$

$$\|\mathbf{V}_{ij}\| \leqslant K_2 B_{ij}, \quad i < j, \tag{7.2.6}$$

where

$$K_1 = \frac{1}{1 - CK_2},$$

$$K_2 = \frac{1 - \sqrt{(1 - 4MC/(1-\delta))}}{2C}$$

and

$$\delta = CMK_1K_2,$$

provided $CMK_1K_2 < 1$, *and* $4CM \leqslant 1 - \delta$.

Proof. The proof follows exactly the same steps as the proof of Theorem 3.1.

Let $\mathbf{U}_i^t$ denote the rows of the i-th row block of $\mathbf{U}$ and let $\mathbf{V}_i$ denote the columns of the i-th column block of $\mathbf{V}$ (so that the first row of $\mathbf{U}_i^t$ corresponds to the $\{\sum_{k=1}^{i-1} v_k + 1\}$ row of $\mathbf{U}$).

It is trivial to show that $\mathbf{U}_1^t$ satisfies (7.2.4), (7.2.5), since

$$\mathbf{L}_{11} = \mathbf{U}_{11}\mathbf{V}_{11},$$

and hence

$$\|\mathbf{U}_{11}\| = \|\mathbf{L}_{11}\| = 1,$$

$$\|\mathbf{U}_{11}^{-1}\| = \|\mathbf{L}_{11}^{-1}\| \leqslant M.$$

Now, assume inductively that $\mathbf{U}_i^t$ satisfies (7.2.3), (7.2.4), (7.2.5) and $\mathbf{V}_i$ satisfies (7.2.6) for $i = 1, 2, \ldots, m - 1 < N$. It then remains to show that $\mathbf{U}_m^t$ and $\mathbf{V}_m$ satisfy (7.2.3), (7.2.4), (7.2.5) and (7.2.6) respectively.

Now,

$$\mathbf{U}_{ij} = \mathbf{L}_{ij} - \sum_{k=1}^{j-1} \mathbf{U}_{ik}\mathbf{V}_{kj}, \qquad i \geqslant j, \tag{7.2.7}$$

and

$$\mathbf{V}_{ij} = \mathbf{U}_{ii}^{-1}\left(\mathbf{L}_{ij} - \sum_{k=1}^{i-1} \mathbf{U}_{ik}\mathbf{V}_{kj}\right), \quad i < j. \tag{7.2.8}$$

Hence,

$$\mathbf{U}_{m1} = \mathbf{L}_{m1} \quad \textit{and} \quad \mathbf{V}_{1m} = \mathbf{U}_{11}^{-1}\mathbf{L}_{1m},$$

and therefore

$$\begin{aligned} \|\mathbf{U}_{m1}\| &\leqslant \|\mathbf{L}_{m1}\| \leqslant A_{m1} \leqslant K_1 A_{m1}, \\ \|\mathbf{V}_{1m}\| &\leqslant \|\mathbf{U}_{11}^{-1}\| \, \|\mathbf{L}_{1m}\| \leqslant MB_{1m} \leqslant K_2 B_{1m}. \end{aligned} \tag{7.2.9}$$

We now assume inductively that

$$\|\mathbf{U}_{mr}\| \leqslant K_1 A_{mr} \quad \text{and} \quad \|\mathbf{V}_{rm}\| \leqslant K_2 B_{rm},$$

for $r = 1, 2, \ldots, j - 1 < m - 1$; then, by (7.2.7), (7.2.8),

$$\begin{aligned} \|\mathbf{U}_{mj}\| &\leqslant \|\mathbf{L}_{mj}\| + \sum_{k=1}^{j-1} \|\mathbf{U}_{mk}\| \, \|\mathbf{V}_{kj}\| \\ &\leqslant A_{mj} + \sum_{k=1}^{j-1} K_1 K_2 A_{mk} B_{kj} \\ &\leqslant (1 + CK_1K_2)A_{mj} \\ &\leqslant K_1 A_{mj}, \end{aligned} \tag{7.2.10}$$

and

$$\begin{aligned} \|\mathbf{V}_{jm}\| &\leqslant \|\mathbf{U}_{jj}^{-1}\| \left(\|\mathbf{L}_{jm}\| + \sum_{k=1}^{i-1} \|\mathbf{U}_{jk}\| \, \|\mathbf{V}_{km}\|\right) \\ &\leqslant \frac{M}{1-\delta}\left(B_{jm} + \sum_{k=1}^{i-1} K_1K_2A_{jk}B_{km}\right) \\ &\leqslant \frac{M}{1-\delta}(1 + CK_1K_2)B_{jm} \\ &\leqslant K_2 B_{jm}. \end{aligned} \tag{7.2.11}$$

Also, from (7.2.7),

$$\mathbf{U}_{mm} = \mathbf{L}_{mm} - \sum_{k=1}^{m-1} \mathbf{U}_{mk}\mathbf{V}_{km}.$$

Hence,

$$\begin{aligned}\|\mathbf{U}_{mm}\| &\leqslant \|\mathbf{L}_{mm}\| + \sum_{k=1}^{m-1} \|\mathbf{U}_{mk}\| \, \|\mathbf{V}_{km}\| \\ &\leqslant 1 + \sum_{k=1}^{m-1} K_1 K_2 A_{mk} B_{km} \\ &\leqslant 1 + CK_1K_2.\end{aligned}$$

Also,

$$\mathbf{U}_{mm}^{-1} = \left(\mathbf{I} - \mathbf{L}_{mm}^{-1} \sum_{k=1}^{m-1} \mathbf{U}_{mk}\mathbf{V}_{km}\right)^{-1} \mathbf{L}_{mm}^{-1},$$

and

$$\left\| \mathbf{L}_{mm}^{-1} \sum_{k=1}^{m-1} \mathbf{U}_{mk}\mathbf{V}_{km} \right\| \leqslant CMK_1K_2 = \delta < 1.$$

Hence,

$$\|\mathbf{U}_{mm}^{-1}\| \leqslant \frac{M}{1-\delta}.$$

It now remains to show that there exist constants K_1, K_2 satisfying (7.2.9), (7.2.10), (7.2.11). We first note that (7.2.9) is immediately satisfied by any K_2 that satisfies (7.2.11). In order to satisfy (7.2.10), we take

$$K_1 = \frac{1}{1 - CK_2}. \tag{7.2.12}$$

If we substitute (7.2.12) into (7.2.11), we obtain

$$\frac{M}{1-\delta}\left(1 + \frac{CK_2}{1 - CK_2}\right) \leqslant K_2,$$

i.e. $CK_2^2 - K_2 + M/(1-\delta) \leqslant 0$, since $CK_2 < 1$ from (7.2.10). This inequality is satisfied by all K_2 lying between the real roots of the equation

$$C\gamma^2 - \gamma + \frac{M}{1-\delta} = 0. \tag{7.2.13}$$

The existence of real roots is guaranteed if

$$1 - \frac{4CM}{1-\delta} \geqslant 0,$$

i.e. $4CM \leqslant 1 - \delta$.

We then take K_2 to be the smaller of the real roots of (7.2.13),

$$K_2 = \frac{1 - \sqrt{(1 - 4CM/(1 - \delta))}}{2C}.$$

It should be noted that the proof of this theorem only requires conditions (7.1.12)–(7.1.15) of Definition 7.1 to be satisfied; these conditions provide the generalisation of a WD matrix of Section 3.2.

7.3 Bounds on the inverse of a BWAD matrix

Lemma 7.1. *Under the conditions of Theorem* 7.1, *the* $N \times N$ *block matrices* $\mathbf{U}, \mathbf{V}$ *have unique inverses.*

Proof. Theorem 7.1 shows that the diagonal blocks $\mathbf{U}_{ii}, \mathbf{V}_{ii}, i = 1, 2, \ldots, N$ are non-singular. Hence,

$$\det \mathbf{U} = \prod_{i=1}^{N} \det \mathbf{U}_{ii} \neq 0,$$

and

$$\det \mathbf{V} = \prod_{i=1}^{N} \det \mathbf{V}_{ii} \neq 0,$$

and therefore $\mathbf{U}$ and $\mathbf{V}$ are non-singular and have unique inverses.

We now proceed to generalise Section 3.3, and provide bounds on the blocks of the inverse matrices $\mathbf{U}^{-1}, \mathbf{V}^{-1}$.

Theorem 7.2. *Under the conditions of Theorem* 7.1, *let*

$$\mathbf{U} = \mathbf{F} - \mathbf{X},$$

and

$$\mathbf{U}^{-1} = \mathbf{F}^{-1} + \mathbf{P},$$

where $\mathbf{F}$ *is a block diagonal matrix with i-th block* $\mathbf{U}_{ii}$. *Similarly, let*

$$\mathbf{V} = \mathbf{I} - \mathbf{Y},$$

and

$$\mathbf{V}^{-1} = \mathbf{I} + \mathbf{R}.$$

Then $\mathbf{P}$ *and* $\mathbf{R}^t$ *are strictly block lower triangular matrices which satisfy*

$$\|\mathbf{P}_{ij}\| \leqslant K_p \bar{A}_{ij}, \quad i > j, \tag{7.3.1}$$

$$\|\mathbf{R}_{ij}\| \leqslant K_r \bar{B}_{ij}, \quad i < j, \tag{7.3.2}$$

$$\|\mathbf{F}_{ii}^{-1}\| \leqslant M', \tag{7.3.3}$$

where

$$K_p = \frac{M'^2 K_1}{1 - M' C_A K_1},$$

$$K_r = \frac{K_2}{1 - C_B K_2},$$

$$M' = \frac{M}{1 - \delta},$$

provided $M' C_A K_1 < 1$ *and* $C_B K_2 < 1$.

Proof. Since $\mathbf{U}$ is block lower triangular, and $\mathbf{F}$ is block diagonal,

$$\mathbf{F}_{ii}^{-1} = (\mathbf{F}^{-1})_{ii} = (\mathbf{U}^{-1})_{ii} = \mathbf{U}_{ii}^{-1},$$

and hence,

$$\|\mathbf{F}_{ii}^{-1}\| = \|\mathbf{U}_{ii}^{-1}\| \leqslant \frac{M}{1 - \delta} = M'.$$

By definition,

$$\mathbf{U}\mathbf{U}^{-1} = \mathbf{I}.$$

Hence,

$$(\mathbf{F} - \mathbf{X})(\mathbf{F}^{-1} + \mathbf{P}) = \mathbf{I} - \mathbf{X}\mathbf{F}^{-1} + \mathbf{F}\mathbf{P} - \mathbf{X}\mathbf{P} = \mathbf{I},$$

$$\mathbf{F}\mathbf{P} = \mathbf{X}\mathbf{F}^{-1} + \mathbf{X}\mathbf{P},$$

and

$$\mathbf{P}_{ij} = (\mathbf{F}^{-1})_{ii} \left\{ \mathbf{X}_{ij} (\mathbf{F}^{-1})_{jj} + \sum_{k=j+1}^{i-1} \mathbf{X}_{ik} \mathbf{P}_{kj} \right\}, \quad i > j. \tag{7.3.4}$$

Let $\mathbf{P}_i^t$ denote the i-th block row of the block matrix $\mathbf{P}$.
From (7.3.4)

$$\mathbf{P}_{21} = (\mathbf{F}^{-1})_{22} \{ \mathbf{X}_{21} (\mathbf{F}^{-1})_{11} \}.$$

Hence

$$\begin{aligned}\|\mathbf{P}_{21}\| &\leqslant M'^2 K_1 A_{21}\\ &\leqslant M'^2 K_1 \bar{A}_{21}\\ &\leqslant K_p \bar{A}_{21}.\end{aligned} \tag{7.3.5}$$

Hence $\mathbf{P}_2^t$ satisfies (7.3.1). Assume inductively that $\mathbf{P}_i^t$ satisfies (7.3.1) for $i = 3, 4, \ldots, k-1$. From (7.3.4),

$$\mathbf{P}_{kj} = (\mathbf{F}^{-1})_{kk}\left\{\mathbf{X}_{kj}(\mathbf{F}^{-1})_{jj} + \sum_{l=j+1}^{k-1} \mathbf{X}_{kl}\mathbf{P}_{lj}\right\}, \; j < k.$$

Hence

$$\begin{aligned}\|\mathbf{P}_{kj}\| &\leqslant \|(\mathbf{F}^{-1})_{kk}\|\left\{\|\mathbf{X}_{kj}\|\,\|(\mathbf{F}^{-1})_{jj}\| + \sum_{l=j+1}^{k-1}\|\mathbf{X}_{kl}\|\,\|\mathbf{P}_{lj}\|\right\}\\ &\leqslant M'\left\{K_1 A_{kj} M' + \sum_{l=j+1}^{k-1} K_1 A_{kl} K_p \bar{A}_{lj}\right\}\\ &\leqslant M'\{K_1 M' \bar{A}_{kj} + C_A K_1 K_p \bar{A}_{kj}\}\\ &\leqslant K_p \bar{A}_{kj}.\end{aligned} \tag{7.3.6}$$

It is easy to show that the choice

$$K_p = \frac{M'^2 K_1}{1 - M' C_A K_1},$$

satisfies (7.3.5), (7.3.6).

In a similar way, by considering the identity $\mathbf{V}\mathbf{V}^{-1} = \mathbf{I}$, we can show that

$$\|\mathbf{R}_{ij}\| \leqslant K_r \bar{B}_{ij}, \quad i < j,$$

where $K_r = K_2/(1 - C_B K_2)$.

We note that the proof of Theorem 7.2 uses conditions (7.1.17) of Definition 7.1. We could, instead, use conditions (7.1.18) by considering the identities $\mathbf{U}^{-1}\mathbf{U} = \mathbf{I}$ and $\mathbf{V}^{-1}\mathbf{V} = \mathbf{I}$.

The second theorem of this section uses the above results to provide bounds on the blocks of the inverse matrix $\mathbf{L}^{(N)-1}$.

Theorem 7.3. *Under the conditions of Theorem* 7.2,

$$\|(\mathbf{L}^{(N)-1})_{ij}\| \leqslant \begin{cases} K_1' \bar{A}_{ij}, & i > j,\\ K_2' \bar{B}_{ij}, & i < j,\\ K_0', & i = j.\end{cases} \tag{7.3.7}$$

where

$$K'_1 = K_p(1 + K_r\Omega)$$
$$K'_2 = K_r(M' + K_p\Omega)$$
$$K'_0 = M' + K_pK_r\Omega.$$

Proof. From Theorem 7.2,

$$\begin{aligned}\mathbf{L}^{(N)-1} &= \mathbf{V}^{-1}\mathbf{U}^{-1}\\ &= (\mathbf{I} + \mathbf{R})(\mathbf{F}^{-1} + \mathbf{P})\\ &= \mathbf{F}^{-1} + \mathbf{R}\mathbf{F}^{-1} + \mathbf{P} + \mathbf{R}\mathbf{P}.\end{aligned}$$

For $i > j$,

$$(\mathbf{L}^{(N)-1})_{ij} = \mathbf{P}_{ij} + \sum_{k=i+1}^{N} \mathbf{R}_{ik}\mathbf{P}_{kj}.$$

Hence,

$$\begin{aligned}\|(\mathbf{L}^{(N)-1})_{ij}\| &\leqslant \|\mathbf{P}_{ij}\| + \sum_{k=i+1}^{N} \|\mathbf{R}_{ik}\|\,\|\mathbf{P}_{kj}\|\\ &\leqslant K_p\bar{A}_{ij} + \sum_{k=i+1}^{N} K_r\bar{B}_{ik}K_p\bar{A}_{kj}\\ &\leqslant (K_p + K_pK_r\Omega)\bar{A}_{ij}\\ &\leqslant K'_1\bar{A}_{ij}.\end{aligned}$$

Similarly, for $i < j$,

$$\begin{aligned}\|(\mathbf{L}^{(N)-1})_{ij}\| &\leqslant \|\mathbf{R}_{ij}\|\,\|(\mathbf{F}^{-1})_{jj}\| + \sum_{k=j+1}^{N} \|\mathbf{R}_{ik}\|\,\|\mathbf{P}_{kj}\|\\ &\leqslant K_r\bar{B}_{ij}M' + \sum_{k=j+1}^{N} K_r\bar{B}_{ik}K_p\bar{A}_{kj}\\ &\leqslant (K_rM' + K_rK_p\Omega)\bar{B}_{ij}\\ &\leqslant K'_2\bar{B}_{ij}.\end{aligned}$$

Finally, for $i = j$,

$$\begin{aligned}\|(\mathbf{L}^{(N)-1})_{ii}\| &\leqslant \|(\mathbf{F}^{-1})_{ii}\| + \sum_{k=i+1}^{N} \|\mathbf{R}_{ik}\|\,\|\mathbf{P}_{ki}\|\\ &\leqslant M' + \sum_{k=i+1}^{N} K_r\bar{B}_{ik}K_p\bar{A}_{ki}\\ &\leqslant M' + K_pK_r\Omega = K'_0.\end{aligned}$$

The results of these theorems are independent of N and are therefore applicable to the infinite matrix $\mathbf{L}$. The details of this limiting procedure are very similar to those of Section 3.4, and we therefore do not repeat them here.

7.4 Behaviour of $\mathbf{a}^{(N)}$ and b

Throughout this section we assume that, for any N, $\mathbf{L}^{(N)}$, the $N \times N$ block principal submatrix of the infinite matrix $\mathbf{L}$, is (normalised) BWAD. We also assume that the matrix $\mathbf{L}$ and the image matrix $\bar{\mathbf{L}} = \mathbf{I} + \bar{\mathbf{A}} + \bar{\mathbf{B}}$ are bounded. These assumptions guarantee that the limiting procedure mentioned at the end of Section 7.3 is valid.

As in Section 3.5 we make use of property MN. Property MN in the context in which we need it will refer to the point $N \times N$ matrices $\mathbf{A}, \mathbf{B}$ of (7.1.13), (7.1.14) (or the image matrices $\bar{\mathbf{A}}$, $\bar{\mathbf{B}}$ of (7.1.16) *et seq.*). The definition is therefore identical with that given in Definition 3.5.

This property then allows some of the conditions of Definition 7.1 to be replaced by rather stronger, but clearer sufficient conditions as the following lemma shows.

Lemma 7.2. *Let the strictly lower triangular* $N \times N$ *matrices* $\mathbf{A}$, $\mathbf{B}^t$ *have property* MN *and satisfy* $A_{i,i-1} < 1$, $B_{i-1,i} < 1, i = 2,3,\ldots,N$. *Then they satisfy* (7.1.15) *and* (7.1.20), *with* $\bar{\mathbf{A}} = \mathbf{A}$, $\bar{\mathbf{B}} = \mathbf{B}$, $C = \max\{\|\mathbf{A}\|_\infty, \|\mathbf{B}\|_1\}$ *and* $\Omega = \max\{\|\mathbf{A}\|_1, \|\mathbf{B}\|_\infty\}$.

The proof of this lemma follows step by step the proof of Lemma 3.1, and we therefore omit it.

Property MN is a somewhat stronger condition than is required by the theorems of this section. However, the results of the theorems are often considerably simplified if we assume that appropriate matrices have property MN. We therefore give these simplified results as corollaries to the theorems.

In order to consider the behaviour of $\mathbf{b}$ and $\mathbf{a}^{(N)}$, we need to know something about the i-convergence rate of $\|\mathbf{g}_i\|$, where $\mathbf{g}_i$ is the i-th partition of the infinite right-hand side vector of (7.1.9). Throughout this section, we therefore assume that there exists a vector $\hat{\mathbf{g}}$, with components $\hat{g}_i$, such that if we partition $\mathbf{g}$ in the same manner as the columns of $\mathbf{L}$, then the i-th partition $\mathbf{g}_i$, of $\mathbf{g}$ satisfies

$$\|\mathbf{g}_i\| \leqslant \hat{g}_i, \quad \forall i, \tag{7.4.1}$$

and $\hat{g}_i$ is semi-monotone decreasing in i.

We are now in a position to bound the partitions $\mathbf{b}_i$ of $\mathbf{b}$, the solution of (7.1.9).

Theorem 7.4. *Under the conditions of this section,*

$$\|\mathbf{b}_i\| \leqslant K_0'\hat{g}_i + K_1'\bar{\mathbf{a}}_i^t\hat{\mathbf{g}} + K_2'\bar{\mathbf{b}}_i^t\hat{\mathbf{g}}, \tag{7.4.2}$$

where $\bar{\mathbf{a}}_i^t$, $\bar{\mathbf{b}}_i^t$ denote the i-th rows of $\bar{A}$ and $\bar{B}$, respectively.

Proof.

$$\mathbf{b} = \mathbf{L}^{-1}\mathbf{g}$$

Hence,

$$\mathbf{b}_i = \sum_{j=1}^{\infty} (\mathbf{L}^{-1})_{ij}\mathbf{g}_j,$$

and

$$\begin{aligned}\|\mathbf{b}_i\| &\leqslant \sum_{j=1}^{i-1} \|(\mathbf{L}^{-1})_{ij}\|\,\|\mathbf{g}_j\| + \|(\mathbf{L}^{-1})_{ii}\|\,\|\mathbf{g}_i\| + \sum_{j=i+1}^{\infty} \|(\mathbf{L}^{-1})_{ij}\|\,\|\mathbf{g}_j\| \\ &\leqslant \sum_{j=1}^{i-1} K_1'\bar{A}_{ij}\hat{g}_j + K_0'\hat{g}_i + \sum_{j=i+1}^{\infty} K_2'\bar{B}_{ij}\hat{g}_j \\ &\leqslant K_0'\hat{g}_i + K_1'\bar{\mathbf{a}}_i^t\hat{\mathbf{g}} + K_2'\bar{\mathbf{b}}_i^t\hat{\mathbf{g}}.\end{aligned}$$

Corollary 7.1. If, additionally, the infinite matrix $\bar{\mathbf{B}}^t$ has property MN, then

$$\|\mathbf{b}_i\| \leqslant (K_0' + \Omega K_2')\hat{g}_i + K_1'\bar{\mathbf{a}}_i^t\hat{\mathbf{g}}. \tag{7.4.3}$$

The proof of this corollary is exactly the same as the proof of Corollary 3.5.

In particular cases, tight bounds on $\|\mathbf{b}_i\|$ are obtained by inserting the relevant forms for $\bar{\mathbf{A}}$ and $\hat{\mathbf{g}}$. However we note that, in general, the result of Corollary 7.1 depends on $\bar{\mathbf{A}}$ and not on $\bar{\mathbf{B}}$, and hence depends only on the behaviour of the block lower triangular part of $\mathbf{L}$.

In practice, it is not possible to solve the infinite system (7.1.9) and, instead, we solve the finite system (7.1.10). The next theorem considers the N-convergence rate of the difference, $\|\mathbf{b}_i - \mathbf{a}_i^{(N)}\|$, of the i-th partitions of $\mathbf{b}$ and $\mathbf{a}^{(N)}$.

Theorem 7.5. *Under the conditions of this section,*

$$\|\mathbf{b}_i - \mathbf{a}_i^{(N)}\| \leqslant K_r K_p \sum_{j=1}^{N} \hat{g}_j \sum_{k=N+1}^{\infty} \bar{B}_{ik}\bar{A}_{kj} + K_2' \sum_{j=N+1}^{\infty} \bar{B}_{ij}\hat{g}_j,$$

for $i \leqslant N$.

Proof.

$$\mathbf{b}_i - \mathbf{a}_i^{(N)} = \sum_{j=1}^{N} \{(\mathbf{L}^{-1})_{ij} - (\mathbf{L}^{(N)-1})_{ij}\}\mathbf{g}_j + \sum_{j=N+1}^{\infty} (\mathbf{L}^{-1})_{ij}\mathbf{g}_j.$$

From the proof of Theorem 7.3, we find

$$(\mathbf{L}^{-1})_{ij} - (\mathbf{L}^{(N)-1})_{ij} = \sum_{k=N+1}^{\infty} \mathbf{R}_{ik}\mathbf{P}_{kj},$$

and hence,

$$\mathbf{b}_i - \mathbf{a}_i^{(N)} = \sum_{j=1}^{N} \sum_{k=N+1}^{\infty} \mathbf{R}_{ik}\mathbf{P}_{kj}\mathbf{g}_j + \sum_{j=N+1}^{\infty} \left\{\mathbf{R}_{ij}(\mathbf{F}^{-1})_{jj} + \sum_{k=j+1}^{\infty} \mathbf{R}_{ik}\mathbf{P}_{kj}\right\}\mathbf{g}_j.$$

Thus,

$$\begin{aligned}\|\mathbf{b}_i - \mathbf{a}_i^{(N)}\| &\leqslant \sum_{j=1}^{N} \sum_{k=N+1}^{\infty} K_r\bar{B}_{ik}K_p\bar{A}_{kj}\hat{g}_j \\ &\quad + \sum_{j=N+1}^{\infty} \left\{K_r\bar{B}_{ij}M' + \sum_{k=j+1}^{\infty} K_r\bar{B}_{ik}K_p\bar{A}_{kj}\right\}\hat{g}_j \\ &\leqslant K_rK_p \sum_{j=1}^{N} \hat{g}_j \sum_{k=N+1}^{\infty} \bar{B}_{ik}\bar{A}_{kj} + (K_rM' + K_rK_p\Omega) \sum_{j=N+1}^{\infty} \bar{B}_{ij}\hat{g}_j,\end{aligned}$$

and $K_2' = K_r(M' + K_p\Omega)$.

Corollary 7.2. If, in addition, $\bar{\mathbf{A}}$ satisfies property MN, then, for $i \leqslant N$,

$$\|\mathbf{b}_i - \mathbf{a}_i^{(N)}\| \leqslant C_{i,\,N+1}\{K_rK_p\bar{\mathbf{a}}_{N+1}^t\hat{\mathbf{g}} + K_2'\bar{g}_{N+1}\},$$

where

$$C_{i,\,m} = \sum_{j=m}^{\infty} \bar{B}_{ij}.$$

The proof of this corollary is the same as that of Corollary 3.6.

As before, tight bounds on $\|\mathbf{b}_i - \mathbf{a}_i^{(N)}\|$ follow on inserting the relevant forms for $\bar{\mathbf{A}}$, $\bar{\mathbf{B}}$ and $\hat{\mathbf{g}}$.

7.5 Convergence of Galerkin calculations

We now consider the case when the linear systems (7.1.9), (7.1.10) arise from a Galerkin or variational method solution of the multi-dimensional operator equation

$$\mathscr{L}f = g. \tag{7.5.1}$$

In particular, we are interested in the convergence of the truncated expansion

$$f_N = \sum_{i=1}^{N} \mathbf{h}_i^t \mathbf{a}_i^{(N)} \tag{7.5.2}$$

to the solution

$$f = \sum_{i=1}^{\infty} \mathbf{h}_i^t \mathbf{b}_i \tag{7.5.3}$$

of (7.5.1). (7.5.2) reflects the practical situation, in which the expansion functions are introduced a block at a time.

We note that the analysis of this section is easily modified to deal with the coupled operator equations

$$\mathscr{L}\mathbf{f} = \mathbf{g}.$$

We first extend Definition 3.6 to the systems under consideration.

Definition 7.3. *Let $\mathscr{L}f = g$ be an inhomogeneous operator equation, and let* $\mathbf{Lb} = \mathbf{g}$ *be the corresponding infinite linear system of equations, with* $\{\mathbf{h}_i, i = 1, 2, \ldots\}$ *as expansion set. We refer to this system as a block weakly nice system (BWNS), if, for the given expansion set, every submatrix* $\mathbf{L}^{(N)}$ *of* $\mathbf{L}$ *satisfies the conditions of Section 7.4, and the vector* $\mathbf{g}$ *satisfies* (7.4.1).

We note that $e_N = f - f_N$ can be written as

$$e_N = \sum_{i=1}^{N} \mathbf{h}_i^t(\mathbf{b}_i - \mathbf{a}_i^{(N)}) + \sum_{i=N+1}^{\infty} \mathbf{h}_i^t \mathbf{b}_i. \tag{7.5.4}$$

We first consider the natural norm of e_N,

$$\|e_N\| = (e_N, e_N)^{1/2},$$

under the assumption that the expansion set $\{\bar{\mathbf{h}}_i, i = 1, 2, \ldots\}$ is orthonormal. This assumption implies that

$$\bar{\mathbf{h}}_i \bar{\mathbf{h}}_j^t = \mathbf{E}_{ij}, \tag{7.5.5}$$

where the $v_i \times v_j$ matrix $\mathbf{E}_{ij}$ satisfies

$$\mathbf{E}_{ii} = \mathbf{I}$$

and

$$\mathbf{E}_{ij} = \mathbf{\Phi}, \text{ the null matrix, for } i \neq j.$$

Then, in terms of this orthonormal expansion set,

$$\begin{aligned}
\|e_N\|^2 &= \left(\left(\sum_{i=1}^{N} \bar{\mathbf{h}}_i^t(\bar{\mathbf{b}}_i - \bar{\mathbf{a}}_i^{(N)}) + \sum_{i=N+1}^{\infty} \bar{\mathbf{h}}_i^t \bar{\mathbf{b}}_i\right), \left(\sum_{j=1}^{N} \bar{\mathbf{h}}_j^t(\bar{\mathbf{b}}_j - \bar{\mathbf{a}}_j^{(N)}) + \sum_{j=N+1}^{\infty} \bar{\mathbf{h}}_j^t \bar{\mathbf{b}}_j\right)\right) \\
&= \sum_{i=1}^{N} (\bar{\mathbf{b}}_i - \bar{\mathbf{a}}_i^{(N)})^t(\bar{\mathbf{b}}_i - \bar{\mathbf{a}}_i^{(N)}) + \sum_{i=N+1}^{\infty} \bar{\mathbf{b}}_i^t \bar{\mathbf{b}}_i, \\
&= \sum_{i=1}^{N} \|\bar{\mathbf{b}}_i - \bar{\mathbf{a}}_i^{(N)}\|_2^2 + \sum_{i=N+1}^{\infty} \|\bar{\mathbf{b}}_i\|_2^2.
\end{aligned} \tag{7.5.6}$$

However, the normalisation of the expansion set $\{\bar{\mathbf{h}}_i, i = 1, 2, \ldots\}$ is different from that assumed in the earlier theorems of this chapter. We therefore introduce the transformation

$$\bar{\mathbf{h}}_i = \gamma_i \mathbf{h}_i, \tag{7.5.7}$$

where γ_i is a $v_i \times v_i$ diagonal matrix, such that $\{\bar{\mathbf{h}}_i, i = 1, 2, \ldots\}$ satisfies (7.5.5) and $\{\mathbf{h}_i, i = 1, 2, \ldots\}$ has the normalisation of Theorems 7.1–7.5.

In terms of the coefficients $\mathbf{a}_i^{(N)}$, $\mathbf{b}_i$ considered in Theorems 7.4 and 7.5, (7.5.6) becomes

$$\|e_N\|^2 \leqslant \sum_{i=1}^{N} \|\gamma_i^{-2}\|_2 \, \|\mathbf{b}_i - \mathbf{a}_i^{(N)}\|_2^2 + \sum_{i=N+1}^{\infty} \|\gamma_i^{-2}\|_2 \, \|\mathbf{b}_i\|_2^2. \tag{7.5.8}$$

It should be noted that (7.5.8) requires $\|\cdot\|_2$ of $\mathbf{b}_i - \mathbf{a}_i^{(N)}$ and $\mathbf{b}_i$, while the results of Theorems 7.4 and 7.5 did not refer to any specific norm. In any particular example, the BWAD property may be derived directly in the norm $\|\cdot\|_2$; then no difficulty arises. If not, we use the fact that for any finite dimensional vector $\mathbf{x}$, and any norm $\|\cdot\|$, there exists some n_i dependent only on v_i, the dimension of $\mathbf{x}$, such that

$$\|\mathbf{x}\|_2^2 \leqslant n_i^2 \|\mathbf{x}\|^2. \tag{7.5.9}$$

Then Theorems 7.4 and 7.5 provide sufficient information to characterise the N-convergence rate of e_N in the natural norm (and also in the energy norm).

This approach is efficient in particular cases although, in general, the resulting bound is not very neat. In addition, the analysis is very similar to that of Section 3.6 and we therefore consider the N-convergence rate of $\|e_N\|$ by a generalisation of Theorems 2.4 and 2.5; this leads to slightly weaker, but more compact, results.

We assume for the remainder of this chapter that $f = \lim_{N\to\infty} f_N$, where f_N is defined by (7.5.2). Then

$$\|e_N\| = \|f - f_N\| \leqslant \sum_{k=N}^{\infty} \|f_{k+1} - f_k\|.$$

Since,

$$f_{k+1} - f_k = \sum_{i=1}^{k} \mathbf{h}_i^t(\mathbf{a}_i^{(k+1)} - \mathbf{a}_i^{(k)}) + \mathbf{h}_{k+1}^t \mathbf{a}_{k+1}^{(k+1)},$$

$$\|f_{k+1} - f_k\|^2 \leqslant \sum_{i=1}^{k} (\mathbf{a}_i^{(k+1)} - \mathbf{a}_i^{(k)})^t \gamma_i^{-2} (\mathbf{a}_i^{(k+1)} - \mathbf{a}_i^{(k)}) + \mathbf{a}_{k+1}^{(k+1)t} \gamma_{k+1}^{-2} \mathbf{a}_{k+1}^{(k+1)},$$

where γ_i is defined by (7.5.7). Hence, using (7.5.9),

$$\|f_{k+1} - f_k\|^2 \leqslant \sum_{i=1}^{k} n_i^2 \|\gamma_i^{-2}\| \, \|\mathbf{a}_i^{(k+1)} - \mathbf{a}_i^{(k)}\|^2 + n_{k+1}^2 \|\gamma_{k+1}^{-2}\| \, \|\mathbf{a}_{k+1}^{(k+1)}\|^2. \tag{7.5.10}$$

In the next lemma, we prove a relationship between $\|\mathbf{a}_i^{(k+1)} - \mathbf{a}_i^{(k)}\|$ and $\|\mathbf{a}_{k+1}^{(k+1)}\|$.

Lemma 7.3. *For a BWNS,*

$$\|\mathbf{a}_i^{(k+1)} - \mathbf{a}_i^{(k)}\| \leqslant N_1 \|\mathbf{a}_{k+1}^{(k+1)}\| \bar{B}_{i,\,k+1},$$

where

$$N_1 = \bar{C}K_1' + K_0' + C_B K_2'.$$

Proof. Consider the $(k + 1)$ block system of equations

$$\mathbf{L}^{(k+1)}\mathbf{a}^{(k+1)} = \mathbf{g}^{(k+1)}, \tag{7.5.11}$$

and the related k block system

$$\mathbf{L}^{(k)}\mathbf{a}^{(k)} = \mathbf{g}^{(k)}. \tag{7.5.12}$$

The first k block equations of (7.5.11) may be written as

$$\tilde{\mathbf{L}}^{(k+1)}\mathbf{a}^{(k+1)} = \mathbf{g}^{(k)}, \tag{7.5.13}$$

where $\tilde{\mathbf{L}}^{(k+1)}$ is a $k \times (k + 1)$ block matrix. Subtracting (7.5.13) from (7.5.12) gives

$$\mathbf{L}^{(k)}\mathbf{a}^{(k)} - \tilde{\mathbf{L}}^{(k+1)}\mathbf{a}^{(k+1)} = \mathbf{0}. \tag{7.5.14}$$

Let

$$\mathbf{a}^{(k+1)t} = (\tilde{\mathbf{a}}^{(k+1)t}, \mathbf{a}_{k+1}^{(k+1)t}).$$

Then

$$\mathbf{L}^{(k)}(\mathbf{a}^{(k)} - \tilde{\mathbf{a}}^{(k+1)}) = \mathbf{Q}\begin{bmatrix} \mathbf{0} \\ \mathbf{a}_{k+1}^{(k+1)} \end{bmatrix},$$

where $\mathbf{0}$ is the k block null vector, and $\mathbf{Q}$ is a $k \times (k + 1)$ block matrix, with (i, j)-th block

$$\begin{aligned} \mathbf{Q}_{ij} &= \mathbf{\Phi}, && \text{for } j < k + 1 \\ &= \mathbf{L}_{i,\,k+1} && \text{for } j = k + 1. \end{aligned}$$

Thus,

$$\mathbf{a}^{(k)} - \tilde{\mathbf{a}}^{(k+1)} = \mathbf{L}^{(k)-1}\mathbf{Q}\begin{bmatrix} \mathbf{0} \\ \mathbf{a}_{k+1}^{(k+1)} \end{bmatrix},$$

and for $i \leqslant k$,

$$\mathbf{a}_i^{(k)} - \mathbf{a}_i^{(k+1)} = \sum_{j=1}^{k} (\mathbf{L}^{(k)-1})_{ij} \mathbf{L}_{j,k+1} \mathbf{a}_{k+1}^{(k+1)}.$$

Hence,

$$\begin{aligned}\|\mathbf{a}_i^{(k)} - \mathbf{a}_i^{(k+1)}\| &\leqslant \left\{ \sum_{j=1}^{i-1} K_1' \bar{A}_{ij} \bar{B}_{j,k+1} + K_0' \bar{B}_{i,k+1} \right.\\ &\quad \left. + \sum_{j=i+1}^{k} K_2' \bar{B}_{ij} B_{j,k+1} \right\} \|\mathbf{a}_{k+1}^{(k+1)}\| \\ &\leqslant \{K_1' \bar{C} \bar{B}_{i,k+1} + K_0' \bar{B}_{i,k+1} + K_2' C_B \bar{B}_{i,k+1}\} \|\mathbf{a}_{k+1}^{(k+1)}\| \\ &\leqslant N_1 \bar{B}_{i,k+1} \|\mathbf{a}_{k+1}^{(k+1)}\|.\end{aligned}$$

It should be noted that the above proof is the first occasion on which condition (7.1.19) of Definition 7.1 is required. Using the result of Lemma 7.3 together with (7.5.10) gives the following result.

Lemma 7.4. *For a BWNS*

$$\|f_{k+1} - f_k\| \leqslant \bar{N}_{k+1} \|\mathbf{a}_{k+1}^{(k+1)}\|,$$

where

$$\bar{N}_{k+1}^2 = N_1^2 \sum_{i=1}^{k} \|\gamma_i^{-2}\| n_i^2 \bar{B}_{i,k+1}^2 + \|\gamma_{k+1}^{-2}\| n_{k+1}^2.$$

Finally, we obtain the following theorem which characterises the N-convergence rate of e_N in the natural norm.

Theorem 7.6. *For a BWNS,*

$$\|e_N\| \leqslant \sum_{k=N+1}^{\infty} \bar{N}_k \|\mathbf{a}_k^{(k)}\|.$$

Proof.

$$\begin{aligned}\|e_N\| &\leqslant \sum_{k=N}^{\infty} \|f_{k+1} - f_k\| \\ &\leqslant \sum_{k=N}^{\infty} \bar{N}_{k+1} \|\mathbf{a}_{k+1}^{(k+1)}\|.\end{aligned}$$

Thus, we see that the rate of convergence of the error to zero in the natural norm is related to the diagonal convergence rate of the blocks of the approximate expansion coefficients $\mathbf{a}_{k+1}^{(k+1)}$.

Having characterised the rate of convergence in the natural norm, we now consider the rate of convergence of the energy norm of the error.

Theorem 7.7. *For a BWNS, with $\mathscr{L}$ positive definite,*

$$\|e_N\|_{\mathscr{L}} \leqslant \sum_{k=N+1}^{\infty} E_k \|\mathbf{a}_k^{(k)}\|,$$

where

$$E_k^2 = N_1 n_k \bar{C} \sum_{j=1}^{k-1} n_j + n_k^2.$$

Proof.

$$\|e_N\|_{\mathscr{L}} = \|f - f_N\|_{\mathscr{L}} \leqslant \sum_{k=N}^{\infty} \|f_{k+1} - f_k\|_{\mathscr{L}}.$$

If we write $u = f_{k+1} - f_k$, then

$$(u, \mathscr{L}u) = \sum_{i=1}^{k} \sum_{j=1}^{k} (\mathbf{a}_i^{(k+1)} - \mathbf{a}_i^{(k)})^t \mathbf{L}_{ij} (\mathbf{a}_j^{(k+1)} - \mathbf{a}_j^{(k)})$$
$$+ \sum_{i=1}^{k} (\mathbf{a}_i^{(k+1)} - \mathbf{a}_i^{(k)})^t \mathbf{L}_{i,k+1} \mathbf{a}_{k+1}^{(k+1)} + \sum_{j=1}^{k} \mathbf{a}_{k+1}^{(k+1)t} \mathbf{L}_{k+1,j} (\mathbf{a}_j^{(k+1)} - \mathbf{a}_j^{(k)})$$
$$+ \mathbf{a}_{k+1}^{(k+1)t} \mathbf{L}_{k+1,k+1} \mathbf{a}_{k+1}^{(k+1)}.$$

Now, from (7.5.14), for $j \leqslant k$,

$$\sum_{j=1}^{k} \mathbf{L}_{ij} (\mathbf{a}_j^{(k+1)} - \mathbf{a}_j^{(k)}) = -\mathbf{L}_{i,k+1} \mathbf{a}_{k+1}^{(k+1)}.$$

Hence,

$$(u, \mathscr{L}u) = \sum_{j=1}^{k} \mathbf{a}_{k+1}^{(k+1)t} \mathbf{L}_{k+1,j} (\mathbf{a}_j^{(k+1)} - \mathbf{a}_j^{(k)}) + \mathbf{a}_{k+1}^{(k+1)t} \mathbf{L}_{k+1,k+1} \mathbf{a}_{k+1}^{(k+1)}.$$

If we let

$$\mathbf{d}_j^{(k+1)} = \mathbf{L}_{k+1,j} (\mathbf{a}_j^{(k+1)} - \mathbf{a}_j^{(k)}) \quad \text{and} \quad \boldsymbol{\delta}_{k+1}^{(k+1)} = \mathbf{L}_{k+1,k+1} \mathbf{a}_{k+1}^{(k+1)},$$

then

$$(u, \mathscr{L}u) \leqslant \sum_{j=1}^{k} \|\mathbf{a}_{k+1}^{(k+1)}\|_2 \|\mathbf{d}_j^{(k+1)}\|_2 + \|\mathbf{a}_{k+1}^{(k+1)}\|_2 \|\boldsymbol{\delta}_{k+1}^{(k+1)}\|_2.$$

Now, using (7.5.9), Lemma 7.3 and the fact that $\|\mathbf{L}_{k+1,\,k+1}\| = 1$, we obtain,

$$\begin{aligned}(u, \mathscr{L}u) &\leqslant \sum_{j=1}^{k} n_{k+1}\|\mathbf{a}_{k+1}^{(k+1)}\| n_j \|\mathbf{L}_{k+1,\,j}\| N_1 \|\mathbf{a}_{k+1}^{(k+1)}\| \bar{B}_{j,\,k+1} + n_{k+1}^2 \|\mathbf{a}_{k+1}^{(k+1)}\|^2 \\ &\leqslant \|\mathbf{a}_{k+1}^{(k+1)}\|^2 \left\{ N_1 n_{k+1} \sum_{j=1}^{k} n_j A_{k+1,\,j} \bar{B}_{j,\,k+1} + n_{k+1}^2 \right\} \\ &\leqslant \|\mathbf{a}_{k+1}^{(k+1)}\|^2 \left\{ N_1 n_{k+1} \bar{C} \sum_{j=1}^{k} n_j + n_{k+1}^2 \right\} \\ &\leqslant E_{k+1}^2 \|\mathbf{a}_{k+1}^{(k+1)}\|^2.\end{aligned}$$

Hence,

$$\|e_N\|_{\mathscr{L}} \leqslant \sum_{k=N}^{\infty} E_{k+1} \|\mathbf{a}_{k+1}^{(k+1)}\|.$$

We have now shown that the error e_N in both the natural and energy norms depends on the vectors $\mathbf{a}_k^{(k)}$, $k = N+1, N+2, \ldots$. In the next Lemma we relate this vector to known quantities.

Lemma 7.5. *For a BWNS, the vector* $\mathbf{a}_k^{(k)}$ *satisfies the bounding inequality*

$$\|\mathbf{a}_k^{(k)}\| \leqslant \alpha \mathbf{a}_k^{*t} \hat{\mathbf{g}},$$

where $\mathbf{a}_k^{*t}$ *is the k-th row of the matrix* $\mathbf{A}^* = \bar{\mathbf{A}} + \mathbf{I}$, *and* $\alpha = K_0' + K_1'$.

Proof.

$$\mathbf{L}^{(k)}\mathbf{a}^{(k)} = \mathbf{g}^{(k)}.$$

Hence,

$$\mathbf{a}_k^{(k)} = \sum_{j=1}^{k} (\mathbf{L}^{(k)-1})_{kj} \mathbf{g}_j,$$

and

$$\begin{aligned}\|\mathbf{a}_k^{(k)}\| &\leqslant \sum_{j=1}^{k-1} \|(\mathbf{L}^{(k)-1})_{kj}\| \, \|\mathbf{g}_j\| + \|(\mathbf{L}^{(k)-1})_{kk}\| \, \|\mathbf{g}_k\| \\ &\leqslant \sum_{j=1}^{k-1} \mathrm{K}_1' \bar{A}_{kj} \hat{g}_j + \mathrm{K}_0' \hat{g}_k \\ &\leqslant (K_1' + K_0') \mathbf{a}_k^{*t} \hat{\mathbf{g}}.\end{aligned}$$

Using this result, we immediately obtain the final theorem of this chapter.

Theorem 7.8. *For a BWNS,*

$$\|e_N\| \leqslant \sum_{k=N+1}^{\infty} \bar{N}_k \alpha \mathbf{a}_k^{*t} \hat{\mathbf{g}},$$

and if $\mathscr{L}$ is positive definite,

$$\|e_N\|_{\mathscr{L}} \leqslant \sum_{k=N+1}^{\infty} E_k \alpha \mathbf{a}_k^{*t} \hat{\mathbf{g}}.$$

We again note the dependence of these bounds on the block lower triangular half of the matrix $\mathbf{L}$.

In conclusion, this chapter has extended the analysis of Chapter 3 to a more general class of block matrices. We have already demonstrated, in Chapter 4, the usefulness of the results of Chapter 3 in analysing the convergence properties of an expansion method solution of an operator equation in a single variable. The assumptions of this chapter are designed to reflect the systems of equations that arise in an expansion method solution of coupled operator equations or multi-dimensional operator equations.

8

Numerical Methods for Weakly Asymptotically Diagonal Matrices

8.1 Introduction

The structure of WAD matrices can be used to advantage in providing efficient algorithms for the solution of sets of linear equations in which they arise. We consider here numerical methods for various linear systems which utilise this structure. The most efficient of these methods are iterative, but we begin by considering direct methods and, in particular, Gauss elimination for WAD systems. To avoid excessive length, no attempt is made to keep the discussion self-contained, and reference is made to the literature for the standard numerical analysis upon which we draw. The matrices considered in this chapter are finite; except where the context was felt to demand the emphasis, the order is not displayed explicitly: thus, $\mathbf{L}$ rather than $\mathbf{L}^{(N)}$.

8.2 Gauss elimination

The number of operations required to solve a full set of $N \times N$ linear equations by a direct method such as Gauss elimination is approximately $\frac{1}{3}N^3$ multiplications and additions ($\frac{1}{6}N^3$ for a symmetric matrix). This compares with $\mathcal{O}(N^2)$ operations for the iterative methods discussed below; however, a direct method may be convenient for small systems. We restrict discussion here to the *stability* of Gauss elimination for WAD systems. The stability properties are determined by the pivoting strategy used (for a general error analysis of this problem, see Wilkinson, 1965, Chapter 4;

and the generalisation considered in Reid, 1971). In the latter reference, it is shown that backward error bounds valid for any pivoting strategy may be derived in terms of the growth $G \equiv G^{(N-1)}$ defined as follows:

Definition 8.1. *Let* $\mathbf{B}^{(k)} = \{B_{ij}^{(k)}\}$ *be the* $(N - k) \times (N - k)$ *matrix resulting from k stages of Gauss elimination, with* $\mathbf{B}^{(0)} \equiv \mathbf{L}$. *Then*

$$G^{(k)} = \max_{i,j=k+1}^{N} |B_{ij}^{(k)}| / \max_{i,j=1}^{N} |L_{ij}|. \tag{8.2.1}$$

Reid's bounds demonstrate clearly that the purpose of any pivotal strategy is to avoid growth in the size of the matrix elements. We show that WAD systems are innately stable against such growth. Under suitable (rather restrictive) conditions the growth is limited automatically so that row interchanges are not necessary. More generally, the first few equations may require interchanges to ensure limited growth; but subsequent equations may then be solved in their 'natural' order. This allows the efficient addition of further rows and columns to a reduced matrix.

8.2.1 Symmetric positive definite systems

The following result is standard in the case that $\mathbf{L}$ is symmetric positive definite.

Theorem 8.1 (*see e.g., Wilkinson* 1965, *p.* 220). *If* $\mathbf{L}$ *is symmetric positive definite, then* $G^{(k)} \leqslant 1$, $k = 1,2,\ldots,N-1$.

This result ensures (see Reid, 1971) that numerical elimination without interchanges is equivalent to the exact solution of a perturbed matrix $\mathbf{L} + \mathbf{E}$, where $\|\mathbf{E}\|$ is small.

8.2.2 Non-positive definite systems

For non-positive definite matrices, Theorem 8.1 does not apply; however, the results below show that growth is unlikely to cause problems in Gauss elimination of WAD matrices, and cannot do so unless C is large. We give first a connection with the theory of diagonally dominant matrices.

Lemma 8.1. *Let* $\mathbf{L}$ *be normalised and BWAD such that* $\mathbf{A}$, $\mathbf{B}^t$ *have property MN. Then if C and* Ω *are sufficiently small,* $\mathbf{L}$ *is strictly block diagonally dominant by rows; that is, for every integer N and for some* γ *independent of N*

$$\sum_{\substack{j=1 \\ j \neq i}}^{N} \|L_{ij}\| \leqslant \gamma < 1. \tag{8.2.2}$$

Proof. Under the stated conditions, we may write $\mathbf{L} = \mathbf{E} + \mathbf{D} + \mathbf{F}$ where $\mathbf{D}$ is a block diagonal matrix with $\|D_{ii}\| = 1$ and $\mathbf{E}, \mathbf{F}^t$ are strictly lower triangular block matrices satisfying the conditions

$$\|\mathbf{E}\|_\infty \leqslant C, \quad \|\mathbf{F}\|_\infty \leqslant \Omega \tag{8.2.3}$$

for some constants C, Ω independent of N. Hence the lemma follows provided that $(C + \Omega) \leqslant \gamma$.

Corollary 8.1. For a WAD matrix with property MN, Gauss elimination without interchanges will yield a growth $G \leqslant 2$. For a BWAD matrix, with property MN, interchanges across blocks are not necessary to limit growth.

Proof. Follows from the standard result for diagonally dominant matrices (see e.g. Reid, 1971).

Lemma 8.1 depends crucially on property MN, which states that the row sums in (8.2.3) are bounded uniformly in N. However, the essential result does not depend on this property, and even WAD matrices with unbounded row sums do not lead to unbounded growth.

Theorem 8.2. *Let* $\mathbf{L}$ *be normalised and WAD. Then for* C, C_1 *sufficiently small, the pivots in Gauss elimination without interchanges are bounded away from zero, and the growth* G *is bounded, uniformly in* N.

Proof. The theorem follows immediately from Theorem 3.1 of Chapter 3, which shows that WAD matrices have a Crout triangular form $\mathbf{L} = \mathbf{UV}$ with $\mathbf{U}, \mathbf{V}$ uniformly bounded in N and $V_{ii} = 1$, $|U_{ii}|$ uniformly bounded away from zero; together with the observation that Gauss elimination with no interchanges produces this triangular reduction.

Comment

It is also shown in Chapter 7, Theorem 7.1 that under similar restrictions on C, normalised BWAD matrices may be reduced to block triangular form with $\mathbf{U}, \mathbf{V}$ bounded uniformly in N, and $\mathbf{V}_{ii} = \mathbf{I}$, $\|\mathbf{U}_{ii}^{-1}\|$ bounded uniformly, provided that the diagonal blocks $\mathbf{L}_{ii}$ of $\mathbf{L}$ have condition number bounded uniformly in N, i.

It also follows from Theorem 7.3 of Chapter 7 that under the given conditions $\|\mathbf{L}^{(N)-1}\|$ is bounded uniformly in N. Thus, block elimination without interchanges forms a stable direct method for solving large systems, for C not too large.

Theorem 8.2 gives sufficient conditions for the sequence of matrices $\mathbf{L}^{(N)}$ to be non-singular, $N = 1,2,\ldots$ and hence no such result can be valid for arbitrary C, C_1. Even so, the theorem suggests that the choice of pivoting strategy is never likely to be important for WAD matrices. If interchanges are needed, they are likely to be confined to the first few rows of $\mathbf{L}$, and to be infrequent. This implies that:

(a) direct interchange of rows will be more efficient than index holding;
(b) given an initial matrix which has been reduced to triangular form with interchanges, further rows and columns can be added efficiently with interchanges switched off or limited to the new entries.

8.3 Iterative solution of sets of equations

For large N, the direct solution of the full $N \times N$ system

$$\mathbf{L}^{(N)}\mathbf{a} = \mathbf{g}^{(N)} \tag{8.3.1}$$

is expensive, and we seek cheaper, iterative methods. One convergent scheme is that given by Mikhlin (1971) for strongly minimal matrices. With notation as in Chapter 5, he proves the following theorem:

Theorem 8.3. *Let $\mathbf{L}^{(N)}$ be symmetric positive definite with eigenvalues λ_i, $i = 1, \ldots, N$, and let α satisfy*

$$1 - \alpha\lambda_1^{(N)} < 1; \quad 1 - \alpha\lambda_N^{(N)} > -1. \tag{8.3.2}$$

Then the iterative scheme

$$\mathbf{a}_{m+1} = [\mathbf{I}^{(N)} - \alpha\mathbf{L}^{(N)}]\,\mathbf{a}_m + \alpha\mathbf{g}^{(N)} \tag{8.3.3}$$

converges to the solution of (8.3.1). *Further, the choice*

$$\alpha = \frac{2}{\lambda_1^{(N)} + \lambda_N^{(N)}} \tag{8.3.4}$$

yields a spectral radius for the iteration matrix of

$$r = \rho(\mathbf{I} - \alpha\mathbf{L}^{(N)}) = (\rho(\mathbf{L}^{(N)}) - 1)/(\rho(\mathbf{L}^{(N)}) + 1) < 1. \tag{8.3.5}$$

For a proof of Theorem 8.3, see Mikhlin, 1971, p. 54.

If $\mathbf{L}^{(N)}$ is strongly minimal but not almost orthonormal, $\lambda_N^{(N)} \underset{N}{\to} \infty$ and for large N, $r \to 1$; then (8.3.3) suffers from the same disadvantage as is typical for corresponding schemes for finite difference methods. If, however, $\mathbf{L}^{(N)}$ is almost orthonormal, the convergence rate of the iterative scheme is essentially independent of N; for we may then replace (8.3.4) by the choice

$$\alpha = 2/(\lambda_0 + \Lambda_0) \tag{8.3.6}$$

where λ_0, Λ_0 are given by Definitions 5.3 and 5.4, and find

$$r \leqslant (\Lambda_0 - \lambda_0)/(\Lambda_0 + \lambda_0). \tag{8.3.7}$$

Then (8.3.3) forms a very attractive computational scheme for large systems, since the number of iterations required to achieve a given accuracy is to first order independent of N, and the total solution time is $\mathcal{O}(N^2)$. The iterative scheme is then much faster for large N than direct elimination.

In practice, the matrix **L** may be neither positive definite nor symmetric. Then the iterative scheme (8.3.3) no longer applies. It is, however, possible to provide simple and rapidly convergent iterative schemes for WAD and for BWAD systems, as the following results show:

Hypothesis H1. Let **L** be almost orthonormal in the extended sense given in Chapter 5, Definition 5.6.

Theorem 8.4. *Let* **L** *satisfy H1, and let* **L** *be split in the form*

$$\mathbf{L} = \mathbf{D} + \mathbf{E} + \mathbf{F} \tag{8.3.8}$$

where **E**, $\mathbf{F}^t$ *are strictly lower block triangular, and* **D** *contains the leading* $M \times M$ *submatrix of* **L** *together with a diagonal strip of half-width* μ:

$$\begin{aligned} D_{ij} &= L_{ij} \qquad i,j \leqslant M \\ &= L_{ij} \qquad |i-j| \leqslant \mu \\ &= 0 \qquad \text{otherwise.} \end{aligned} \tag{8.3.9}$$

Then there exist integers M, μ, independent of N, such that block Jacobi and block Gauss Seidel iterations on the equations (8.3.1) *are convergent. Moreover, for any positive δ independent of N, (M, μ) may be chosen so that the spectral radius r of the iteration matrix satisfies*

$$r < \delta. \tag{8.3.10}$$

Proof. Now from Hypothesis H1, there exists a constant γ such that for M, μ sufficiently large

$$\|\mathbf{D}^{-1}\| \leqslant \gamma. \tag{8.3.11}$$

Further, $\lim_{\mu,M\to\infty} \|\mathbf{E}\|, \|\mathbf{F}\| = 0$ and hence we may choose M, μ so that uniformly in N and for any $\varepsilon > 0$,

$$\|\mathbf{E}\|, \|\mathbf{F}\| \leqslant \varepsilon. \tag{8.3.12}$$

We then have:

(*a*) *Jacobi*

The iteration matrix $\mathbf{P} = -\mathbf{D}^{-1}(\mathbf{E} + \mathbf{F})$ and

$$r(\mathbf{P}) \leqslant \|\mathbf{P}\| \leqslant 2\gamma\varepsilon. \tag{8.3.13}$$

(*b*) *Gauss-Seidel:*

$$\mathbf{P} = -(\mathbf{D} + \mathbf{E})^{-1}\,\mathbf{F}$$

and for $\gamma\varepsilon < 1$

$$r(\mathbf{P}) \leqslant \|\mathbf{P}\| \leqslant \gamma\varepsilon/(1 - \gamma\varepsilon). \tag{8.3.14}$$

Hence for any $\delta > 0$ we may satisfy (8.3.10).

Comments
Hypothesis H1 is not invariant under a congruence transformation of $\mathbf{L}$; hence, if $\mathbf{L}$ is a Galerkin matrix, H1 may hold for one normalisation of the expansion set $\{h_i\}$ but not for another. However, the convergence properties of the iterative schemes are invariant under the choice of normalisation. For the rescaling $h_i \to \Lambda_{ii}^{-1} h_i$ induces the transformation (5.2.1), (5.2.3) on equation (8.3.1) and under the iterative schemes discussed, at each iteration m:

$$\mathbf{a}_{(m)} \to \Lambda \mathbf{a}_{(m)}.$$

This, in turn, implies that the sequence of Galerkin solutions $f_N^m = \sum_{i=1}^{N} (a_m)_i h_i$ is invariant under the congruence transformation so that the iterations in any two normalisations converge or diverge together. We therefore have:

Corollary 8.2. Let there exist some non-singular diagonal matrix Λ such that under the transformation (5.2.1), (5.2.3), Hypothesis H1 holds, then Theorem 8.4 follows.

We also have a connection with WAD matrices:

Hypothesis H2. Let the system (8.3.1) be a WADS (Definition 5.2), and let $\mathbf{A}$ and $\mathbf{B}^t$ have property *MN*.

Corollary 8.3. Let Hypothesis H2 hold, then Theorem 8.4 follows.

Proof. Under this hypothesis, $\mathbf{L}$ may be normalised and then satisfies Hypothesis H1, almost orthonormality following from property *MN*.

An almost identical result follows for BWAD systems under the following hypothesis, which is an obvious generalisation of Hypothesis H2:

Hypothesis H3.

(a) $\mathbf{L}$ is BWAD and $\mathbf{A}$ and $\mathbf{B}^t$ have property MN.

(b) For some Q, independent of M, and for i sufficiently large, the diagonal blocks $\mathbf{L}_{ii}$ satisfy the inequality

$$\|\mathbf{L}_{ii}\| \, \|\mathbf{L}_{ii}^{-1}\| \leqslant Q.$$

(c) Let $\overline{\mathbf{L}}_{ij} = \mathbf{L}_{ij}/\{\|\mathbf{L}_{ii}\| \, \|\mathbf{L}_{jj}\|\}^{1/2}$ so that $\|\overline{\mathbf{L}}_{ii}\| = 1, \forall i$. Then $\overline{\mathbf{L}}$ is strongly minimal in the extended sense of Chapter 5.

Theorem 8.5. *Let* $\mathbf{L}$ *satisfy Hypothesis H3 and let* $\mathbf{L}$ *be split as in Theorem* 8.4. *Then there exist integers* M, μ *independent of* N, *such that block Jacobi and block Gauss-Seidel iterations on the equations* (8.3.1) *are convergent. Moreover, for any positive* δ *independent of* N, (M, μ) *may be chosen so that the spectral radius of the iteration matrix satisfies* (8.3.10).

Proof. Hypothesis H3 is independent of the normalisation of $\mathbf{L}$. We may therefore consider $\mathbf{L}$ normalised, and then the proof follows closely that of Theorem 8.4.

8.4 Practical applications

Theorems 8.4 and 8.5 may be used to yield a rapid iterative alogrithm with guaranteed convergence for almost orthonormal or for WAD or BWAD systems. The constants M, μ are not in practice known in advance, but it is possible to choose these initially so that (8.3.12) is satisfied with $\varepsilon = 0.1$ (say) and to increase M during the course of iterations if convergence is not rapid enough. Similar convergence results may be given for SOR, but the use of acceleration techniques (other than increasing M) is hardly warranted.

For WAD systems, it is usually possible to set $\mu = 0$; then the matrix of Theorem 8.4 contains the leading $M \times M$ submatrix of $\mathbf{L}$, plus the remaining diagonal terms; that is,

$$\begin{aligned} D_{ij} &= L_{ij} \qquad i, j \leqslant M \\ &= L_{ii}\delta_{ij} \quad i > M \quad \text{or} \quad j > M. \end{aligned} \tag{8.4.1}$$

The iterative schemes discussed have the form

$$\mathbf{D}\mathbf{a}^{(k+1)} = \mathbf{g}^{(N)} - (\mathbf{E} + \mathbf{F})\,\mathbf{a}^{(k)} \quad \text{(Jacobi)} \tag{8.4.2}$$

$$[\mathbf{D} + \mathbf{E}]\mathbf{a}^{(k+1)} = \mathbf{g}^{(N)} - \mathbf{F}\mathbf{a}^{(k)} \quad \text{(Gauss-Seidel)} \tag{8.4.3}$$

and with $\mathbf{D}$ of the form (8.4.1) these equations may be more simply written in terms of the alternative partitioning of $\mathbf{L}^{(N)}$:

$$\begin{bmatrix} \mathbf{L}^{(M)} & \mathbf{L}_{12} \\ \mathbf{L}_{21} & \mathbf{L}_{22} \end{bmatrix} \begin{bmatrix} \mathbf{a}_{(1)} \\ \mathbf{a}_{(2)} \end{bmatrix} = \begin{bmatrix} \mathbf{g}_{(1)} \\ \mathbf{g}_{(2)} \end{bmatrix}. \tag{8.4.4}$$

In terms of (8.4.4) a partitioned Gauss-Seidel iterative scheme has the form

$$\begin{aligned} \mathbf{L}^{(M)}\,\mathbf{a}_{(1)}^{(k+1)} &= \mathbf{g}_{(1)} - \mathbf{L}_{12}\,\mathbf{a}_{(2)}^{(k)} \\ (\mathbf{D}_{22} + \mathbf{E}_{22})\,(\mathbf{a}_{(2)}^{(k+1)} - \mathbf{a}_{(2)}^{(k)}) &= \mathbf{g}_{(2)} - \mathbf{L}_{21}\,\mathbf{a}_{(1)}^{(k+1)} - \mathbf{L}_{22}\,\mathbf{a}_{(2)}^{(k)} \end{aligned} \tag{8.4.5}$$

where we have partitioned $\mathbf{L}_{22} = \mathbf{D}_{22} + \mathbf{E}_{22} + \mathbf{F}_{22}$ with $\mathbf{D}_{22}$ diagonal and $\mathbf{E}_{22}$, $\mathbf{F}_{22}^t$ strictly lower triangular matrices.

This scheme is rapidly convergent under the same conditions as (8.4.2), (8.4.3), and is particularly simple to implement; a procedure, including the automatic choice of M, is given in Delves (1977b).

There are however situations where we may not set $\mu = 0$. For example, we recall the bounding inequalities for a symmetric, normalised AD matrix of type $\mathrm{B}(p, r; C)$:

$$\begin{aligned} |L_{ii}| &= 1 \\ |L_{ij}| &\leqslant Ci^{-p}(i-j)^{-r}, \quad i > j. \end{aligned}$$

Matrices of this type occur in the Galerkin solution of second order differential equations (see Chapter 13), but possibly with $p = 0$. For this value of p, it is necessary to take $\mu > 0$ to ensure convergence of the iterative schemes discussed here.

However, in either case we emphasize the most important consequences of the simple Theorems 8.4 and 8.5—matrices with this structure can be solved in $\mathcal{O}(N^2)$ operations. This result has far-reaching effects on the competitiveness of "global" Galerkin procedures. As an example, we quote the experience of Davies and Hendry (1977) in which a global Galerkin solution of the nonlinear two-dimensional channel design problem for compressible fluid flow is discussed. A direct solution of the (linearised) equations dominated the computing time for this problem; resort to the iterative solution discussed here led to an overall time saving of a factor of more than 5, and this factor increases with the size of the problem. A second application of the theorems, to the solution of Fredholm integral equations, is discussed in Chapter 12. There, it is shown that the Galerkin solution of Fredholm integral equations with a Chebyshev polynomial expansion set yields matrices which are AD of type $A(p; C)$. Table 8.1 shows timings obtained using the iterative scheme (8.4.5), iterating to machine accuracy (about 11 digits), compared with a standard Gauss elimination routine. The matrix in this table comes from the solution of the integral equation

$$f(x) = g(x) + \int_0^1 e^{\beta xy} f(y)\,dy \tag{8.4.6}$$

with

$$g(x) = e^x - [e^{1+\beta x} - 1]/[1 + \beta x]$$

and exact solution $f(x) = e^x$. The constant C for this system increases with β,

TABLE 8.1

Solution of the linear equations arising from the integral equation (8.4.6). All times are in milliseconds and were obtained on an ICL 1906S using the RRE Algol 68R compiler with scheme (8.4.5) and the initial block size M as shown. Times for a direct solution are given for comparison.

	N =	5	9	17	33	65
(Gauss elimination)	T =	15	36	125	616	3780
$\beta = 1$	M =	3	3	3	3	3
	T =	59	51	68	137	320
$\beta = 5$	M =	5	8	10	15	15
	T =	18	44	101	144	443
$\beta = 20$	M =	5	9	17	22	22
	T =	19	53	128	630	745

and we see from Table 8.1 that the value of M required in (8.4.5) also increases with β; for fixed β and $N \gg M$, however, M stabilises at a fixed value as predicted by Theorem 8.4 and substantial savings are made over direct elimination.

8.5 Asymptotic estimates of the solution vector

The iterative algorithms of Section 8.4 depend only weakly for their success on the structure of WAD matrices. We show now that, under stronger conditions, a simple pointwise iteration leads to accurate prediction of the components a_i, for i large.

We consider the *infinite* system

$$\mathbf{Lb} = \mathbf{g} \tag{8.5.1}$$

Setting $\mu = 0$, $M = 1$ in Theorem 8.5, a simple Jacobi or Gauss-Seidel iteration yields the following approximations (prediction coefficients)

D_n, d_n to b_n:

$$D_n = g_n/L_{nn} \qquad \text{(Jacobi)} \tag{8.5.2}$$

$$d_n = \frac{1}{L_{nn}}\left(g_n - \sum_{j=1}^{n-1} L_{nj}\, d_j\right) \qquad \text{(Gauss-Seidel)} \tag{8.5.3}$$

Then the following theorem shows that, for symmetric systems of type **B** and under rather strong conditions, $\lim_{n\to\infty} D_n/b_n, d_n/b_n = 1$.

Theorem 8.6. *Let* **b** *satisfy* (8.5.1) *and* **L** *be symmetric and normalised AD of type* $B(p, r; C)$ *and let* $|b_n| \leqslant \kappa n^{-q}$ *with* $q, r > 1$ *and* $p \geqslant 0$. *Then there exists a constant* $\mathscr{C}_1$ *such that for every* n

(*i*)
$$|D_n - b_n| \leqslant \mathscr{C}_1\, n^{-\nu}, \tag{8.5.4}$$

where $\nu = p + \min[r, q]$. *If, in addition,*

$$C < \left[2^r\left(\frac{\nu}{\nu - 1}\right) + 2^\nu\left(\frac{r}{r-1}\right)\right]^{-1} \tag{8.5.5}$$

then there exists a constant $\mathscr{C}_2$ *such that for every* n

(*ii*)
$$|d_n - b_n| \leqslant \mathscr{C}_2\, n^{-\nu}. \tag{8.5.6}$$

Corollary 8.4. If in addition for some subsequence $\{m\}$ of the integers $|b_m| \geqslant km^{-q'}$ and $q' < \nu$ then

$$\left|\frac{D_m}{b_m} - 1\right| \leqslant \frac{\mathscr{C}_1}{k} m^{-\varepsilon}$$

and

$$\left|\frac{d_m}{b_m} - 1\right| \leqslant \frac{\mathscr{C}_2}{k} m^{-\varepsilon} \tag{8.5.7}$$

where $\varepsilon = (\nu - q')$. Note that we may usually identify $q' = q$, but that the subsequence $\{m\}$ cannot in general be extended to include all of the integers; consider for example an odd-even effect which yields $b_{2j} = 0, \forall j$.

The conditions under which the relative error in the prediction coefficients can be expected to tend to zero, can be identified more closely. For C sufficiently small, and with $|g_n| < C_g n^{-s}, s > 1$, it follows from (4.2.1a) that we may take $q = \min(p + r, s)$. It then follows that, if we can identify $q' = q$, the exponent ε will be positive provided that $p + r > s$.

Proof. For brevity we give the proof for part (i) only, i.e. for the validity of the coefficients D_n. For the proof of part (ii) see Mead (1971).

We have from (8.5.2) and the normalisation condition on $\mathbf{L}$,

$$L_{nn} D_n = g_n = \sum_{i=1}^{\infty} b_i L_{ni}$$

$$= L_{nn} b_n + \sum_{i=1}^{n-1} b_i L_{ni} + \sum_{i=n+1}^{\infty} b_i L_{ni}. \quad (8.5.8)$$

The conditions of the theorem yield the inequalities

$$|L_{ni}| \leqslant Cn^{-p} (n-i)^{-r}, \quad \text{for } n > i,$$

and

$$|b_n| \leqslant \kappa n^{-q}.$$

Inserting these into equation (8.5.8) we have

$$|D_n - b_n| \leqslant C\kappa \sum_{i=1}^{n-1} i^{-q} n^{-p} (n-i)^{-r} + C\kappa \sum_{i=n+1}^{\infty} i^{-p-q} (i-n)^{-r},$$

$$\leqslant C\kappa n^{-p} \sum_{i=1}^{n-1} i^{-q} (n-i)^{-r} + C\kappa n^{-(p+q)} \sum_{j=1}^{\infty} j^{-r}.$$

Now from Lemma A9 of the Appendix

$$\sum_{i=1}^{n-1} i^{-q} (n-i)^{-r} \leqslant \frac{2t}{t-1} \left(\frac{n}{2}\right)^{-t}, \quad t = \min(q, r).$$

Thus

$$|D_n - b_n| \leqslant C\kappa \frac{t}{t-1} 2^{t+1} n^{-(p+t)} + C\kappa \left(\frac{r}{r-1}\right) n^{-(p+q)}$$

$$\leqslant \mathscr{C} n^{-\nu}$$

provided that

$$\nu = p + \min[r, q]$$

and

$$\mathscr{C} \geqslant C\kappa \left[\frac{2^{t+1} t}{t-1} + \left(\frac{r}{r-1}\right)\right].$$

This can always be satisfied since $r,q > 1$; and this completes the proof of part (i).

We now comment that the ratios $D_n : b_n$ and $d_n : b_n$ are invariant under a diagonal congruence transformation of **L**, and hence the results are essentially independent of the scaling used.

Theorem 8.6 yields results also for type A systems (set $r = 0$). We quote without proof (for which see Mead, 1971) the similar results for type C systems.

Theorem 8.7. *Let* **b** *satisfy* (8.5.1); *let* **L** *be symmetric and normalised AD of type* $C(p, r; C)$; *and let* $|b_n| \leqslant \kappa n^{-q}$. *Then there exists a constant* $\mathscr{C}_3$ *such that for every n,*

$$|D_n - b_n| \leqslant \mathscr{C}_3\, n^{-\nu},$$

where

$$\nu = \begin{cases} p - 1, & q \geqslant r, \\ p + q - r - 1, & q < r. \end{cases}$$

If in addition $p > r + 1, t > r + 1$ *and* $C < (t - r - 1)/(t - r)$, *where* $t = \min\{p, p + q - r - 1\}$, *then there exists a constant* $\mathscr{C}_4$ *such that for every n,*

$$|d_n - b_n| \leqslant \mathscr{C}_4\, n^{-t}.$$

Similar corollaries follow as for Theorem 8.6.

Comment

If for some s, C_g and all i, $|g_i| < C_g i^{-s}$, it follows from (4.3.2) that with a suitable restriction on the constant C, we may take $q = \min(p, s)$. Hence, as before, the sufficiently large p the relative error in the prediction coefficients tend to zero.

8.5.1 Discussion and example

The foregoing theorems discuss under various conditions on the matrix **L** the extent to which the prediction coefficients D_n, d_n can be expected to approximate the expansion coefficients b_n. The existence of simple approximations to b_n for large n is clearly of some interest, and the prediction coefficients defined have two possible uses. The first is as a means of giving *a priori* estimates of the vertical convergence rate of a proposed calculation; for this purpose they give rather more detail than is available from the bounding theorems of Chapters 2–4. The second use is to extrapolate a given variational calculation, when we might choose to augment the computed vector of expansion coefficients $\mathbf{a}^{(N)}$, equation (1.2.4), by the prediction coefficients $D_{N+1}, \ldots, D_{N+M}$.

As an example of their use as predictors for the b_n, we return to the one-dimensional example (1.5.1) described in Chapter 1.

$$\left[\frac{d^2}{dx^2} - \lambda x\right] f(x) = g(x), \quad f(0) = f(\pi) = 0,$$

$$g(x) = x^4 - (1 + \pi)\, x^3 + \pi x^2 - 6x + 2(1 + \pi).$$

For $\lambda = 1$ the exact solution is

$$f(x) = x(x - 1)(\pi - x).$$

With the expansion set as in Chapter 1

$$h_i(x) = (2/\pi)^{1/2} \sin ix, \quad i = 1,2,3,\ldots,$$

we find that

$$\begin{aligned} b_i &= \left(\frac{2}{\pi}\right)^{1/2} \cdot \frac{2}{i^3}(\pi - 2), \quad i \text{ odd}, \\ &= -\left(\frac{2}{\pi}\right)^{1/2} \cdot \frac{6\pi}{i^3}, \qquad i \text{ even}. \end{aligned} \tag{8.5.9}$$

From (1.5.9) we recall that the matrix $\mathbf{L}$ is AD of type B(2,2; $8\lambda/\pi$). Hence from Corollary 8.4 we predict that

$$\left|\frac{D_n}{b_n} - 1\right| \quad \text{and} \quad \left|\frac{d_n}{b_n} - 1\right|$$

are both $\mathcal{O}(n^{-\varepsilon})$ where $\varepsilon = (v - q') = 4 - 2 = 2$.

We may easily compute the non-recursive coefficients D_n analytically in this case. We find for n odd

$$\begin{aligned} D_n = \frac{g_n}{L_{nn}} &= \left[\left(\frac{2}{\pi}\right)^{1/2} \frac{2}{n}(\pi - 2)\left[1 + \mathcal{O}(n^{-2})\right]\right] \Big/ \left(n^2 + \frac{\pi}{2}\right), \\ &= \left(\frac{2}{\pi}\right)^{1/2} \frac{2}{n^3}(\pi - 2)\left[1 + \mathcal{O}(n^{-2})\right], \end{aligned}$$

and for n even

$$D_n = -\left(\frac{2}{\pi}\right)^{1/2} \frac{6\pi}{n^3}\left[1 + \mathcal{O}(n^{-2})\right],$$

so that in both cases

$$\left|\frac{D_n}{b_n} - 1\right| = \mathcal{O}(n^{-2}).$$

TABLE 8.2

n	b_n	$\|D_n/b_n - 1\|$	$\|d_n/b_n - 1\|$
1	$0{\cdot}1821718 \times 10$	$0{\cdot}2$	$0{\cdot}2$
2	$-0{\cdot}1879971 \times 10$	$0{\cdot}1$	$0{\cdot}2 \times 10^{-1}$
3	$0{\cdot}6747105 \times 10^{-1}$	$0{\cdot}2 \times 10$	$0{\cdot}3$
4	$-0{\cdot}2349964$	$0{\cdot}3 \times 10^{-1}$	$0{\cdot}5 \times 10^{-2}$
5	$0{\cdot}1457375 \times 10^{-1}$	$0{\cdot}8$	$0{\cdot}1$
6	$-0{\cdot}6962856 \times 10^{-1}$	$0{\cdot}2 \times 10^{-1}$	$0{\cdot}1 \times 10^{-2}$
7	$0{\cdot}5311132 \times 10^{-2}$	$0{\cdot}4$	$0{\cdot}8 \times 10^{-1}$
8	$-0{\cdot}2937455 \times 10^{-1}$	$0{\cdot}9 \times 10^{-2}$	$0{\cdot}6 \times 10^{-3}$
9	$0{\cdot}2498928 \times 10^{-2}$	$0{\cdot}3$	$0{\cdot}5 \times 10^{-1}$
10	$-0{\cdot}1503977 \times 10^{-1}$	$0{\cdot}6 \times 10^{-2}$	$0{\cdot}3 \times 10^{-3}$
11	$0{\cdot}1368684 \times 10^{-2}$	$0{\cdot}2$	$0{\cdot}4 \times 10^{-1}$
12	$-0{\cdot}8703570 \times 10^{-2}$	$0{\cdot}4 \times 10^{-2}$	$0{\cdot}2 \times 10^{-3}$
13	$0{\cdot}8291845 \times 10^{-3}$	$0{\cdot}1$	$0{\cdot}3 \times 10^{-1}$
14	$-0{\cdot}5480966 \times 10^{-2}$	$0{\cdot}3 \times 10^{-2}$	$0{\cdot}9 \times 10^{-4}$
15	$0{\cdot}5397684 \times 10^{-3}$	$0{\cdot}9 \times 10^{-1}$	$0{\cdot}2 \times 10^{-1}$
16	$-0{\cdot}3671819 \times 10^{-2}$	$0{\cdot}2 \times 10^{-2}$	$0{\cdot}6 \times 10^{-4}$
17	$0{\cdot}3707955 \times 10^{-3}$	$0{\cdot}7 \times 10^{-1}$	$0{\cdot}2 \times 10^{-1}$
18	$-0{\cdot}2578836 \times 10^{-2}$	$0{\cdot}2 \times 10^{-2}$	$0{\cdot}4 \times 10^{-4}$
19	$0{\cdot}2655953 \times 10^{-3}$	$0{\cdot}6 \times 10^{-1}$	$0{\cdot}1 \times 10^{-1}$
20	$-0{\cdot}1879971 \times 10^{-2}$	$0{\cdot}1 \times 10^{-2}$	$0{\cdot}2 \times 10^{-4}$
21	$0{\cdot}1967086 \times 10^{-3}$	$0{\cdot}5 \times 10^{-1}$	$0{\cdot}1 \times 10^{-1}$
22	$-0{\cdot}1412450 \times 10^{-2}$	$0{\cdot}1 \times 10^{-2}$	$0{\cdot}2 \times 10^{-4}$
23	$0{\cdot}1497262 \times 10^{-3}$	$0{\cdot}4 \times 10^{-1}$	$0{\cdot}1 \times 10^{-1}$
24	$-0{\cdot}1087946 \times 10^{-2}$	$0{\cdot}1 \times 10^{-2}$	$0{\cdot}1 \times 10^{-4}$
25	$0{\cdot}1165900 \times 10^{-3}$	$0{\cdot}3 \times 10^{-1}$	$0{\cdot}9 \times 10^{-2}$
26	$-0{\cdot}8556992 \times 10^{-3}$	$0{\cdot}8 \times 10^{-3}$	$0{\cdot}7 \times 10^{-5}$
27	$0{\cdot}9255288 \times 10^{-4}$	$0{\cdot}3 \times 10^{-1}$	$0{\cdot}8 \times 10^{-2}$
28	$-0{\cdot}6851207 \times 10^{-3}$	$0{\cdot}7 \times 10^{-3}$	$0{\cdot}5 \times 10^{-5}$
29	$0{\cdot}7469426 \times 10^{-4}$	$0{\cdot}2 \times 10^{-1}$	$0{\cdot}7 \times 10^{-2}$
30	$-0{\cdot}5570285 \times 10^{-3}$	$0{\cdot}6 \times 10^{-3}$	$0{\cdot}3 \times 10^{-5}$

This is precisely the convergence rate predicted above using the results of Corollary 8.4.

The results of a full numerical computation of both sets of coefficients, $\{D_n\}$ and $\{d_n\}$, are given in Table 8.2. While both sets give the predicted rate of convergence to the true Fourier coefficients, the Gauss-Seidel set $\{d_n\}$ is in general an order of magnitude better than the Jacobi set $\{D_n\}$. This superiority appears to be maintained in other cases too.

8.6 The least squares solution of overdetermined systems

In applications that involve the imposition of boundary conditions or side conditions on the solution, some methods yield overdetermined systems of equations which may be written in the partitioned form

$$\begin{bmatrix} \mathbf{L} \\ \mathbf{B} \end{bmatrix} \mathbf{a} = \begin{bmatrix} \mathbf{g} \\ \mathbf{k} \end{bmatrix} \tag{8.6.1}$$

where $\mathbf{L}$ is $N \times N$ and $\mathbf{B}$ is $P \times N$; $\mathbf{a}$, $\mathbf{g}$ are N-vectors and $\mathbf{k}$ is a P-vector. In these applications P is typically much less than N; and we consider the efficient solution of (8.6.1) when in addition $\mathbf{L}$ is WAD. The method described uses only the fact that a square WAD system can be solved in $\mathcal{O}(N^2)$ operations. We assume here that $\mathbf{L}$ is non-singular, so that the solution to (8.6.1) is unique; for the case when $\begin{pmatrix} \mathbf{L} \\ \mathbf{B} \end{pmatrix}$ or $\mathbf{L}$ is rank deficient, see Delves and Barrodale (1979).

Let $\mathbf{C}$ be any non-singular matrix, and let

$$\mathbf{Ca} = \mathbf{y}. \tag{8.6.2}$$

Then, in any norm, system (8.6.1) is equivalent via (8.6.2) to the modified system (8.6.1′):

$$\begin{bmatrix} \mathbf{LC}^{-1} \\ \mathbf{BC}^{-1} \end{bmatrix} \mathbf{y} = \begin{bmatrix} \mathbf{g} \\ \mathbf{k} \end{bmatrix}. \tag{8.6.1′}$$

If $\mathbf{L}$ is non-singular we may choose $\mathbf{C} = \mathbf{L}$. Then (8.6.1′) takes the form

$$\begin{bmatrix} \mathbf{I} \\ \mathbf{Q} \end{bmatrix} \mathbf{y} = \begin{bmatrix} \mathbf{g} \\ \mathbf{k} \end{bmatrix}, \quad \mathbf{Q} = \mathbf{BL}^{-1}, \tag{8.6.1″}$$

and the normal equations associated with (8.6.1″) are

$$[\mathbf{I} + \mathbf{Q}^{*t}\mathbf{Q}]\,\mathbf{y} = \mathbf{g} + \mathbf{Q}^{*t}\mathbf{k} = \mathbf{c} \quad \text{(say)}, \tag{8.6.3}$$

where $\mathbf{Q}^{*t}$ is the complex conjugate transpose of $\mathbf{Q}$.

The use of the transformed equations (8.6.3) appears to have been suggested first by Noble (1969), who shows that the modified normal equations are likely to be better conditioned than the unmodified equations. Noble discusses only the direct solution of (8.6.3) by elimination. However, it is neither necessary nor efficient (if $P << N$) to solve (8.6.3) directly.

Let $\mathbf{Q}_k$ be the k-th column of $\mathbf{Q}^t$. Then (8.6.3) may be written as

$$\left[\mathbf{I} + \sum_{k=1}^{P} \mathbf{Q}_k^* \mathbf{Q}_k^t\right] \mathbf{y} = \mathbf{c} \tag{8.6.3′}$$

where now the second term is displayed as a rank P modification to the matrix $\mathbf{I}$; standard techniques then yield the expansion for $\mathbf{y}$:

$$\mathbf{y} = \mathbf{c} - \sum_{k=1}^{P} \gamma_k \mathbf{Q}_k^* \tag{8.6.4}$$

where $\gamma = (\gamma_k)$ satisfies the $P \times P$ linear system

$$\begin{aligned}[\mathbf{I} + \mathbf{Q}\mathbf{Q}^{*t}]\,\gamma &= \mathbf{Qc} \\ &= \mathbf{Qg} + \mathbf{Q}\mathbf{Q}^{*t}\mathbf{k}.\end{aligned} \tag{8.6.5}$$

The direct solution of (8.6.5) has negligible cost if $P << N$. We can now summarise the complete least squares procedure for a non-singular WAD matrix $\mathbf{L}$ as follows:

(a) Form $\mathbf{Q}$ as the solution of the system

$$\mathbf{L}^t\mathbf{Q}^t = \mathbf{B}^t.$$

This can be done at a cost of $\mathcal{O}(PN^2)$ operations.

(b) Set up and solve (8.6.5): cost $\frac{1}{2}P^2N + \mathcal{O}(P^3)$ operations.

(c) Form $\mathbf{a}$ as the solution of $\mathbf{La} = \mathbf{y}$: cost $\mathcal{O}(N^2)$ operations.

The method is applicable also to weighted least squares problems. Let $\begin{bmatrix} \mathbf{W}_1 & \mathbf{0} \\ \mathbf{0} & \mathbf{W}_2 \end{bmatrix}$ be the partitioned $(N + P) \times (N + P)$ positive definite diagonal weighting matrix. The unweighted normal equations (8.6.3) become in the weighted case:

$$[\mathbf{W}_1 + \mathbf{Q}^{*t}\mathbf{W}_2\,\mathbf{Q}]\,\mathbf{y} = \mathbf{W}_1\mathbf{g} + \mathbf{Q}^{*t}\,\mathbf{W}_2\mathbf{k}. \tag{8.6.6}$$

Equation (8.6.6) displays the normal matrix as a symmetric, rank P modification to the diagonal matrix $\mathbf{W}_1$, and can hence be solved by the same techniques as before.

8.7 Eigenvalue problems

We next consider briefly the numerical solution of the eigenvalue problem

$$[\mathbf{L} - \lambda\,\mathbf{M}]\,\mathbf{a} = \mathbf{0} \tag{8.7.1}$$

where $\mathbf{L}, \mathbf{M}$ are $N \times N$, WAD matrices.

8.7.1 Calculation of a single eigensolution

Commonly, only a single eigensolution $(\lambda, \mathbf{a})$ of (8.7.1) is required. Provided that an initial approximation λ_0 is available, and that λ is a simple eigenvalue, this can be computed using inverse iteration:

$$\begin{aligned}
[\mathbf{L} - \lambda_0 \mathbf{M}]\,\mathbf{b}^{(m+1)} &= \mathbf{M}\mathbf{a}^{(m)} \\
\mathbf{a}^{(m+1)} &= \mathbf{b}^{(m+1)}/\|\mathbf{b}^{(m+1)}\| \qquad\qquad (8.7.2) \\
\lambda^{(m+1)} &= \lambda_0 + \operatorname{sign}(b_1^{(m+1)})/\|\mathbf{b}^{(m+1)}\|.
\end{aligned}$$

The cost of each iteration in this scheme is $\mathcal{O}(N^2)$ operations. The number of iterations P required depends on the closeness of λ_0 to λ. However, it follows from the analysis of Chapter 6 that if $\mathbf{L}, \mathbf{M}$ are AD of type A or B, the sequence of ordered eigenvalues $\lambda_i^{(N)}$ converges to λ_i for fixed i as $N \to \infty$, provided that certain restrictions are satisfied,

$$\lambda_i^{(N)} \underset{N \to \infty}{\to} \lambda_i$$

and hence for a fixed λ_0 and N sufficiently large, the number of iterations required to achieve a given accuracy is essentially independent of N.

8.7.2 Calculation of a group of eigensolutions

More generally, several eigensolutions may be of interest and in that case, provided that initial approximate eigenvalues $\lambda_{0_1}, \ldots, \lambda_{0_r}$ are available, simultaneous inverse iteration (see, e.g. Stewart, 1976) also leads to an $\mathcal{O}(N^2)$ calculation of each eigensolution; this approach also handles the case when one or more eigenvalues are not simple.

8.7.3 Calculation of initial estimates

If either the largest or the smallest (group of) eigenvalues are of interest, the problem of providing initial estimates does not arise. For the largest eigenvalues, direct simultaneous iteration and for the smallest, inverse iteration with initial eigenvalue estimates zero, will converge with usually a satisfactory rate of convergence. Typically, the smallest eigenvalues are physically meaningful for a system that arises from differential equations, the largest for systems that arise from integral equations.

If an intermediate group of eigenvalues is required, or if we seek the rapid convergence associated with a good initial estimate, then we may proceed as follows:

Case 1: $\mathbf{M} = \mathbf{I}$
In this case we seek an approximate eigensolution to the equation

$$[\mathbf{L} - \lambda\mathbf{I}]\mathbf{a} = \mathbf{0}. \tag{8.7.3}$$

Now, if we split $\mathbf{L}$ as in (8.3.8):

$$\mathbf{L} = \mathbf{D} + \mathbf{E} + \mathbf{F} = \mathbf{D} + \mathbf{B} \quad \text{(say)}, \tag{8.7.4}$$

it follows from Lemma 8.2 (see below) that, at least if $\mathbf{L}$ is symmetric, the complete eigensystem of $\mathbf{D}$ can be computed in $\mathcal{O}(N^2)$ operations. It further follows from Theorem 6.1 that for AD matrices of types A or B the eigenvalues of $\mathbf{D}$ may be kept arbitrarily close to those of $\mathbf{L}$ by a suitabe choice of bandwidth, and that the required bandwidth is independent of N. An identical procedure in fact applies to non-symmetric $\mathbf{L}$, save that in stage (i) of Lemma 8.2 the reduction using Given's method yields a band Hessenberg matrix.

Case 2: *The generalised eigenvalue problem*
When $\mathbf{M} \neq \mathbf{I}$ we split $\mathbf{L}$ as in (8.7.4) and introduce a similar splitting for $\mathbf{M}$:

$$\mathbf{M} = \mathbf{P} + \mathbf{Q} \tag{8.8.5}$$

where $\mathbf{P}$ is a band matrix. Then the eigensolutions of

$$[\mathbf{D} - \bar{\lambda}\mathbf{P}]\mathbf{c} = \mathbf{0} \tag{8.7.6}$$

are close to those of (8.7.1), and we seek an efficient method for solving (8.7.6). If $\mathbf{D}$ is symmetric and $\mathbf{P}$ positive definite, we may use the algorithm of Crawford (1973) to reduce (8.7.6) to the form covered by case 1:

$$[\overline{\mathbf{D}} - \bar{\lambda}\mathbf{I}]\bar{\mathbf{c}} = \mathbf{0} \tag{8.7.7}$$

where $\overline{\mathbf{D}}$ is again a band matrix. In the more difficult general case, we resort to the QZ algorithm (Moler and Stewart, 1973) for a solution of (8.7.6).

8.7.4 Calculation of the complete eigensystem

Finally, we consider the calculation of the complete eigensystem of a WAD matrix $\mathbf{L}$. For simplicity, we take $\mathbf{L}$ to be symmetric; then the standard procedure is to introduce the orthogonal $N \times N$ matrix $\mathbf{U}$ of eigenvectors of $\mathbf{L}$

$$\begin{aligned} \mathbf{L}\mathbf{U} &= \mathbf{U}\mathbf{\Lambda}, \\ \mathbf{U}^t &= \mathbf{U}^{-1}, \\ \mathbf{\Lambda} &= \mathbf{diag}(\lambda_i). \end{aligned} \tag{8.7.8}$$

Here, all of the eigenvalues λ_i of $\mathbf{L}$ are real. The numerical solution of (8.7.8) to yield the eigensystem $(\mathbf{U}, \mathbf{\Lambda})$ takes $\mathcal{O}(N^3)$ operations. We therefore seek a more efficient method for WAD matrices.

We approach the problem in two stages, and utilise the splitting (8.3.8) of $\mathbf{L}$:

$$\mathbf{L} = \mathbf{D} + \mathbf{E} + \mathbf{F} = \mathbf{D} + \mathbf{B} \quad \text{(say)} \tag{8.7.9}$$

where $\mathbf{D}$ is defined by (8.3.9) and in stage (i), we consider the calculation of the eigensystem $(\mathbf{U}_0, \mathbf{\Lambda}_0)$ of $\mathbf{D}$:

$$\begin{aligned} \mathbf{D}\mathbf{U}_0 &= \mathbf{U}_0\mathbf{\Lambda}_0, \\ \mathbf{U}_0^t &= \mathbf{U}_0^{-1}, \\ \mathbf{\Lambda}_0 &= \mathbf{diag}(\lambda_{0i}). \end{aligned} \tag{8.7.10}$$

Lemma 8.2. $(\mathbf{U}_0, \mathbf{\Lambda}_0)$ *may be computed in* $\mathcal{O}(N^2)$ *operations.*

Proof. Condition (8.3.9) displays $\mathbf{D}$ as a (symmetric) band matrix of non-constant bandwidth. For simplicity, we treat it as a constant band matrix of half width $m = \max(M, \mu)$. Then the steps required to compute $(\mathbf{U}_0, \mathbf{\Lambda}_0)$ are the following:

(1) Reduce $\mathbf{D}$ to trigdiagonal form using (for example) Given's method: $\mathcal{O}(mN^2)$ operations.

(2) Find λ_{0i}, $i = 1, \ldots, N$ usinf the QR algorithm:
$\mathcal{O}(N)$ operations per iteration; $\mathcal{O}(2)$ iterations per eigenvalue;
$\mathcal{O}(N^2)$ operations in all.

(3) Generate the columns of $\mathbf{U}_0$ (the eigenvectors) using inverse iteration. For each eigenvalue this requires:
Triangular decomposition of $\mathbf{D} - \lambda_{0i}\mathbf{I}$: $\mathcal{O}(m^2N)$ operations. Solution of triangular system for inverse iteration: $\mathcal{O}(mN)$ operations per iteration. Number of iterations per eigenvector: $\mathcal{O}(2)$.
Hence in total, this step requires $\mathcal{O}(m^2N^2)$ operations.

Stage (ii) consider the effect of the "perturbation" $\mathbf{B}$. Now for a symmetric matrix

$$\|\mathbf{\Lambda} - \mathbf{\Lambda}_0\| \leqslant \|\mathbf{B}\|. \tag{8.7.11}$$

We assume that the eigenvalues are required to an accuracy ε, and seek a value of the half bandwidth m such that

$$\|\mathbf{B}\|_\infty \leqslant \varepsilon. \tag{8.7.12}$$

We consider only the two most commonly occurring systems, those of type A and of type B.

Systems of type A(p; C)
For these systems, which arise typically from integral equations of the second kind, we assume that the matrix is approximately normalised:

$$|L_{ii}| < \delta, \qquad \forall i. \tag{8.7.13}$$

Then

$$|L_{ij}| \leqslant C\delta i^{-p}, \quad i > j. \tag{8.7.14}$$

But for all N, and for $p > 1$,

$$\begin{aligned}
\|\mathbf{B}\|_{\infty} &= \max_i \left[\sum_{j=1}^{i-m-1} |L_{ij}| + \sum_{j=i+m+1}^{\infty} |L_{ij}| \right] \\
&\leqslant C\delta \max_i \left[\sum_{j=1}^{i-m-1} i^{-p} + \sum_{j=i+m+1}^{\infty} j^{-p} \right] \\
&\leqslant C\delta \max_i \left[\max(0, i-m-1)\, i^{-p} + \frac{(i+m)^{-p+1}}{p-1} \right] \\
&\leqslant \frac{2C\delta}{p-1} m^{-p+1}.
\end{aligned} \tag{8.7.15}$$

Hence (8.7.12) is satisfied provided that

$$\frac{2C\delta}{p-1} m^{-p+1} \leqslant \varepsilon \tag{8.7.16}$$

and this defines a bandwidth m which is *independent of* N. It then follows from Lemma 8.2 that the complete eigensystem can be calculated to any required accuracy ε in $\mathcal{O}(N^2)$ operations.

Systems of type B(p, r; C)
These systems arise typically from differential equations; we make the assumption, characteristic of such systems:

$$|L_{ii}| < \delta i^{2t}, \quad t > 0. \tag{8.7.17}$$

Then for $i > j$

$$|L_{ij}| \leqslant C\delta i^{(t-p)} j^{t} (i-j)^{-r}. \tag{8.7.18}$$

We make the strong assumptions

$$2t - p + 1 < 0, r > 1. \tag{8.7.19}$$

Then

$$\|\mathbf{B}\|_\infty = \max_i \left[\sum_{j=1}^{i-m-1} + \sum_{j=i+m+1}^{\infty} \right] |L_{ij}|$$

$$\leqslant C\delta \max_i \left[\frac{(m+1)^{-r}}{t+1} i^{(2t-p+1)} + (i+m+1)^{2t-p} \frac{m^{-r+1}}{r-1} \right]$$

$$\leqslant C\delta m^{-r} \left[\frac{1}{t+1} + \frac{m^{2t-p+1}}{r-1} \right]. \tag{8.7.20}$$

This is again bounded uniformly in N; hence the bandwidth required to satisfy (8.7.12) is independent of N, and the operation count for the calculation of the complete eigensystem is $\mathcal{O}(N^2)$. Note that because of assumption (8.7.19), this result does not apply to the important practical case $t > 0$, $p = 0$.

8.8 Calculation of $e^{\mathbf{L}t}$

8.8.1 Use of the eigensystem of L

Finally, we discuss briefly numerical methods for generating the matrix exponential $e^{\mathbf{L}t}$, where $\mathbf{L}$ is an $N \times N$ WAD matrix and t is a scalar. This problem arises for example in the solution of parabolic partial differential equations, and for general $\mathbf{L}$ has been extensively discussed, with a review by Van Loan (1975). We again take $\mathbf{L}$ to be symmetric; then probably the best standard procedure is to compute the eigensystem (8.7.8) of $\mathbf{L}$, and to generate $e^{\mathbf{L}t}$ as

$$e^{\mathbf{L}t} = \mathbf{U}\, e^{\mathbf{\Lambda}t}\, \mathbf{U}', \tag{8.8.1}$$

where

$$e^{\mathbf{\Lambda}t} = \mathbf{diag}(e^{\lambda_i t}). \tag{8.8.2}$$

Under the assumptions of Section 8.7.4, the eigensystem $(\mathbf{U}, \mathbf{\Lambda})$ can be computed in $\mathcal{O}(N^2)$ operations. The matrix multiplication in (8.8.1) takes $\mathcal{O}(N^3)$ operations; but very often all that is required is the vector $e^{\mathbf{L}t}\mathbf{y}$, which can be computed in $\mathcal{O}(N^2)$ operations by the three-stage process

$$\begin{aligned} \mathbf{z}_1 &= \mathbf{U}'\mathbf{y}, \\ \mathbf{z}_2 &= e^{\mathbf{\Lambda}t}\mathbf{z}_1 \\ e^{\mathbf{L}t}\mathbf{y} &= \mathbf{U}\mathbf{z}_2 \end{aligned} \tag{8.8.3}$$

8.8.2 Iterative methods

The use of the eigensystem is not wholly satisfactory, for two reasons. The first is that the $\mathcal{O}(N^2)$ method of Section 8.7.4 fails for an important practical class of systems; and the second is the aesthetic feeling that the method, which completely ignores elements of **L** outside the band of width m, savours rather too much of brute force (and ignorance?). Other approaches have therefore been sought for the calculation of $e^{\mathbf{L}t}\mathbf{y}$ but again not wholly successfully. The simplest is based on the observation that for any $N \times N$ matrix **L**, the Taylor series

$$e^{\mathbf{L}t} = \sum_{i=0}^{\infty} \frac{\mathbf{L}^i t^i}{i!} \tag{8.8.4}$$

is convergent. If we now define the sequence

$$\mathbf{y}_0 = \mathbf{y}; \quad \mathbf{z}_0 = \mathbf{y},$$

$$\mathbf{y}_{i+1} = \frac{t\mathbf{L}}{(i+1)}\mathbf{y}_i, \quad \mathbf{z}_{i+1} = \mathbf{z}_i + \mathbf{y}_{i+1}, \tag{8.8.5}$$

then, in the absence of round-off errors, $\lim_{i\to\infty} \mathbf{z}_i = e^{\;\;t}\mathbf{y}$. Moreover,

$$e^{\mathbf{L}t}\mathbf{y} - \mathbf{z}_n = \sum_{i=n+1}^{\infty} \frac{\mathbf{L}^i t^i}{i!y} \|\mathbf{y}\|, \tag{8.8.6}$$

and hence

$$\begin{aligned} \|e^{\mathbf{L}t}\mathbf{y} - \mathbf{z}_n\| &\leqslant \sum_{i=n+1}^{\infty} \frac{\|\mathbf{L}\|^i |t|^i}{i!} \|\mathbf{y}\| \\ &= \frac{\|\mathbf{L}\|^{n+1} |t|^{n+1}}{(n+1)!} \left\{1 + \sum_{i=n+2}^{\infty} \frac{\|\mathbf{L}\|^{i-n-1} |t|^{i-n-1}}{(n+2)(n+3)\ldots i}\right\} \|\mathbf{y}\| \\ &\leqslant \frac{\|\mathbf{L}\|^{n+1} |t|^{n+1}}{(n+1)!} \sum_{i=0}^{\infty} \frac{\|\mathbf{L}\|^i |t|^i}{i!} \|\mathbf{y}\| \\ &= \frac{\|\mathbf{L}\|^{n+1} |t|^{n+1}}{(n+1)!} e^{\|\mathbf{L}\|\,|t|} \|\mathbf{y}\|. \end{aligned} \tag{8.8.7}$$

It then follows that, provided $\|\mathbf{L}\|$ is uniformly bounded in N, the number of iterations that are required to achieve a given accuracy ε in (8.8.5) is independent of N, and hence the iterative scheme has a total count of $\mathcal{O}(N^2)$.

However, the use of the Taylor series in this way has a fatal flaw: although the series (8.8.4) converges eventually very fast, it is very common in practice for the terms in the series to increase rapidly initially (a phenomenon known as "the hump"). The result is that many terms may be needed (sometimes hundreds of terms) so that the method is in fact very slow. It is also disastrously inaccurate, because of the resulting severe cancellation errors.

A more complicated iterative scheme, which attempts to avoid "the hump" by calculating $e^{\mathbf{D}t}$ directly (see (8.7.9)) and then forming $e^{(\mathbf{D}+\mathbf{B})t}$ iteratively, has been devised (Richmond, 1981). However, this also suffers in practice from "the hump", so that no generally satisfactory iterative scheme is in fact known.

Section 2

Orthogonal Expansions

9

The Convergence of Orthogonal Expansions on the Finite Interval

9.1 Introduction

So far we have covered the theory and numerical analysis of WAD systems of equations, together with the formal application of the theory of variational/Galerkin calculations, on the assumption that these lead to WAD systems. In succeeding chapters we discuss a number of application areas in more detail. We show that certain classes of expansion methods for integral and for boundary value differential equations indeed lead to WAD systems of equations; and we also discuss the ways in which this influences the structure of the algorithms, leading in many cases to large gains in efficiency.

The expansion methods under discussion can be typified loosely but descriptively as those based on orthogonal expansions in one or more variables. The demonstration that these lead to WAD systems, as well as the accompanying error analysis and error estimates, depends on an analysis of the convergence of the orthonormal expansion

$$f(\mathbf{x}) = \sum_{i=1}^{\infty} b_i h_i(\mathbf{x}) \tag{9.1.1}$$

$$(h_i, h_j) = \delta_{ij}$$

where $f(\mathbf{x})$ is a known function of one or more independent variables. In particular, the theory depends upon asymptotic estimates of the expansion coefficients b_i. We discuss in this chapter the problem of estimating b_i under various assumptions on $f(x)$, and for various choices of expansion functions $h_i(x)$ in one space dimension. Our interest lies partly in the form of the results, which determine directly the structure of the Galerkin matrix for a given application area, and partly in a comparison of the i-convergence rates of b_i achieved by alternative expansions for a given class of functions; these comparisons together with the theory developed in the preceding chapters allow an *a priori* estimation of the convergence rate achievable for a given problem with a given expansion set, and hence to an efficient choice of expansion. The connection between convergence of a variational calculation with an orthogonal basis and the (generalised) Fourier series convergence studied in this chapter is clear from (1.4.5), which expresses the total variational error as the sum of two parts S_1 and S_2. The term S_2 is

$$S_2 = \sum_{i=N+1}^{\infty} b_i^2 \equiv \|f - f_{(N)}\|_2^2 \tag{9.1.2}$$

where $f_{(N)} = \sum_{i=1}^{N} b_i h_i$ is the truncated generalised Fourier series for f: rapid convergence of the expansion (9.1.1) is therefore a necessary condition for the rapid convergence of a Galerkin calculation with exact solution f.

Asymptotic estimates of the coefficients b_n can be found by the straightforward process of repeatedly integrating by parts the defining equation for b_n. We illustrate the general process first in one dimension; the analogous process in higher dimensions is considered in Chapter 11.

Let $f(x)$ be defined on the interval (a, b). The set of expansion functions $\{h_i(x)\}$ is assumed to be continuous and orthonormal with respect to a weight function $w(x)$:

$$(h_i, h_j) = \int_a^b h_i(x) h_j(x) w(x)\, dx = \delta_{ij}.$$

The Fourier coefficient of f with respect to h_n is then

$$b_n = (f, h_n) = \int_a^b f h_n w\, dx. \tag{9.1.3}$$

We consider now the process of integration by parts.

Lemma 9.1. *Let $I = (a, b)$ be a finite, semi-infinite or infinite interval. Also let G be a real- or complex-valued function defined on I which has a derivative almost everywhere in I; let H be a real- or complex-valued function continuous almost everywhere in I, and suppose that H has a primitive $\int H\, dx$ on I such that $G \int H\, dx$ is continuous on I. Then*

$$\int_a^b G(x)H(x)\, dx = \left[G(x) \int H(t)\, dt \right]_a^b - \int_a^b G'(x) \left(\int H(t)\, dt \right) dx \tag{9.1.4}$$

whenever the limits and the integral on the right exist.

Proof. (see Flett, 1966, p. 229). The primitive $\int H(t)\, dt$ may be chosen to be

$\int_c^x H(t)\,\mathrm{d}t$ for any constant c. Equation (9.1.4) is valid for any choice of c; however, the terms on the right and left are individually different for different choices.

We may apply (9.1.4) to (9.1.3) with either of the obvious choices

$$G = f(x)w(x) \quad H = h_n(x) \tag{9.1.5a}$$

or

$$G = f(x) \qquad H = h_n(x)w(x). \tag{9.1.5b}$$

Both alternatives prove useful; if we use the non-committal notation

$$G = \bar{g}(x), \quad H = \bar{h}_n(x) \tag{9.1.5c}$$

we may integrate (9.1.3) by parts $p + 1$ times provided the above conditions are satisfied for

$$G = \bar{g}^{(i)}(x),$$

$$H = \int \dots \int \bar{h}_n(x)\,\mathrm{d}x^i \equiv \int_{c_i}^{x} \dots \int_{c_1}^{x_2} \bar{h}_n(x_1)\,\mathrm{d}x_1 \dots \mathrm{d}x_i$$

for all n and $i = 1, 2, \dots, p$. Proceeding in this way we obtain

$$b_n = \sum_{i=0}^{p} \left[(-1)^i \bar{g}^{(i)}(x) \int \dots \int \bar{h}_n(x)\,\mathrm{d}x^{i+1} \right]_a^b$$
$$+ (-1)^{p+1} \int_a^b \bar{g}^{(p+1)}(x) \left[\int \dots \int \bar{h}_n(x)\,\mathrm{d}x^{p+1} \right] \mathrm{d}x. \tag{9.1.6}$$

The choice $c_1, \dots, c_i$ is arbitrary, different choices yielding different expansions for b_n. We choose the c_j so that the sequence of indefinite multiple integrals $\int \dots \int h_n(x)\,\mathrm{d}x^i$ is of successively increasing order in n^{-1}. The required order of the coefficient b_n is then given by the first non-vanishing term of the expansion (9.1.6) together with any additional terms that result from finite discontinuities in $\bar{g}$ or its derivatives.

For given properties of f, we choose p, which governs the number of terms in (9.1.6), to be as large as possible without invalidating the expansion. This number is limited by two factors: the behaviour of f and its derivatives at the interval boundaries, and its behaviour in the interior of the interval. We characterize the interior behaviour by an integer p^*, defined as the largest integer such that

(s) $f^{(p^*-1)}$ is absolutely continuous on any closed subinterval of (a, b); and
(b) $f^{(p^*)}$ is absolutely continuous on any closed subinterval of (a, b), except possibly at a finite number of finite discontinuities.

It is clear that, in general, we must take $p \leqslant p^*$. A second bound may be derived by considering the boundary behaviour of f; however, this bound depends also on the weight function $w(x)$ being employed and we consider it separately in each case.

9.2 Classical Fourier expansions

As an example of the procedure we consider the straightforward problem of expansion of a function f using the set

$$h_n(x) = (2/\pi)^{1/2} \sin nx \quad \text{on } [0, \pi] \text{ with unit weight function.}$$

With the choice of integration constants

$$c_i = \pi/2n, \quad i \text{ odd}, \qquad c_i = 0, \quad i \text{ even},$$

we have

$$\int \cdots \int \sin nx \, \mathrm{d}x^i = \begin{cases} \dfrac{(-1)^{i/2}}{n^i} \sin nx, & i \text{ even} \\ \dfrac{(-1)^{(i+1)/2}}{n^i} \cos nx, & i \text{ odd} \end{cases}$$

and hence

$$\left| \int \cdots \int h_n(x) \, \mathrm{d}x^i \right| \leqslant (2/\pi)^{1/2} n^{-i} \tag{9.2.1}$$

and the sequence of multiple integrals converges to zero. With unit weight (9.1.5a, b) are equivalent and we substitute into (9.1.6) to obtain

$$(\pi/2)^{1/2} b_n = \begin{cases} \displaystyle\sum_{j=0}^{p/2} \left[\frac{(-1)^{j+1}}{n^{2j+1}} (f^{(2j)}(\pi)(-1)^n - f^{(2j)}(0)) \right] \\ \displaystyle + \frac{(-1)^{p/2}}{n^{p+1}} \int_a^b f^{(p+1)}(x) \cos nx \, \mathrm{d}x, \qquad \text{for } p \text{ even}, \\ \displaystyle\sum_{j=0}^{(p-1)/2} \left[\frac{(-1)^{j+1}}{n^{2j+1}} (f^{(2j)}(\pi)(-1)^n - f^{(2j)}(0)) \right] \\ \displaystyle + \frac{(-1)^{(p+1)/2}}{n^{p+1}} \int_a^b f^{(p+1)}(x) \sin nx \, \mathrm{d}x, \quad \text{for } p \text{ odd}. \end{cases} \tag{9.2.2}$$

The number of partial integrations p will be restricted by singularities of f or its derivatives. We may take p to be the largest integer such that:

(a) f has $(p-1)$ continuous derivatives on (a, b);
(b) $f^{(p)}$ is bounded on $[a, b]$;
(c) the remainder integral of (9.2.2) is finite.

The remainder integral (and any terms which result from discontinuities of $f^{(p)}$) will then be $\mathcal{O}(n^{-p-1})$. Next, we look at the "surface terms". If $f^{(2i)}$ is non-zero at either $x = 0$ or $x = \pi$, then in general $[f^{(2i)}(\pi)(-1)^n - f^{(2i)}(0)]$ will also be non-zero. We define a parameter γ to be the largest integer such that

$$f^{(2i)}(0) = f^{(2i)}(\pi) = 0 \quad \text{for } i = 0, 1, \ldots, \gamma - 1,$$

and the predominant surface term will be $\mathcal{O}(n^{-2\gamma-1})$. It follows that

$$b_n^2 = \mathcal{O}(n^{-Q})$$

where

$$Q = \min[4\gamma + 2, 2p + 2].$$

We illustrate the analysis for the function

$$f(x) = x^{7/2}(\pi - x).$$

The fourth derivative of this function is singular at the origin; hence we may perform only four partial integrations, i.e. we take $p = 3$. f itself clearly vanishes at both $x = 0$ and $x = \pi$ but its first derivative does not, so we have $\gamma = 1$ for this example. From above we predict

$$b_n^2 = \mathcal{O}(n^{-Q})$$

with

$$Q = \min[6, 8] = 6$$

and we check this by direct integration by parts.

$$\begin{aligned} b_n = &\left[(\pi x^{7/2} - x^{9/2})\left(-\frac{\cos nx}{n}\right)\right]_0^\pi - \left[\left(\frac{7}{2}\pi x^{5/2} - \frac{9}{2}x^{7/2}\right)\left(-\frac{\sin nx}{n^2}\right)\right]_0^\pi \\ &+ \left[\left(\frac{35}{4}\pi x^{3/2} - \frac{63}{4}x^{5/2}\right)\left(\frac{\cos nx}{n^3}\right)\right]_0^\pi - \left[\left(\frac{105}{8}\pi x^{1/2} - \frac{315}{8}x^{3/2}\right)\left(\frac{\sin nx}{n^4}\right)\right]_0^\pi \\ &+ \int_0^\pi \left(\frac{105}{16}\pi x^{-1/2} - \frac{945}{16}x^{1/2}\right)\left(\frac{\sin nx}{n^4}\right) dx \\ = {} & 0 - 0 + \left(-\frac{28}{4}\pi^{5/2}\right)\frac{(-1)^n}{n^3} - 0 \\ &+ \int_0^\pi \left(\frac{105}{16}\pi x^{-1/2} - \frac{945}{16}x^{1/2}\right)\left(\frac{\sin nx}{n^4}\right) dx. \end{aligned}$$

The integrand in this last integral is bounded on $[0, \pi]$, and hence the integral is at most $\mathcal{O}(n^{-4})$. The first non-vanishing term therefore characterises the n-convergence rate of b_n and, as predicted, for this problem

$$b_n^2 = \mathcal{O}(n^{-6}).$$

For the half-range sine expansion, it is apparent that to ensure rapid convergence of b_n to zero the function f must satisfy stringent conditions, namely the vanishing of a large number of low-order derivatives. The n-convergence rates of b_n for various classes of function, together with the corresponding ones arising from the similar analysis for half-range cosine and for full-range (classical) Fourier expansions are given in Table 9.1. It will be seen that the classes that lead to fast convergence are very restricted; in particular, one of the few non-trivial classes for which n-exponential (faster than any power rate) convergence of b_n may be obtained is that of analytic, completely periodic functions, when expanded in a full range Fourier series.

In subsequent paragraphs we repeat the above analysis for a number of alternative expansion sets. For brevity, only the results are given; for details the reader is referred to Mead (1971) and Bain (1974).

9.3 Sturm-Liouville eigenfunction expansion sets on a finite interval $[a,b]$

The classical Fourier series are the simplest members of the class of Sturm-Liouville eigenfunction expansions, and it might be thought that other Sturm-Liouville expansions would converge more rapidly. We consider here a restricted class of such expansions and take as our expansion set the functions $\psi_n(x)$ satisfying

$$\psi_n'' + \{\lambda_n - q(x)\}\psi_n = 0, \tag{9.3.1}$$

and logarithmic boundary conditions at a, b, of the form

$$\psi_n'(a)/\psi_n(a) = \cot\alpha \tag{9.3.1a}$$

at $x = a$, with a similar condition at $x = b$. We assume that $q(x) \in C^{(r)}[a, b]$. It is well known that these functions are orthogonal with unit weight, and we may require them to be normalised, so that

$$\int_a \psi_n\psi_m \,\mathrm{d}x = \delta_{nm}.$$

These functions yield one natural generalisation of the classical Fourier expansions. For the process described above, we need more explicit information concerning these eigenfunctions and their indefinite multiple integrals,

TABLE 9.1

Convergence rates Q of the expansion coefficients $b_n^2 = \mathcal{O}(n^{-Q})$ for a function f on a finite interval

Boundary properties of f on $[a, b]$	*Half-range Fourier*		*General Sturm-Liouville eigenfunction expansion*		*Full-range Fourier expansion*
	Sine expansion	*Cosine expansion*	$\sin \alpha = 0$ (*cf. sine set*)	$\sin \alpha \neq 0$ (*cf. cosine set*)	
Asymmetric $f(a) \neq f(b)$	2	$\underline{4}$	2	$\underline{4}$	2
$f^{(i)}(a) = f^{(i)}(b)$ for $i = 0, 1, \ldots, m-1$	2	4	2	4	$\underline{2m+2}$
$f^{(2i)}(a) = f^{(2i)}(b) = 0$ for $i = 0, 1, \ldots, m-1$	$\underline{4m+2}$	4	6	4	4
$f^{(2i+1)}(a) = f^{(2i+1)}(b) = 0$ for $i = 0, 1, \ldots, m-1$	2	$\underline{4m+4}$	2	4	2
$f^{(i)}(a) = f^{(i)}(b) = 0$ for $i = 0, 1, \ldots, m-1$; m odd	$\underline{2m+4}$	$2m+2$	$\underline{2m+4}$	$2m+2$	$2m+2$
m even	$2m+2$	$\underline{2m+4}$	$2m+2$	$\underline{2m+4}$	

The values of Q are listed in the Table under various assumptions on f. In each case it is assumed that mid-interval singularities of f do not contribute to the convergence rate; that is, that $f \in C^p [a, b]$ with $p > m$. The underlined entries are those which yield the fastest convergence rates under the given assumptions.

and to obtain this we consider the unnormalised eigenfunctions $\phi_n(x)$ satisfying boundary conditions

$$\phi_n(a) = \sin\alpha, \quad \phi_n'(a) = \cos\alpha.$$

These functions satisfy the integral equation

$$\phi_n(x) = \cos\{s_n(x-a)\}\sin\alpha - \sin\{s_n(x-a)\}\cos\alpha/s_n$$
$$+ \frac{1}{s_n}\int_a^x \sin\{s_n(x-y)\}q(y)\phi_n(y)\,dy, \tag{9.3.2}$$

where (see Titchmarsh, 1946, p. 13)

$$s_n = \lambda_n^{1/2} = \frac{n\pi}{(b-a)} + \delta_n \quad \text{and} \quad \delta_n = \mathcal{O}(n^{-1}). \tag{9.3.3}$$

Iterating (9.3.2) produces a series expansion for ϕ_n which may then be substituted in (9.1.6) to yield an estimate of b_n. We omit the rather involved details (Mead, 1971) and describe only the results obtained.

Assuming $q(x)$ to be free from singularities, we may choose p as for the case of classical Fourier expansion, and may show also that the integral remainder of (9.1.6) and any terms which result from a discontinuity of $f^{(p)}$ are of order n^{-p-1}.

Turning to the "surface terms", we again define a boundary parameter γ to be the greatest integer such that

$$f^{(i)}(a) = f^{(i)}(b) = 0 \quad \text{for } i = 1, 2, \ldots, \gamma - 1. \tag{9.3.4}$$

The term involving $\int\ldots\int\psi_n(x)\,dx^{\gamma+1}$ at the endpoints will in general be non-zero and of order $n^{-\gamma-1}$ or $n^{-\gamma-2}$, depending both on whether γ is odd or even and the choice of α.

Lastly, we consider the effect of singularities of $q(x)$. If q has only r derivatives, no multiple integrals can be shown to be of order less than $\mathcal{O}(n^{-r-1})$. This places a further bound on the n-convergence rate of b_n. Putting these factors together, we obtain

$$b_n^2 = \mathcal{O}(n^{-Q})$$

$$\begin{aligned} \text{where } Q &= \min[2\gamma+2, 2p+2, 2r+2] && \text{for } \sin\alpha \neq 0, \gamma \text{ odd} \\ &&& \text{or } \sin\alpha = 0, \gamma \text{ even} \\ &= \min[2\gamma+4, 2p+2, 2r+2] && \text{for } \sin\alpha \neq 0, \gamma \text{ even} \\ &&& \text{or } \sin\alpha = 0, \gamma \text{ odd.} \end{aligned}$$

In practice, the most useful results are those that relate to functions for which p and r are comparatively large and the boundary behaviour dominates. These results are contained in Table 9.1 alongside the corresponding ones for Fourier half range series, which are special cases of the eigenfunction expansion (in our notation, $q(x) \equiv 0$ and $\alpha = \pi/2$ or π). It is clear, in general, that they do no better than the classical expansion and that computationally they will be more difficult to use.

9.4 Expansions in Jacobi polynomials

We now turn to the case of most practical importance: that of expansion in the various classical orthogonal polynomials. Their importance stems from the results we find: the n-convergence rates of the expansion coefficients b_n can be characterised loosely by saying that they depend on the smoothness of the function $f(x)$, but not on rather irritating and restrictive details of its behaviour at the boundary, as did the results above. This makes the orthogonal polynomials particularly suitable as the basis for a Galerkin calculation; to achieve tight error estimates for such a calculation we need to know the weakest possible conditions under which convergence of a given rate is achieved, and we therefore consider these expansions with some care.

Denoting the orthogonal polynomial by $p_n(x)$, the required expansion set in its orthonormal form is

$$h_n(x) = A_n p_n(x) \tag{9.4.1}$$

where A_n is the normalising factor given by

$$\int_a^b w(x) p_n(x)\, p_m(x)\, \mathrm{d}x = A_n^{-2} \delta_{nm}. \tag{9.4.2}$$

Now all the classical orthogonal polynomials satisfy a generalised Rodrigues formula (Erdélyi *et al.*, 1953, p. 164)

$$p_n(x) = \frac{1}{\mu_n w(x)} \frac{\mathrm{d}^n}{\mathrm{d}x^n} [w(x) H^n(x)]. \tag{9.4.3}$$

Table 9.2 lists the values of A_n, μ_n and $H(x)$ for a number of polynomial bases. Equation (9.4.3) suggests the use of the split (9.1.5b) in (9.1.3); (9.1.3) then takes the very simple form

$$b_n = \frac{A_n}{\mu_n} \int_a^b f(x) \frac{\mathrm{d}^n}{\mathrm{d}x^n} [w(x) H^n(x)]\, \mathrm{d}x. \tag{9.4.4}$$

We now integrate (9.4.4) repeatedly by parts as before. Then, if the function

TABLE 9.2
The classical orthogonal polynomials

Name of polynomial	*Range* *a*	*b*	*Weight function* $w(x)$	*Normalising factor* A_n *(see (9.4.2))*	μ_n	*Rodrigue's formula (see (9.4.3))* $H(x)$
Legendre or spherical $P_n(x)$	-1	1	1	$\left(\frac{2n+1}{2}\right)^{1/2}$	$(-1)^n 2^n (n!)$	$(1-x^2)$
Chebyshev $T_n(x)$	-1	1	$(1-x^2)^{-1/2}$	$\sqrt{\frac{2}{\pi}}\left(\sqrt{\frac{1}{\pi}} \text{ if } n=0\right)$	$\frac{(-1)^n 2^n \Gamma(n+\frac{1}{2})}{\Gamma(\frac{1}{2})}$	$(1-x^2)$
Gegenbauer or ultraspherical $P_n^{(\lambda)}(x)(\lambda > -\frac{1}{2}, \neq 0)$	-1	1	$(1-x^2)^{\lambda-1/2}$	$\left\{\frac{\Gamma(\lambda)^2(n+\lambda)\Gamma(n+1)}{2^{1-2\lambda}\pi\Gamma(n+2\lambda)}\right\}^{1/2}$	$\frac{n!\,\Gamma(\lambda)\,\Gamma(2n+2\lambda)}{(-2)^n\,\Gamma(n+\lambda)\,\Gamma(n+2\lambda)}$	$(1-x^2)$
Jacobi or hypergeometric $P_n^{(\alpha,\beta)}(x)(\alpha, \beta > -1)$	-1	1	$(1-x)^\alpha(1+x)^\beta$	$\left\{\frac{(2n+\alpha+\beta+1)\Gamma(n+1)\Gamma(n+\alpha+\beta+1)}{2^{\alpha+\beta+1}\Gamma(n+\alpha+1)\Gamma(n+\beta+1)}\right\}^{1/2}$	$(-1)^n 2^n n!$	$(1-x^2)$
(Generalised) Laguerre $L_n^{(\alpha)}(x)$ $(\alpha > -1)$	0	∞	$x^\alpha e^{-x}$	$\left\{\frac{n!}{\Gamma(n+\alpha+1)}\right\}^{1/2}$	$n!$	x
Hermite $H_n(x)$	$-\infty$	∞	e^{-x^2}	$\{\pi^{1/2}2^n n!\}^{-1/2}$	$(-1)^n$	1

f is such that (9.4.4) may be integrated by parts $(t + 1)$ times, we have:

$$b_n = \frac{A_n}{\mu_n}\left\{\sum_{r=0}^{t}\left[(-1)^r f^{(r)}(x)\frac{d^{n-r-1}}{dx^{n-r-1}}[w(x)H^n(x)]\right]_a^b + (-1)^{t+1}\int_a^b f^{(t+1)}(x)\frac{d^{n-t-1}}{dx^{n-t-1}}[w(x)H^n(x)]\,dx\right\} \quad (9.4.5)$$

where we have assumed $n \geqslant t + 1$.

We restrict our attention now to the case when the expansion functions are the Jacobi polynomials.

We let $h_n(x) = A_n P_n^{(\alpha,\beta)}(x)$, $(a, b) = (-1,1)$ and the normalising factor

$$A_n = \left\{\frac{n!\Gamma(n + \alpha + \beta + 1)(2n + \alpha + \beta + 1)}{2^{\alpha+\beta+1}\Gamma(n + \alpha + 1)\Gamma(n + \beta + 1)}\right\}^{1/2}$$

Further in (9.4.3)

$$w(x) = (1 - x)^\alpha(1 + x)^\beta, \quad \alpha, \beta > -1,$$
$$\mu_n = (-2)^n n!,$$

and

$$H(x) = (1 - x^2).$$

It is clear that if $f \in C^{(p)}[-1, 1]$ then (9.4.5) will be valid with $t = p$ and using Lemma 9.2, below, the integrated terms can be shown to vanish, the remainder integral being estimated using Lemma 9.3, below. This condition is, however, stronger than necessary and we can impose weaker conditions as follows.

Suppose that $f(x)$ belongs to the class $C_{(v,\mu)}^{(p,s)}(a, b)$, that is:

(a) f has p continuous derivatives on the open interval (a, b);
(b) $f^{(p+1)}(x)$ is continuous on (a, b) except for possibly a finite number of finite discontinuities;
(c) the integer s is such that $\lim_{x\to b_-} f^{(r)}(x)$ and $\lim_{x\to a_+} f^{(r)}(x)$ exist for $r < s$;
(d) if $s > p + 1$ we define $v = \mu = 0$. Otherwise:
Let $\bar{v} = \inf\{v'\}$ and $\bar{\mu} = \inf\{\mu'\}$ such that

$$\lim_{x\to b_-}(b - x)^{v'+r}f^{(s+r)}(x) \quad\text{and}\quad \lim_{x\to a_+}(x - a)^{\mu'+r}f^{(s+r)}(x)$$

exist for $r = 0, 1, \ldots, p - s$. Then v and μ satisfy

(i) $v \geqslant \bar{v}, \mu \geqslant \bar{\mu}$,
(ii) $\lim_{x\to b_-}(b - x)^{v+p+1-s}f^{(p+1)}(x)$

and

$$\lim_{x \to a+} (x - a)^{\mu + p + 1 - s} f^{(p+1)}(x)$$

exist.

Assume also that

$$\alpha - \nu + s > -1 \quad \text{and} \quad \beta - \mu + s > -1, \tag{9.4.6}$$

which ensures finiteness of b_n; then substitution of (9.4.3) into (9.4.5) yields with $(a, b) = (-1, 1)$:

$$b_n = \frac{A_n}{\mu_n} \sum_{r=0}^{t} [(-1)^r f^{(r)}(x) \mu_{n-r-1} (1-x)^{\alpha+r+1} (1-x)^{\beta+r+1} P_{n-r-1}^{(\alpha+r+1, \beta+r+1)}(x)]_{-1}^{1}$$
$$+ (-1)^{t+1} \mu_{n-t-1} \int_{-1}^{1} f^{(t+1)}(x)(1-x)^{\alpha+t+1}(1-x)^{\beta+t+1} P_{n-t-1}^{(\alpha+t+1, \beta+t+1)}(x)\, dx. \tag{9.4.7}$$

We let $t = p$; then the integrated terms are all zero and the remainder integral exists. This is clear for $s > p + 1$, and follows from conditions (c) and (d) above for $s \leqslant p + 1$, together with (9.4.6).

Thus it only remains to bound the remainder integral of (9.4.7). Now for fixed p and $n > p$ we have,

$$|b_n| \leqslant A_n \left| \frac{\mu_{n-p-1}}{\mu_n} \right| M(p+1) \int_{-1}^{1} (1-x)^{\alpha-\nu+s}(1+x)^{\beta-\mu+s} |P_{n-p-1}^{(\alpha+p+1, \beta+p+1)}(x)|\, dx \tag{9.4.8}$$

where

$$M(p+1) = \max_{-1 \leqslant x \leqslant 1} |(1-x)^{\nu+p+1-s}(1+x)^{\mu+p+1-s} f^{(p+1)}(x)| \tag{9.4.9}$$

exists and is independent of n.

We introduce the following lemmas:

Lemma 9.2. *If $\alpha, \beta > -1$ then*

$$\max_{-1 \leqslant x \leqslant 1} |P_n^{(\alpha, \beta)}(x)| \leqslant K n^q$$

where $q = \max(\alpha, \beta, -\frac{1}{2})$ and K is a positive constant.

For the proof see Szego (1939), p. 163.

Lemma 9.3. *If α, β and σ are real numbers > -1, then as $n \to \infty$*

$$\int_0^1 (1-x)^\sigma |P_n^{(\alpha,\beta)}(x)|\, dx = \begin{cases} \mathcal{O}(n^{(\alpha-2\sigma-2)}), & \text{if } 2\sigma < \alpha - \frac{3}{2} \\ \mathcal{O}(n^{-1/2}\log_e n), & \text{if } 2\sigma = \alpha - \frac{3}{2} \\ \mathcal{O}(n^{-1/2}), & \text{if } 2\sigma > \alpha - \frac{3}{2}. \end{cases}$$

For the proof see Szego (1939), pp. 167–169.

Lemma 9.4. *For* $\alpha, \beta > -1$ there exists $K(\alpha, \beta)$ *such that for every positive integer n*

$$\frac{\Gamma(n+\alpha+\beta+1)}{\Gamma(n+\alpha+1)} < K(\alpha,\beta) n^\beta$$

For the proof see Natanson (1965), pp. 111–113.

Now

$$\begin{aligned} &\int_{-1}^1 (1-x)^{\alpha-\nu+s}(1+x)^{\beta-\mu+s}|P_{n-p-1}^{(\alpha+p+1,\beta+p+1)}(x)|\, dx \\ &= \int_0^1 (1-x)^{\alpha-\nu+s}(1+x)^{\beta-\mu+s}|P_{n-p-1}^{(\alpha+p+1,\beta+p+1)}(x)|\, dx \\ &\quad + \int_{-1}^0 (1-x)^{\alpha-\nu+s}(1+x)^{\beta-\mu+s}|P_{n-p-1}^{(\alpha+p+1,\beta+p+1)}(x)|\, dx \\ &\leqslant 2^{\beta-\mu+s}\int_0^1 (1-x)^{\alpha-\nu+s}|P_{n-p-1}^{(\alpha+p+1,\beta+p+1)}(x)|\, dx \\ &\quad + 2^{\alpha-\nu+s}\int_0^1 (1-x)^{\beta-\mu+s}|P_{n-p-1}^{(\beta+p+1,\alpha+p+1)}(x)|\, dx. \end{aligned}$$

Now the conditions of Lemma 9.3 are satisfied for both of the above integrals as

$$\alpha-\nu+s,\ \beta-\mu+s,\ \alpha+p+1,\ \beta+p+1 > -1.$$

So

$$\int_0^1 (1-x)^{\alpha-\nu+s}|P_{n-p-1}^{(\alpha+p+1,\beta+p+1)}(x)|\, dx$$

$$\sim \begin{cases} n^{p-\alpha+2\nu-2s-1}, & \text{if } \alpha-2\nu+2s < p-\frac{1}{2}, \\ n^{-1/2}\log_e n, & \text{if } \alpha-2\nu+2s = p-\frac{1}{2}, \\ n^{-1/2}, & \text{if } \alpha-2\nu+2s > p-\frac{1}{2}, \end{cases}$$

as $n \to \infty$ (p finite).

We find a similar estimate for

$$\int_0^1 (1-x)^{\beta-\mu+s} \left| P_{n-p-1}^{(\beta+p+1,\alpha+p+1)}(x) \right| dx.$$

Moreover by Lemma 9.4,

$$\left| \frac{A_n \mu_{n-p-1}}{\mu_n} \right| < \text{const. } n^{-p-1/2}.$$

Substituting these results into (9.4.8) we now find the following:

(i) Let $f(x) \in C_{(\nu,\mu)}^{(p,s)}(-1, 1)$, where p is finite. Then

$$|b_n| = \mathcal{O}(n^{-l-3/2}), \quad \text{if } q \neq p - \tfrac{1}{2}, \tag{9.4.10a}$$

where

$$l = \min(q, p - \tfrac{1}{2})$$

and

$$q = \min[(\alpha - 2\nu), (\beta - 2\mu)] + 2s;$$

and in the special case $q = p - \frac{1}{2}$,

$$|b_n| = \mathcal{O}(n^{-p-1} \log_e n). \tag{9.4.10b}$$

One immediate consequence of this result is that to achieve a Fourier convergence rate of order p, the number of derivatives of the function f that are required to be bounded at the endpoints is roughly $p/2$ for small α, β. As α and β are increased the required order is reduced.

(ii) Suppose $f(x)$ is analytic over $[-1, 1]$. Then in (9.4.7) we may set $t = p = s - 2 = n - 1$ to obtain

$$|b_n| = \frac{A_n}{n!} M(n) 2^{n+\alpha+\beta+1} \frac{\Gamma(n+\alpha+1)\Gamma(n+\beta+1)}{\Gamma(2n+\alpha+\beta+2)}.$$

This result can be further simplified if $f(x)$ is, in addition, analytic in a region that encloses the interval $[-1, 1]$, for then we can use the following lemma.

Lemma 9.5. (*Cauchy's estimate*). *Let ρ be the shortest distance between $[-1, 1]$ and the nearest singularity of $f(z)$. It follows then that for any fixed $\varepsilon > 0$, a number $\overline{M}(\varepsilon)$ exists such that*

$$\left| \frac{f^{(n)}(x)}{n!} \right| \leqslant \frac{\overline{M}(\varepsilon)}{(\rho - \varepsilon)^n}, \quad -1 \leqslant x \leqslant 1.$$

For the proof see Ahlfors (1953), p. 122.

Using this lemma we find immediately that

$$|b_n| = \mathcal{O}\left[\frac{\overline{M}(\varepsilon)}{[2(\rho - \varepsilon)]^n}\right], \tag{9.4.11}$$

Thus if $\rho > +\frac{1}{2}$ the expansion coefficients converge to zero exponentially fast.

9.4.1 Examples

We test these results by applying the bound (9.4.10) to several functions for which the exact expansions are known (Erdélyi *et al.*, 1953, pp. 212–214).

(i) $f(x) = (1 - x)^\rho$

$$= 2^\rho\Gamma(\alpha + \rho + 1)$$
$$\times \sum_{n=0}^{\infty} \frac{\Gamma(2n+\alpha+\beta+1)\Gamma(n+\alpha+\beta+1)\Gamma(n-\rho)}{\Gamma(n+\alpha+1)\Gamma(n+\alpha+\beta+\rho+2)\Gamma(-\rho)} P_n^{(\alpha,\beta)}(x),$$

valid for $\quad -\rho < \min(\alpha + 1, \alpha/2 + \frac{3}{4}), \quad -1 < x < 1.$ (9.4.12)

For ρ integer this expansion terminates; otherwise it follows from the exact expansion that the normalised expansion coefficients b_n satisfy

$$|b_n| = \mathcal{O}(n^{-\alpha-2\rho-3/2}).$$

It is easy to verify that this is exactly the estimate predicted by (9.4.10).

$$\text{(ii)}\quad f(x) = \log_e\left[1 + \left(\frac{2}{1 - x}\right)^{1/2}\right] = \sum_{n=0}^{\infty} \frac{1}{(n+1)} P_n(x). \tag{9.4.13}$$

From this the normalised expansion coefficients satisfy

$$|b_n| = \mathcal{O}(n^{-3/2}).$$

Once again $f(x)$ is infinitely differentiable on $(-1, 1)$ but is not continuous on $[-1, 1]$; we may let $p \to \infty$, $s = 0$. We satisfy condition (*d*) with $\nu = \mu = 0$, and again (9.4.10) reproduces the exact asymptotic convergence rate.

$$\text{(iii)}\quad f(x) = |x|^\rho = \sum_{m=0}^{\infty} (-1)^m \frac{(2m + \frac{1}{2})(-\rho/2)_m}{(\rho/2 + \frac{1}{2})_{m+1}} P_{2m}(x), \quad \rho > -1. \tag{9.4.14}$$

The normalised expansion coefficients can be estimated from this to be

$$|b_n| = \mathcal{O}(n^{-\rho-1}), \quad \rho \neq 0, 1, 2, 3, \ldots.$$

We consider the case $\rho > 0$ and non-integer. Let

$$\rho = \rho_0 + \delta, \quad \text{where } 0 < \delta < 1, \quad \rho_0 \text{ an integer},$$

then we may take

$$p = \rho_0, \quad s = \infty, \quad v = \mu = 0$$

and determine from (9.4.10) that

$$|b_n| = \mathcal{O}(n^{-\rho_0 - 1}).$$

Comments

Examples (i) and (ii) typify functions with endpoint singularities and show that our estimates of the Fourier convergence rates handle these singularities rather well. Example (iii) typifies functions with mid-interval singularities. For these our treatment is less refined and our estimates may be pessimistic by as much as a factor n^{-1}; however, the dependence on ρ is given correctly, if somewhat coarsely.

Example (i) also typifies functions for which bad behaviour near the boundary can be compensated for by a suitable choice of weight function. For functions such as these our estimates depend explicitly upon α and/or β, and we use these estimates in Chapter 13 to discuss the appropriate choice of weight functions for a class of Ritz-Galerkin calculations.

In conclusion we can make a number of remarks. Firstly, it is apparent that the convergence properties of each of the polynomial expansions is dominated by singularities in the function on the interval under consideration or at its boundaries. In this context we note that much of the complication in the analysis of this section could have been removed if assumptions had been made concerning the continuity of the function and its derivatives at the boundaries; however, by not doing so it has been possible to demonstrate that when it is singular behaviour at a finite boundary that characterises the Fourier convergence rate, much can be done to reduce this effect by a careful choice of the weight function (corresponding to a reasoned choice of the polynomial parameters α and/or β). For the case of expansion as a series of Jacobi polynomials this simply means increasing α and/or β as required.

Another point which emerges from the careful treatment of the boundaries is that in order to achieve a Fourier convergence rate of order p, in an expansion by Jacobi polynomials, the number of derivatives of the function that are required to be bounded as $x \to +1_-, -1_+$ is roughly $p/2$. As α and β are increased, the required order is reduced.

Finally, we remark that the results for functions analytic in a region that encloses the interval of integration indicate that convergence is exponentially fast as long as any singularities are bounded away from the interval of integration.

9.5 Convergence in the uniform norm

The foregoing results yield for the expansion sets treated, estimates of the order of convergence of the Fourier coefficients b_n. We can extend these results to estimate the pointwise convergence rate of the expansions by noting that

$$e_N(x) = |f(x) - f_{(N)}(x)| \leqslant \sum_{i=N+1}^{\infty} |b_i||h_i(x)|,$$

and inserting in this relationship the bounds obtained for the coefficients b_i, and the uniform bounds used in the text on the expansion functions $|h_i(x)|$. We should, however, note that the results obtained depend directly for their usefulness on these latter bounds, while in most cases our estimates for the Fourier coefficients depended only on obtaining quite weak bounds for the $|h_i|$.

We present the results in a uniform notation by relating them to the estimates obtained for the b_i. For given assumptions on $f(x)$, these have the form

$$b_n = \mathcal{O}(n^{-C}).$$

In most cases C is finite; those cases which lead to exponentially fast convergence in the L_2 norm lead also to exponentially rapid convergence in the L_∞ norm and, as above, our methods do not characterise the convergence more closely. For finite C we obtain under the same assumptions on $f(x)$ the following results for the error in the uniform norm.

Classical half-range or full-range Fourier expansions:

$$e_N(x) = \mathcal{O}(N^{-C+1});$$

Jacobi polynomial $P_N^{(\alpha,\beta)}(x)$:

$$e_N(x) = \mathcal{O}(N^{-C+q'+3/2}) \quad \text{where } q' = \max(\alpha, \beta, -\tfrac{1}{2}).$$

Some of these bounds can be improved by a more direct approach to the L_∞ approximation problem; see for example, Bain (1974), Chapter 3 and Bain (1978).

9.6 Transformation of variables

In many practical problems the Fourier convergence rate achieved is limited by the behaviour of $f(x)$ at the boundaries. This is so for the orthogonal polynomial expansion set if a low order derivative of $f(x)$ is singular at a boundary; it is certainly often so for the classical Fourier and Sturm-Liouville expansion sets, for which rapid convergence is obtained only under rather strict conditions on the behaviour of f at the boundaries. We notice, however, that rapid convergence is obtained when f and its first few derivatives vanish at the boundaries of the interval. A feasible method of accelerating convergence is therefore to transform the independent variable so as to ensure that in the transformed variables $f(x)$ has this boundary behaviour.

Let us take as an example the expansion of a function $f(x)$ on $[0, 1]$, subject to $f(0) = f(1) = 0$. In order to ensure that

$$f'(0) = f'(1) = 0$$

we make the transformation $T:[0, 1] \to [0, 1]$ defined by

$$y = T(x)$$

and require that

$$\begin{aligned} T(0) &= 0, \quad \left[\frac{\mathrm{d}x}{\mathrm{d}y}\right]_{y=0} = \left[\frac{\mathrm{d}x}{\mathrm{d}y}\right]_{y=1} = 0 \\ T(1) &= 1, \quad \frac{\mathrm{d}x}{\mathrm{d}y} > 0 \text{ for } y \in (0, 1). \end{aligned} \tag{9.6.1}$$

But since

$$\frac{\mathrm{d}f}{\mathrm{d}y} = \frac{\mathrm{d}f}{\mathrm{d}x} \cdot \frac{\mathrm{d}x}{\mathrm{d}y}$$

we have the first derivative of f vanishing at both ends of the interval as well as f itself. Similarly, using a higher order transformation, we may ensure that higher derivatives of f also vanish on the boundary.

One example of a simple transformation which satisfies the conditions (9.6.1) is

$$y = T(x) = \tfrac{1}{2}[\tan \pi(x - \tfrac{1}{2}) + 1],$$

for we have

$$\frac{\mathrm{d}x}{\mathrm{d}y} = \frac{2}{\pi} \cos^2 \pi(x - \tfrac{1}{2}),$$

which is clearly non-negative on $[0, 1]$ and vanishes only at the boundaries of the interval. The inverse transformation is readily available for this case but unfortunately it does not generalise easily to deal with higher order derivatives.

As a more flexible example we consider T defined implicitly by a minimum order polynomial. The derivative conditions in (9.6.1) will be met if

$$\frac{\mathrm{d}x}{\mathrm{d}y} = k_1 y(1-y), \quad k_1 \text{ constant},$$

that is, if

$$x = k_1 \frac{y^2}{2} - k_1 \frac{y^3}{3} + k_2, \quad k_2 \text{ constant}.$$

The remaining conditions in (9.6.1) give $k_1 = 6$, $k_2 = 0$. Hence the required transformation is defined implicitly by

$$x = 3y^2 - 2y^3.$$

In general, if we require the first m derivatives of $f(y)$ to vanish at the boundaries of the interval we must ensure that

$$\left[\frac{\mathrm{d}x}{\mathrm{d}y}\right]_{y=0} = \left[\frac{\mathrm{d}x}{\mathrm{d}y}\right]_{y=1} = \left[\frac{\mathrm{d}^2x}{\mathrm{d}y^2}\right]_{y=0} = \left[\frac{\mathrm{d}^2x}{\mathrm{d}y^2}\right]_{y=1} = \dots$$
$$= \left[\frac{\mathrm{d}^mx}{\mathrm{d}y^m}\right]_{y=0} = \left[\frac{\mathrm{d}^mx}{\mathrm{d}y^m}\right]_{y=1} = 0$$

and so we choose

$$\frac{\mathrm{d}x}{\mathrm{d}y} = k_1 y^m (1-y)^m.$$

On integrating and requiring that $T(0) = 0$, $T(1) = 1$ we obtain the general transformation

$$x = \left[\sum_{i=0}^{m} \binom{m}{i} \frac{(-1)^i}{(m+i+1)} y^{m+i+1}\right] \Big/ \left(\sum_{i=0}^{m} \binom{m}{i} \frac{(-1)^i}{(m+i+1)}\right].$$

These techniques are equally applicable to the Galerkin solution of an operator equation $\mathscr{L}f = g$; given a suitable transformation $T: x \to y$ we first transform the equation and then perform a variational calculation in the transformed space to obtain Fourier coefficients of f with respect to the expansion set $h_i(y)$. Required values of $f(x)$ may then be computed by inverting the above polynomial expression.

10

Orthogonal Expansions on Infinite Intervals

We now turn to the problem of expansions valid over semi-infinite and infinite intervals. We find that the analysis of the previous chapter can be extended to cover this case, but the additional complications which arise make sharp bounds more difficult to obtain.

10.1 Eigenfunction expansions on the semi-infinite interval

To compare with the results of Chapter 9 we begin by considering the analogue of the classical Fourier or Sturm-Liouville expansions on $[a,b]$. On the interval $[0, \infty)$, natural counterparts of the Sturm-Liouville eigenfunctions are the solutions u of the equation

$$u'' + [\lambda g(x) - k^2]u = 0 \quad \text{on } [0, \infty),$$

with the boundary conditions

$$u(0) = 0, \tag{10.1.1a}$$

$$u(x) \text{ finite as } x \to \infty. \tag{10.1.1b}$$

These are the set of Sturmian functions for the weight $g(x)$. The eigenfunctions are orthogonal with respect to the weight function $g(x)$ and, unlike the Sturm-Liouville equation, this equation has a discrete spectrum when $g(x)$ tends to zero with increasing x. We assume this is so here, and write

$$u(x) = e^{-kx}\, w(x),$$

$$w'' - 2kw' + \lambda g(x)w = 0. \tag{10.1.2}$$

For $g(x)$ which decay sufficiently rapidly (for example, exponentially) it is straightforward to show that $w(x) \to \text{constant}$ as $x \to \infty$.

Equation (10.1.2) proves difficult to handle in general, and we therefore restrict discussion to a particular set of such eigenfunctions which are typical in that they exhibit the above asymptotic behaviour. For the choice $g(x) = \mathrm{e}^{-4kx}$ the normalised eigenfunctions of (10.1.2), satisfying boundary conditions (10.1.1a) and (10.1.1b), may be written

$$u_n = 2k^{1/2}\left[\frac{\sin(n\pi\mathrm{e}^{-2kx})}{\mathrm{e}^{-2kx}}\right]\mathrm{e}^{-kx} \quad \text{on } [0,\infty),$$

$$\lambda_n = 4n^2k^2\pi^2.$$

The convergence properties of the Fourier expansion of a function f with respect to this set are easily determined by transformation to a finite interval, using the transformation $y = \pi\mathrm{e}^{-2kx}$ and the results for the classical half-range sine expansion. For sufficiently smooth functions f we find a convergence rate $b_n^2 = \mathcal{O}(n^{-4m+2})$ provided that

$$f^{(l)}(0) \equiv \frac{\mathrm{d}^l f}{\mathrm{d}x^l}(0) = 0, \quad \text{for } l = 0,1,\ldots,2m-2,$$

and that, for some K_2 and all $x > 0$,

$$|f| \leqslant K_2\mathrm{e}^{-\alpha x}, \quad \text{with } K_2 \text{ constant and } \alpha > 4m-5.$$

As in the case of the Sturm-Liouville eigenfunction expansion on the finite interval, the conditions under which rapid convergence occurs are extremely restrictive. More interesting, therefore, are extensions of the orthogonal polynomial bases to infinite intervals, and in Sections 10.2 and 10.3 we consider the classical orthogonal polynomials on $[0,\infty)$ and on $(-\infty,\infty)$.

10.2 Expansions on [0,∞): Laguerre polynomials

The Laguerre polynomials $L_n^{(\alpha)}(x)$ (Szego, 1939, p. 98) are orthogonal on $[0,\infty)$ with weight $x^\alpha\mathrm{e}^{-x}$; in most applications it is convenient to absorb the factor e^{-x} into the expansion set, and we discuss here the convergence properties of the set

$$h_n(x) = A_n\mathrm{e}^{-x/2}L_n^{(\alpha)}(x) \tag{10.2.1}$$

which is orthonormal with weight x^α on $[0,\infty)$:

$$\int_0^\infty x^\alpha h_n(x)\,h_m(x)\,\mathrm{d}x = \delta_{nm}. \tag{10.2.2}$$

The normalisation constant A_n is given in Chapter 9, Table 9.2.

Now using (9.4.4), b_n satisfies

$$b_n = \frac{A_n}{\mu_n} \int_0^\infty g(x) \frac{d^n}{dx^n} [x^{n+\alpha} e^{-x}]\, dx \tag{10.2.3}$$

where $g(x) = f(x)\, e^{x/2}$.

We may now proceed as before, using the following lemmas:

Lemma 10.1. *If* $\alpha \geqslant 0$ *then* $|L_n^{(\alpha)}(x)| \leqslant \binom{n+\alpha}{n} e^{x/2}$. *If* $-1 < \alpha < 0$ *then*

$$|L_n^{(\alpha)}(x)| \leqslant \left[2 - \binom{n+\alpha}{n}\right] e^{x/2}.$$

For the proof see Erdélyi *et al.*, 1953, p. 207.

Lemma 10.2. *Let* α *and* λ *be arbitrary and real,* $A > 0$. *Then for* $n \to \infty$

$$\max_{x \geqslant A} e^{-x/2} x^\lambda |L_n^{(\alpha)}(x)| = \mathcal{O}(n^Q)$$

where

$$Q = \max(\lambda - \tfrac{1}{3}, \alpha/2 - \tfrac{1}{4}).$$

For the proof see Erdélyi *et al.*, 1953, p. 207.

Definition 10.1. $f(x) \in C_{(\nu,\mu)}^{(p,s)}(0, \infty)$ *iff* f *is such that:*

(a) $f(x)$ *has* p *continuous derivatives on* $(0, \infty)$;

(b) $f^{(p+1)}(x)$ *is continuous on* $(0, \infty)$ *except for possibly at a finite number of finite discontinuities*;

(c) $\lim_{x \to 0+} f^{(r)}(x)$ *exists and is finite for* $r = 0, 1, \ldots, s-1$;

(d) *if* $s > p + 1$ *we define* $\nu = 0$ *otherwise* $\nu = \inf\{\nu'\}$ *such that* $\lim_{x+0+} x^{\nu'} f^{(s)}(x)$ *exists and is finite*;

(e) $\lim_{x \to 0+} x^{\nu+r} f^{(s+r)}(x)$ *exists and is finite for* $r = 0, 1, \ldots, p+1-s$;

(f) $\mu \leqslant \sup\{\mu'\}$ *such that* $\lim_{x \to +\infty} x^{\mu'} f(x)$ *exists and is finite*;

(g) $\lim_{x \to +\infty} x^{\mu+r} f^{(r)}(x)$ *exists and is finite for* $r = 0, 1, \ldots, p+1$.

Again, the parameter s characterises the boundedness of the derivatives at the finite limit; a similar parameter could be introduced to characterise more closely the behaviour of $f(x)$ at infinity, at the cost of complicating the analysis.

Now, integrating (10.2.3) by parts $(t+1)$ times we obtain

$$b_n = \frac{A_n}{\mu_n}\left\{\sum_{r=0}^{t}[(-1)^r g^{(r)}(x)\mu_{n-r-1}\, e^{-x}\, x^{\alpha+r+1}\, L_{n-r-1}^{(\alpha+r+1)}(x)]_0^\infty \right.$$
$$\left. +(-1)^{t+1}\,\mu_{n-t-1}\int_0^\infty g^{(t+1)}(x) e^{-x} x^{\alpha+t+1}\, L_{n-t-1}^{(\alpha+t+1)}(x)\,dx\right\}. \tag{10.2.4}$$

It is assumed, as before, that $t \leqslant p < n$. Now from the definition of $g(x)$, (10.2.3),

$$g^{(r)}(x) = \sum_{k=0}^{r} 2^{-r+k} e^{x/2} f^{(k)}(x). \tag{10.2.5}$$

Substituting (10.2.5) into (10.2.4) yields

$$b_n = \frac{A_n}{\mu_n}\left\{\sum_{r=0}^{t}\sum_{k=0}^{r}[(-1)^r 2^{-r+k} f^{(k)}(x)\mu_{n-r-1} e^{-x/2} x^{\alpha+r+1} L_{n-r-1}^{(\alpha+r+1)}(x)]_0^\infty \right.$$
$$\left. +(-1)^{t+1}\,\mu_{n-t-1}\int_0^\infty \sum_{k=0}^{t+1} 2^{-t-1+k} f^{(k)}(x) e^{-x/2} x^{\alpha+t+1}\, L_{n-t-1}^{(\alpha+t+1)}(x)dx\right\} \tag{10.2.6}$$

Now if $s \geqslant p+1$ then by Definition 10.1 and Lemma 10.1 the limits as $x \to 0_+$ are zero; also by Definition 10.1, for $0 \leqslant s < p+1$ the limits as $x \to 0_+$ again exist and are zero if

$$\alpha > \nu - 1 - s. \tag{10.2.7}$$

Because this is a necessary condition for finiteness of b_n, (10.2.7) is assumed to hold and we have therefore shown that the integrated terms as $x \to 0_+$ exist and are zero for all t. As $x \to +\infty$ it can be seen, using Lemma 10.2 (with $\lambda = r$), Definition 10.1 and (10.2.6) that if

$$-\mu + \alpha + 1 < 0 \tag{10.2.8}$$

then all integrated terms vanish at this end of the range; once again this is a necessary condition for b_n to be finite and so it is assumed to hold and we proceed.

We consider now the two cases.

Case 1: p finite

It only remains now to fix t finally and to bound the remainder integral of (10.2.6), which now reads

$$b_n = (-1)^{t+1}\frac{A_n\mu_{n-t-1}}{\mu_n}\int_0^\infty \sum_{k=0}^{t+1} 2^{-t-1+k} f^{(k)}(x) e^{-x/2} x^{\alpha+t+1} L_{n-t-1}^{(\alpha+t+1)}(x)dx. \tag{10.2.9}$$

Using Table 9.2 and Lemma 9.4, (10.2.9) can be bounded thus:

$$|b_n| < C_0 n^{-(t+1+(\alpha/2))} \sum_{k=0}^{t+1} \left\{ \int_0^A |f^{(k)}(x)|\, e^{-x/2} x^{\alpha+t+1} \, |L_{n-t-1}^{(\alpha+t+1)}(x)| \, dx \right.$$

$$\left. + \int_A^{\infty} |f^{(k)}(x)| e^{-x/2} x^{\alpha+t+1} \, |L_{n-t-1}^{(\alpha+t+1)}(x)| \, dx \right\} \tag{10.2.10}$$

where A is a fixed positive real number. Consider

$$\int_0^A |f^{(k)}(x)\,| e^{-x/2} x^{\alpha+t+1} L_{n-t-1}^{(\alpha+t+1)}(x) \, dx = \tilde{A}. \tag{10.2.11}$$

If $s > t + 1$ this is bounded by

$$\tilde{A} \leqslant M(k) \int_0^A e^{-x/2} x^{\alpha+t+1} \, |L_{n-t-1}^{(\alpha+t+1)}(x)| \, dx \tag{10.2.11a}$$

where $M(k)$ is a constant independent of n. But by the Cauchy-Schwartz inequality we have

$$\tilde{A} \leqslant M(k) \left\{ \int_0^A (e^{-x} x^{\alpha+t+1}) \, |L_{n-t-1}^{(\alpha+t+1)}(x)|^2 \, dx \right\}^{1/2} \left\{ \int_0^A x^{\alpha+t+1} \, dx \right\}^{1/2}$$

$$\leqslant M(k) A^{\alpha+t+2} \left\{ \int_0^{\infty} e^{-x} x^{\alpha+t+1} \, |L_{n-t-1}^{(\alpha+t+1)}(x)|^2 \, dx \right\}^{1/2} \quad (\text{since } \alpha + t > -2)$$

$$= M(k) A^{\alpha+t+2} \left\{ \binom{n+\alpha}{n-t-1} \right\}^{1/2}$$

i.e.

$$\tilde{A} \leqslant C_1 n^{(t+1+\alpha)/2} \quad \text{using Lemma 9.4.} \tag{10.2.12}$$

If $0 \leqslant k < s \leqslant t + 1$, (10.2.12) follows straightforwardly. Consider then $k \geqslant s$. By virtue of Definition 10.1 there exists a constant $M(k, s)$ independent of n such that

$$|x^{\nu+i} f^{(s+i)}(x)| \leqslant M(k, s) \quad \text{for } i = 0, 1, \ldots, t + 1 - s \text{ on } (0, A].$$

Thus (10.2.11) becomes

$$\tilde{A} \leqslant M(k, s) \int_0^A e^{-x/2} x^{\alpha+t+1-\nu-k+s} \, |L_{n-t-1}^{(\alpha+t+1)}(x)| \, dx$$

$$\leqslant M(k, s) \left\{ \int_0^A e^{-x} x^{\alpha+t+1} |L_{n-t-1}^{(\alpha+t+1)}(x)|^2 \, dx \right\}^{1/2} \left\{ \int_0^A x^{\alpha+t+1-2\nu-2k+2s} dx \right\}^{1/2} \tag{10.2.13}$$

by the Cauchy-Schwartz inequality. The second of these integrals is finite if and only if

$$\alpha + t + 1 - 2v - 2k + 2s > -1.$$

But $k = 0,1,\ldots,t+1$, so we must have

$$t < \alpha - 2v + 2s. \tag{10.2.14}$$

Given (10.2.14) (10.2.13) becomes

$$\begin{aligned}\tilde{A} &\leqslant M(k,s)\int_0^\infty e^{-x}x^{\alpha+t+1}\,|L_{n-t-1}^{(\alpha+t+1)}(x)|^2\,dx\cdot A^{\alpha+t+1-2v+2s-2k}\\ &\leqslant C_2 n^{(t+1+\alpha)/2}\end{aligned} \tag{10.2.15}$$

from Table 9.2 and Lemma 9.4.

The constraints upon t now are (from (10.2.4) and (10.2.14))

$$t \leqslant \min\{p, (\alpha - 2v + 2s - \delta)\} \quad \text{where } \delta > 0. \tag{10.2.16}$$

Then

$$\tilde{A} \leqslant C_3 n^{(t+1+\alpha)/2} \quad \text{from (10.2.13) and (10.2.16).} \tag{10.2.17}$$

Let

$$\tilde{B} = \int_A^\infty |f^{(k)}(x)|\,e^{-x/2}x^{\alpha+t+1}\,|L_{n-t-1}^{(\alpha+t+1)}(x)|\,dx.$$

Using Lemma 10.2 we find

$$\tilde{B} \leqslant n^{(\alpha+t+1)/2}\int_A^\infty C_4\,|f^{(k)}(x)|\,x^{[(\alpha+t)/2+1/6]}\,dx,$$

the above integral being finite if its integrand tends to zero as $x \to \infty$ faster than x^{-1}. Using Definition 10.1, this is satisfied if

$$-\mu - k + (\alpha + t)/2 < -7/6.$$

Hence

$$\tilde{B} \leqslant C_5 n^{(\alpha+t+1)/2} \tag{10.2.18}$$

provided that $t < 2\mu - \alpha - 7/3$.

Combining (10.2.17) and (10.2.18), equation (10.2.10) becomes finally

$$|b_n| \leqslant C_6 n^{-(t+1)/2}, \quad \text{as } n \to \infty, \tag{10.2.19}$$

where

$$t = \min\{p, (\alpha - 2v + 2s - \delta), 2\mu - \alpha - 7/3 - \varepsilon\}, \tag{10.2.20}$$

and where ε, δ are any positive real numbers.

Equation (10.2.19) describes the Fourier convergence rate expected with the expansion set (10.2.1), for functions in the class $C^{(p,s)}_{(v,\mu)}(0, \infty)$. Similar results for the expansion set $R_n = A_n L_n^{(\alpha)}(x)$ can be read off directly from (10.2.20) be setting $g(x) = f(x)$ in (10.2.3). We obtain again the estimate (10.2.19) where now

$$t = \min\{p, (\alpha - 2v + 2s - \delta)\} \tag{10.2.21}$$

provided that f does not increase exponentially as $x \to \infty$.

Case 2: p infinite

Here in fact we consider functions analytic in some region which encloses the interval $[0, \infty)$.

We may let $t = n - 1$ in (10.2.6). Then it is necessary only to bound the remainder integral; the result, analogous to (10.2.10), is then

$$|b_n| < C_7 \left|\frac{A_n}{\mu_n}\right| \sum_{k=0}^{n} \left\{2^{-n-k}\left[\int_0^A |f^{(k)}(x)|\, e^{-x/2} x^{n+\alpha}\, dx + \int_A^\infty |f^{(k)}(x)|\, e^{-x/2} x^{n+\alpha}\, dx\right]\right\}. \tag{10.2.22}$$

The second integral of (10.2.22) can be analysed as before; it yields a contribution to b_n of $\mathcal{O}(d^n)$ for some d, $0 < d < 1$, provided that $f(x)$ decays exponentially at the infinite boundary. The first integral cannot be satisfactorily treated as before, however, because the bound obtained depended upon a factor $M(k)$ (see (10.2.11a)), which in this case may now depend upon n also. We consequently use Cauchy's estimate as in Section 9.4.

Let ρ_1 be the shortest distance between the interval $[0, A]$ and the nearest singularity of $f(z)$; it follows that for any fixed $\varepsilon_1 > 0$ there exists a number $M_1(\varepsilon_1)$ such that

$$\left|\frac{f^{(n)}(x)}{n!}\right| < \frac{M_1(\varepsilon_1)}{(\rho_1 - \varepsilon_1)^n}. \tag{10.2.23}$$

Whence

$$\left|\frac{A_n}{\mu_n}\right| \int_0^A |f^{(n)}(x)|\, e^{-x/2} x^{n+\alpha}\, dx \leqslant C_8 \left|\frac{A_n}{\mu_n}\right| \frac{M_1(\varepsilon_1) n!}{(\rho_1 - \varepsilon_1)^n} \int_0^A e^{-x/2} x^{n+\alpha}\, dx$$
$$\leqslant C_9 n^{-\alpha/2} \left\{\frac{A}{\rho_1 - \varepsilon_1}\right\}^n. \tag{10.2.24}$$

(10.2.24) shows that Fourier convergence is at a rate faster than any fixed power rate if $f(x)$ decays exponentially as $x \to \infty$ and if

$$\rho_1 > A. \tag{10.2.25}$$

10.3 Expansions on $(-\infty, \infty)$. Hermite polynomials

On the range $(-\infty, \infty)$ the simplest orthogonal polynomials are the Hermite polynomials $H_n(x)$, which are orthogonal with weight e^{-x^2}. Possible choices of expansion set are therefore

$$h_n(x) = A_n H_n(x) \tag{10.3.1}$$

and

$$h_n(x) = A_n e^{-x^2/2} H_n(x) \tag{10.3.2}$$

where A_n is given in Table 9.2.

We consider expansions in terms of the expansion set (10.3.2); then, as before, we can simply read off the results for expansions in terms of (10.3.1). With the choice (10.3.2), (9.1.3) becomes

$$b_n = A_n \int_{-\infty}^{\infty} g(x)\, e^{-x^2} H_n(x)\, dx \tag{10.3.3}$$

where $g(x) = f(x)\, e^{x^2/2}$. It is now possible to proceed as before, using Rodrigues' formula:

$$b_n = \frac{A_n}{\mu_n} \int_{-\infty}^{\infty} g(x) \frac{d^n}{dx^n}[e^{-x^2}]\, dx. \tag{10.3.4}$$

We introduce Lemma 10.3:

Lemma 10.3 *For some K_2 independent of n and all x*

$$|H_n(x)| \leqslant K_2\, e^{x^2/2}\, 2^{n/2}(n!)^{1/2}$$

See Erdélyi *et al*, 1953, p. 208.

Lemma 10.4. *For all real λ, and for $|x| > a > 0$,*

$$\max_x e^{-x^2/2} x^{\lambda}\, |H_n(x)| = \mathcal{O}(n^s(2^n n!)^{1/2}),$$

where $s = \max(\lambda/2 - 1/12, -1/4)$.

For the proof see Szego, 1939, p. 236.

We also assume $f(x)$ belongs to the following class of functions:

Definition 10.2. $f(x) \in C^{(p,\delta)}(-\infty, \infty)$ *iff f is such that:*

(a) $f(x)$ *has p continuous derivatives on* $(-\infty, \infty)$;

(b) $f^{(p+1)}(x)$ *is continuous on* $(-\infty, \infty)$ *except for possibly at a finite number of discontinuities;*

(c) $\delta = \sup\{\delta'\}$ *such that* $\lim_{x\to-\infty} x^{\delta'} f(x)$ *and* $\lim_{x\to+\infty} x^{\delta'} f(x)$ *exist and are finite*;

(d) *for* $r = 0,1,2,\ldots,p+1$ $\lim_{x\to-\infty} x^{\delta+r} f^{(r)}(x)$ *and* $\lim_{x\to+\infty} x^{\delta+r} f^{(r)}(x)$ *exist and are finite*

Given that $f(x) \in C^{(p,\delta)}(-\infty,\infty)$ we can integrate (10.3.4) by parts $(t+1)$ times to give

$$b_n = \frac{A_n}{\mu_n}\left\{\sum_{r=0}^{t} [(-1)^r g^{(r)}(x)\mu_{n-r-1}\, e^{-x^2} H_{n-r-1}(x)]_{-\infty}^{\infty} + (-1)^{t+1}\mu_{n-t-1}\int_{-\infty}^{\infty} g^{(t+1)}(x)\, e^{-x^2} H_{n-t-1}(x)\,dx\right\} \tag{10.3.5}$$

where, as before, $t \leqslant p < n$.

Again, we must investigate the boundary behaviour of f to determine t. Now,

$$\begin{aligned} g^{(r)}(x) &= \frac{d^r}{dx^r}[f(x)e^{x^2/2}] \\ &= e^{x^2/2}\{f^{(r)}(x) + \ldots + x^r f(x)\} \end{aligned} \tag{10.3.6}$$

and, as before, we consider the conditions under which the integrated terms of (10.3.5) vanish.

Substituting (10.3.6) into (10.3.5) and using Lemma 10.4 and Definition 10.2 it is seen that the condition $\delta > 1$ ensures that all of the integrated terms of (10.3.5) are zero. This is a necessary condition for b_n to be finite; we assume it to hold and proceed to bound the remainder integral of (10.3.5). We suppose that p is finite. Now,

$$\begin{aligned} b_n &= (-1)^{p+1}\frac{A_n\mu_{n-p-1}}{\mu_n}\left\{\int_{-\infty}^{\infty} g^{(p+1)}(x)\, e^{-x^2} H_{n-p-1}(x)\,dx\right\} \\ &= (-1)^{p+1}\frac{A_n\mu_{n-p-1}}{\mu_n}\left\{\int_{A}^{\infty} I\,dx + \int_{-A}^{A} I\,dx + \int_{-\infty}^{-A} I\,dx\right\} \end{aligned} \tag{10.3.7}$$

where

$$I = g^{(p+1)}(x)\, e^{-x^2} H_{n-p-1}(x).$$

We set

$$b_n = (-1)^{p+1}\frac{A_n\mu_{n-p-1}}{\mu_n}\{I_1 + I_2 + I_3\}.$$

Now

$$I_1 \leqslant \int_A^\infty |g^{(p+1)}(x)|\, e^{-x^2}\, |H_{n-p-1}(x)|\, dx,$$

and by Lemma 10.4 we have

$$I_1 \leqslant C_{10} n^s \{2^{n-p-1}(n-p-1)!\}^{1/2} \int_A^\infty |g^{(p+1)}(x)|\, x^{-\lambda} e^{-x^2/2}\, dx,$$

where

$$s = \max\{\lambda/2 - 1/12, -1/4\}.$$

The above integral is finite if the integrand decays faster than x^{-1} as $x \to \infty$. Using (10.3.6) and Definition 10.2 this implies

$$\lambda > p + 2 - \delta.$$

Hence

$$I_1 \leqslant C_{11}\, \{2^n(n-p-1)!\}^{1/2}\, n^s, \tag{10.3.8}$$

where

$$s = \max\left\{\frac{p+(11/6)-\delta+\varepsilon_1}{2}, -\tfrac{1}{4}\right\} \quad (\varepsilon_1 > 0).$$

Similarly,

$$I_3 \leqslant C_{12}\{2^n(n-p-1)!\}^{1/2}\, n^s. \tag{10.3.9}$$

Also,

$$I_2 = \int_{-A}^{A} |g^{(p+1)}(x)|\, e^{-x^2}\, |H_{n-p-1}(x)|\, dx$$

and using Lemma 10.3 we have

$$I_2 \leqslant C_{13}\{2^n(n-p-1)!\}^{1/2}. \tag{10.3.10}$$

Moreover,

$$\left|\frac{A_n \mu_{n-p-1}}{\mu_n}\right| \leqslant C_{14}\{2^n n!\}^{-1/2} \quad \text{using Table 9.2,}$$

and hence, from this, and from (10.3.8)–(10.3.10), equation (10.3.7) yields

$$|b_n| \leqslant C_{15} n^{-(p+1)/2+s},$$

where

$$s = \max\left\{0, \frac{p+(11/6)-\delta+\varepsilon_1}{2}\right\}, \quad \varepsilon_1 > 0.$$

i.e.

$$|b_n| \leqslant C_{15} n^{-q/2}, \tag{10.3.11}$$

where

$$q = \min\{p + 1, \delta - 5/6 - \varepsilon_1\}.$$

Results for expansion set (10.3.1) can now be read off from (10.3.11); we find

$$|b_n| \leqslant C_{16} n^{-(p+1)/2}$$

as long as f does not increase at an exponential rate as $|x| \to \infty$.

10.4 Examples

10.4.1 Results for uniform convergence

As in Chapter 9, the bounds of Lemmas 10.1 and 10.3 can be combined with the bounds on the coefficients b_n derived above to yield uniform convergence rate bounds for the error $e_N(x) = f(x) - f_N(x)$. We quote the results obtained:

Let $f(x)$ be such that b_n is $\mathcal{O}(n^{-c})$; and let ε be positive. Then

(a) Laguerre: $h_n(x) = A_n P_n^{\alpha}(x)\, e^{-x/2}$:

$$|e_n(x)| \leqslant \mathcal{O}(n^{-c+\alpha/2+1}), \qquad 0 \leqslant x < \infty. \tag{10.4.1}$$

(b) Hermite: $h_n(x) = A_n H_n(x) e^{-x^2/2}$

$$|e_n(x)| \leqslant \mathcal{O}(n^{-c+5/2+\varepsilon}), \qquad -\infty < x < \infty. \tag{10.4.2}$$

10.4.2 Comparisons with exact results

As for the finite interval, the results of Sections 10.2 and 10.3 yield *bounds* on the Fourier convergence rates; the expansions may in particular cases converge faster than these bounds indicate. We test their effectiveness with two examples for which the exact expansions are known.

Laguerre

From Erdélyi *et al.* (1953), p. 214, we find that

$$\text{for } \alpha > -1 \quad \text{and} \quad -\rho < 1 + \min(\alpha, \tfrac{1}{2}\alpha - \tfrac{1}{4}): \tag{10.4.3}$$

$$x^{\rho} e^{-x/2} = \frac{\Gamma(\alpha + \rho + 1)}{\Gamma(-\rho)} \sum_{n=0}^{\infty} \frac{\Gamma(n - \rho)}{[n!\Gamma(n + \alpha + 1)]^{1/2}} h_n(x) \tag{10.4.4}$$

where $h_n(x)$ is given by (10.2.1). From this result we find the exact order relationship

$$|b_n| = \mathcal{O}(n^{-(2\rho+\alpha+2)/2}) \quad \rho \neq 0, 1, 2, \tag{10.4.5}$$

and we note that (10.4.3) implies that

$$\text{for } \alpha \geqslant -\tfrac{1}{2}, \qquad \rho > -\tfrac{1}{2}\alpha - \tfrac{3}{4},$$
$$\text{for } -1 < \alpha < -\tfrac{1}{2}, \quad \rho > -\alpha - 1.$$

Equations (10.2.19) and (10.2.20) yield the estimate

$$|b_n| = \mathcal{O}(n^{-(t+1)/2}),$$

$t = \min[p, (\alpha - 2\nu + 2s - \delta), 2\mu - \alpha - \tfrac{7}{3} - \varepsilon], \delta, \varepsilon > 0.$
For $f(x) = x^\rho \, \mathrm{e}^{-x/2}$ we may set

$$p = \mu = \infty, \quad \nu = -\rho, \quad s = 0,$$

and hence we find

$$|b_n| = \mathcal{O}(n^{-(2\rho+\alpha+1-\delta)/2}), \tag{10.4.6}$$

which is pessimistic by a factor $n^{1/2+\delta}$.

Hermite
From Erdélyi *et al.* (1953), p. 216, we find

$$\mathrm{e}^{-x^2/2}\,|x|^\rho = \frac{\Gamma(\tfrac{1}{2}+\rho/2)}{\pi^{1/4}\Gamma(-\rho/2)} \sum_{m=0}^{\infty} (-1)^m \frac{\Gamma(m-\rho/2)2^m}{[(2m)!]^{1/2}} h_{2m}(x) \tag{10.4.7}$$

where $h_m(x)$ is given by (10.3.2). From (10.4.7) we derive the exact order relationship

$$|b_{2n}| = \mathcal{O}(n^{-\rho/2-1}), \quad \rho \neq 0,1,2, \tag{10.4.8}$$

while the theory above yields

$$b_n = \mathcal{O}(n^{-q/2}), \quad q = \min(p + \tfrac{3}{2}, \delta + \varepsilon - \tfrac{5}{6}), \quad \varepsilon > 0.$$

We consider $\rho > 0$ and non-integer. Let $\rho = \rho_0 + \gamma$, ρ_0 integer. Then we may set $p = \rho_0$, $\delta = \infty$, and estimate

$$|b_{2n}| = \mathcal{O}(n^{-\rho_0/2-3/4}), \tag{10.4.9}$$

which is pessimistic by a factor $n^{(\gamma/2)+(1/4)}$.

These examples indicate that the Fourier convergence rate bounds which we achieve are fairly tight.

10.5 Alternative expansion sets

In Section 9.6 we discussed the use of non-linear mappings to generate new expansion sets from those already analysed, with possibly better convergence properties for some interesting class of functions. This technique is particularly important for the semi-infinite and infinite interval because of the variety of ways in which the function to be expanded may behave at infinity; this behaviour may or may not be well described by the exponential decay associated with both the Laguerre and the Hermite sets. In particular, we may make use of the results given in Chapter 9 for the Jacobi polynomials by mapping the interval involved in any convenient (and hopefully serviceable) way onto $[-1,1]$. We illustrate the possibilities by considering a particular mapping for the interval $[0, \infty)$. A convenient transformation is given by

$$T:[-1,1] \to [0,\infty)$$

$$x = T(y) = \ln\left(\frac{2}{y+1}\right)$$

with the inverse

$$y = (2e^{-x} - 1).$$

Let $f(x)$ be the function to be expanded on $[0, \infty)$ and define

$$g(y) = f(x)$$

where $y \in [-1, 1]$ is defined as above. The Jacobi expansion coefficients of g are given by

$$b_n = A_n \int_{-1}^{1} g(y) P_n^{(\alpha,\beta)}(y)\, w(y)\, dy,$$

where $w(y) = (1-y)^\alpha (1+y)^\beta$. Transforming to the interval $[0, \infty)$ we obtain:

$$b_n = -2^{\alpha+\beta+1} A_n \int_0^\infty f(x)\, P_n^{(\alpha,\beta)}(2e^{-x} - 1)\,(1 - e^{-x})^\alpha e^{-(\beta+1)x}\, dx.$$

The expansion set which we have defined in this way is a set of polynomials in e^{-x}, orthogonal with respect to the rather cumbersome weight function

$$w(x) = -2^{\alpha+\beta+1}(1 - e^{-x})^\alpha e^{-(\beta+1)x}.$$

In the special case of the transformed Legendre polynomials which have $\alpha = \beta = 0$, the expansion functions are

$$h_n(x) = P_n(2e^{-x} - 1), \tag{10.5.1}$$

and the weight function reduces to the simple exponential

$$w(x) = -4e^{-x}.$$

To determine the Fourier convergence rate for an expansion in this class we need only transform the relevant properties of $f(x)$ into properties of the equivalent function $g(y)$ and read the results off from those given in the preceding chapter. In particular, we note that if f is free from mid-interval singularities, then g will be also. It is less easy to generalise as to the effect of the transformation on boundary behaviour; usually, however, exponential decay of f as $x \to \infty$ will become simple power decay of g at the boundary $y = -1$, and no singularities will be introduced by the process.

We turn now to the use of Hermite expansions for functions on the semi-infinite interval. The classical Hermite functions which have already been discussed are defined on the whole real axis so we look for a related expansion set on half this range.

The first method which we discuss is exactly analogous to the use of sine and cosine functions in classical half-range Fourier expansions. As above, let $f(x)$ be the function to be expanded on the interval $[0, \infty)$. We then define the function $g(x)$ on $(-\infty, \infty)$ either by

$$g(x) \equiv \begin{cases} f(x) & \text{for } x \geqslant 0 \\ f(-x) & \text{for } x < 0, \end{cases} \tag{10.5.2}$$

or by

$$g(x) \equiv \begin{cases} f(x) & \text{for } x \geqslant 0. \\ -f(-x) & \text{for } x > 0, \end{cases} \tag{10.5.3}$$

i.e. $g(x)$ is either the even or the odd extension of $f(x)$ to the whole range. The function $g(x)$ may be now be expanded by Hermite functions on this range but, because of its symmetry, only half of these functions will be involved in the expansion. Thus with the even extension (10.5.2) we use

$$h_n(x) = A_{2n} e^{-x^2/2} H_{2n}(x),$$

and with the odd extension (10.5.3)

$$h_n(x) = A_{2n-1} e^{-x^2/2} H_{2n-1}(x),$$

both expansions having unit weight on $(-\infty, \infty)$. Since $g(x)$ is always identically equal to $f(x)$ on the half-axis, both series will provide an expansion of $f(x)$ on this interval.

Convergence rates for these series are determined in a very straightforward way from the results of Section 10.3. Firstly, the parameter which describes the behaviour of f at the infinite boundary will describe the behaviour of its extension g at both infinite boundaries. In particular, if f decays exponentially for large x, g will also do so for large positive and negative x. Secondly, if f has no mid-interval singularities, g will also be free of them except, in general, at the origin $x = 0$; this origin singularity, introduced by the reflection process, will generally dominate in determining the overall convergence properties of the expansion. Let us suppose that $f(x)$ is continuous in the neighbourhood of $x = 0$, and that $f(0)$ is finite. Then (10.5.2) will be the more useful extension if $f(0) = 0$, and (10.5.3) if $f'(0) = 0$, since in both these cases $g(x)$ and $g'(x)$ will be continuous at $x = 0$.

Although simple, the effectiveness of this type of expansion is limited by the induced "origin singularity" effect. We consider now a natural generalisation which is used considerably in practice because of its computational convenience. We refer to expansion by "harmonic oscillator" functions.

These expansion functions originally arose in physics, being eigenfunctions of the Schrödinger equation with harmonic oscillator (i.e. inverse square) potential (Morse and Feshbach, 1953, pp. 1662 ff). They are given by

$$R_n^{(l)}(r) = N_n e^{-\beta r^2/2} L_n^{(l+1/2)}(\beta r^2),$$

where β is a scale constant, and satisfy the orthogonality relationship

$$\int_0^\infty R_n^{(l)}(r) R_m^{(l)}(r)\, r^{2l+2}\, dr = \delta_{nm}$$

for all l provided

$$N_n = \left[\frac{2\beta^{(l+3/2)} n!}{\Gamma(n + l + \frac{3}{2})}\right]^{1/2}$$

For a function $f(r)$ defined on $[0, \infty)$, the n-th Fourier coefficient will be given by

$$\begin{aligned} b_n &= \int_0^\infty f(r)\, R_n^{(l)}(r)\, r^{2l+2}\, dr \\ &= N_n \int_0^\infty f(r)\, e^{-\beta r^2/2} L_n^{(l+1/2)}(\beta r^2)\, r^{2l+2}\, dr. \end{aligned}$$

The transformation $x = \beta r^2$ reduces the problem to a Laguerre expansion which we have already considered in some detail. We obtain

$$\begin{aligned} b_n &= \frac{1}{(2\beta^{\alpha+1})^{1/2}} \left[\frac{n!}{\Gamma(n + \alpha + 1)}\right] \int_0^\infty f(x) e^{-\alpha/2} L_n^{(\alpha)}(x)\, x^\alpha\, dx, \quad \alpha = l + \tfrac{1}{2} \\ &= b_n' \end{aligned}$$

where b_n' is the Fourier coefficient of $(2\beta^{\alpha+1})^{-1/2} f(x)$ with respect to the polynomial $L_n^{(\alpha)}(x)$. It then follows that, if $f(x)$ is well behaved on $[0, \infty)$, rapid Fourier convergence is obtained for integer α, that is, $\alpha + l$ half integral. However, the more widely employed expansion set is that which arises when l takes integer values, the chief advantage of this case being that it is frequently possible to calculate the relevant integrals analytically and hence to avoid numerical integration in the course of computation. We therefore look further at this case.

Firstly, we look at the choice $l = -1$, which gives rise to a unit weight function on the interval and which is the least integral value of l consistent with an integrable weight function. We have

$$R_n^{(-1)}(r) = N_n e^{-\beta r^2/2} L_n^{(-1/2)}(\beta r^2)$$
$$= \left[\frac{2\beta^{1/2} n!}{\Gamma(n+\frac{1}{2})}\right]^{1/2} e^{-x^2/2} L_n^{(-1/2)}(x^2), \quad \text{where } x^2 = \beta r^2.$$

Szego (1939) gives the relationship

$$H_{2m}(x) = (-1)^m 2^{2m} m!\, L_m^{(-1/2)}(x^2) \tag{10.5.4}$$

which we substitute above to obtain

$$R_n^{(-1)}(r) = (-1)^n 2^{1/2} \beta^{1/4} A_{2n} e^{-x^2/2} H_{2n}(x)$$

where A_n is the normalizing factor for $H_n(x)$ (Table 9.2). The Fourier coefficient of $f(r)$ will be

$$b_n = \int_0^\infty f(r) R_n^{(-1)}(r)\, dr$$
$$= (-1)^n \left(\frac{2}{\beta^{1/2}}\right)^{1/2} A_{2n} \int_0^\infty f(x) e^{-x^2/2} H_{2n}(x)\, dx$$
$$= b'_n,$$

where b'_n is the Fourier coefficient of $(2/\beta^{1/2})^{1/2} f(x)$ with respect to the polynomial $(-1)^n H_{2n}(x)$ on $[0, \infty)$. This is precisely equivalent to making an even extension of $f(x)$ to $(-\infty, \infty)$, (10.5.2), and performing an ordinary Hermite expansion, as described earlier in this section.

Similarly, for the case $l = 0$, we have the relationship

$$H_{2m+1}(x) = (-1)^m 2^{2m+1} m! x L^{(1/2)}(x^2) \tag{10.5.5}$$

and hence we find, putting $x^2 = \beta r^2$,

$$b_n = (-1)^n \left(\frac{2}{\beta^{3/2}}\right)^{1/2} A_{2n+1} \int_0^\infty [xf(x)]\, e^{-x^2/2} H_{2n+1}(x)\, dx$$
$$= b'_n,$$

where b'_n is the Fourier coefficient of $(2/\beta^{3/2})^{1/2}\, xf(x)$ with respect to the function $(-1)^n H_{2n+1}(x)$ on $[0, \infty)$. This corresponds to Hermite expansion of $[xf(x)]$ by even extension on $(-\infty, \infty)$, also described above.

For positive integral values of l we make use of the fact that

$$L_n^{(l+1/2)}(x) = (-1)^l \frac{d^l}{dx^l} L_{n+1}^{(1/2)}(x)$$

$$= (-1)^{l+1} \frac{\mathrm{d}^{l+1}}{\mathrm{d}x^{l+1}} L_{n+1}^{(-1/2)}(x)$$

together with either (10.5.4) or (10.5.5) to reduce the expansion functions to a linear combination of Hermite polynomials (in general, both odd and even); Fourier convergence rates are easily extracted from preceding results.

We conclude by comparing the Fourier convergence rates achieved in a typical case, by each of the expansion types which we have considered. We make the following assumptions:

(a) $f(x)$ analytic on $[0, \infty)$:
(b) $f(0)$ finite;
(c) $f(x) \sim \mathrm{e}^{-\gamma x}$ as $x \to \infty$ for some constant γ.

The obvious choice of expansion set, that of Laguerre functions with integer α, can easily be seen to give exponential Fourier convergence for such a function. This follows since the origin parameter $\mu = 0$ and exponential decay at infinity implies $\nu = \infty$; in addition, there are no mid-interval singularities.

Making the transformation $y = \beta x^2$ we see that $f(y)$ will remain analytic on $(0, \infty)$ and have the same boundary parameters, namely $\mu = 0$, $\nu = \infty$. Expansion by harmonic oscillator functions with half-integral values of l can therefore also be expected to lead to exponential Fourier convergence.

Next, we turn to the transformed Jacobi set, and in particular the set derived from Legendre polynomials having $\alpha = \beta = 0$. It is useful to define $F(x)$ by

$$f(x) = F(x)\,\mathrm{e}^{-\gamma x}$$

so that $F(x)$ will have properties (a) and (b) in common with $f(x)$. Then we make the transformation

$$y = 2\,\mathrm{e}^{-x} - 1$$

onto the finite interval $[-1, 1]$ giving

$$f(y) = F(y)\left[\frac{y+1}{2}\right]^{\gamma}.$$

This is the function to be expanded by Legendre polynomials, and has boundary parameters $\mu = 0$, $\nu = -\gamma$. For convergence to occur at all, it is necessary to have $\gamma > -1$, and under this condition the Fourier convergence will again be exponentially fast.

Lastly, we consider the expansion sets derived from Hermite polynomials. Since $f(0) \neq 0$, the odd extension of $f(x)$ will be discontinuous at the origin, so we look first at the expansion of the even extension (10.5.2). Only the even polynomials will be involved in this expansion and the predominant term will be that which results from the discontinuity in the first derivative of $g(x)$ at the origin. By reference to Section 10.3 we find

$$b_n \approx |f'(0)| \cdot \mathcal{O}(n^{-5/4}).$$

The convergence rate $\mathcal{O}(n^{-5/4})$ is also to be expected when using harmonic oscillator functions with unit weight, i.e. $l = -1$, since the function $(2/\beta^{1/2}) f(x)$ has the same singularity properties as $f(x)$.

For harmonic oscillator expansion with weight $x^2 (l = 0)$ we form the odd extension (10.5.3) of $[xf(x)]$. The dominant term in $|b_n|$ will then be that which results from the discontinuous second derivative of $g(x)$ at the origin, and we find from Section 10.3

$$|b_n| \approx |f'(0)| \cdot \mathcal{O}(n^{-7/4}).$$

Thus, expansions of the "half-range" Hermite type will have relatively slow power convergence as a result of the origin singularity which the process introduces.

11

The Convergence of Multi-dimensional Expansions

11.1 Introduction

We now face the realities of life: most problems of interest involve an independent variable x in an s-dimensional Euclidian space E^s, for $s > 1$.

Expansion methods are particularly suited to high-dimensional problems provided that rapid convergence can be obtained. In one dimension, an orthogonal expansion method with ten terms may yield the same accuracy as a finite difference method with 100 points; but the cost of both is negligible and the simplicity of the finite difference method then very appealing. In three dimensions, the same comparison would yield a 1000 term orthogonal expansion, and a finite difference (or finite element) mesh of 10^6 points; a ratio of this magnitude has undeniable advantages. Now, the size of the matrix equation involved is not the only factor to affect the cost of a calculation, but we treat the argument as sufficient apologia (should one be needed) for seeking an extension to s-dimensional expansions of the analysis of Chapters 9 and 10.

In s-dimensions, we assume that for a given region R in E^s, a set of expansion functions $[h_p(x)]$ defined on R is available and such that

$$\int_R h_p(x)\, h_q(x)\, w(x)\, \mathrm{d}x = \delta_{pq}. \tag{11.1.1}$$

Then (9.1.1) takes the form

$$f(x) = \sum_{p=1}^{\infty} b_p h_p(x) \tag{11.1.2}$$

$$b_p = \int_R f(x)\, h_p(x)\, w(x)\, \mathrm{d}x. \tag{11.1.3}$$

The results in Chapters 9 and 10 are obtained by repeated integration by parts of this integral form for b_p; there is no unique extension of this process to more than one dimension, and we discuss a number of extensions which yield progressively more detailed information with, regrettably, progressively less elegance.

Most of the extensions discussed assume a product expansion set; for simplicity, we restrict the discussion to two dimensions, although it is (we hope) clear how to extend the analysis to s dimensions. For a product expansion we take

$$h_p(x, y) = r_m(x)\, s_n(y), \qquad (11.1.4)$$
$$b_p = b_{m,n},$$

where $\{r_m\}$, $\{s_n\}$ are orthonormal on (a, b), (A, B) with weight functions w, W respectively and

$$R(x, y) = (a, b) \times (A, B).$$

Then (11.1.2) takes the form

$$f(x, y) = \sum_{m,n=1}^{\infty} b_{m,n} h_{m,n}(x, y), \qquad (11.1.5)$$

where now

$$b_{m,n} = \int_A^B \int_a^b w(x)\, W(y)\, f(x, y)\, r_m(x)\, s_n(y)\, \mathrm{d}x\, \mathrm{d}y. \qquad (11.1.6)$$

In practice, one has some choice in constructing the bijection $p \Leftrightarrow (m, n)$; the choices used in practice yield $m, n = \mathcal{O}(p^{1/2})$.

11.2 Eigenfunction expansions

A common, if usually misguided, approach in the context of multi-dimensional expansion methods is to choose the functions $h_p(x)$ as eigenfunctions of an Hermitian operator appearing naturally in the problem at hand. We analyse one class of such eigenfunction expansions, which yield a natural generalisation of the Sturm-Liouville expansions considered in Section 9.3. Let R be finite with smooth boundary C, and let $h_p(x)$ be the p-th ordered eigenfunction of the equation

$$\nabla^2 h_p(x) + (\lambda_p - V(x))h_p(x) = 0, \quad x \in R, \qquad (11.2.1)$$
$$h_p(x) = 0, \quad x \in C,$$

where $V(x)$ is continuous in R. The eigenfunctions h_p are orthogonal on R, and may be assumed normalised. Then

$$b_p = \int_R f h_p \, dR$$

$$= \lambda_p^{-1} \int_R f(\nabla^2 - V) h_p \, dR.$$

Now, if f satisfies the well-known conditions of Green's theorem we have

$$\int_R (f \nabla^2 h_p - h_p \nabla^2 f) \, dR = \int_C (f \nabla h_p - h_p \nabla f) \, dC$$

and so

$$b_p = (\lambda_p)^{-1} \left\{ \int_C (f \nabla h_p - h_p \nabla f) \, dC + \int_R ((\nabla^2 - V) f) h_p \, dR \right\}.$$

This result corresponds to one stage of integration by parts.

If we can perform this operation a total of s times, we find

$$b_p = \sum_{r=0}^{s-1} (\lambda_p)^{-(r+1)} \int_C (F_r \nabla h_p - h_p \nabla F_r) \, dC + \lambda_p^{-s} \int_R F_s h_p \, dR, \qquad (11.2.2)$$

where $F_r = (\nabla^2 - V)^r f$.

We now continue by attempting to treat the surface integrals in a similar manner and bounding the individual terms when this process terminates. The dominant term then characterises the convergence rate. We illustrate this process in a little more detail by considering the case when R is the square of side π and the expansion set is chosen to be the product set

$$h_p(x) = h_{m,n}(x, y) = (2/\pi) \sin mx \, . \sin ny, \quad m, n = 1, 2, \ldots,$$

where there is defined a bijection of pairs of integers (m, n) onto the integers p. Clearly, this set is orthonormal over R.

Equation (11.2.1) takes the form

$$\nabla^2 h_{m,n} + (m^2 + n^2) h_{m,n} = 0, \quad x \in R,$$

$$h_{m,n} = 0, \quad x \in C,$$

and similarly (11.2.2) yields

$$b_{m,n} = \frac{2}{\pi} \sum_{r=0}^{s-1} (m^2 + n^2)^{-(r+1)} \int_C (\nabla^{2r} f \nabla h_{m,n} - h_{m,n} \nabla(\nabla^{2r} f)) \, dC$$

$$+ (m^2 + n^2)^{-s} \int_R (\nabla^{2s} f) h_{n,m} \, dR. \qquad (11.2.3)$$

Now the line integrals around C consist of four parts, each on the interval $[0, \pi]$; it is therefore straightforward, although messy, to treat these using the results of Chapter 9. The remainder term (the last term in (11.2.3)) can be bounded directly given suitable assumptions on f. We omit details and pass to the final result, which is embodied in the following theorem.

Theorem 11.1. *Let* $f \in D^{q_1, q_2}_{t_1, t_2}[R]$; *that is:*

(i) *the r-th partial derivative of f with respect to* $\begin{Bmatrix} x \\ y \end{Bmatrix}$ *is continuous in R and on its boundary for* $r = 0, 1, \ldots, \begin{Bmatrix} q_1 \\ q_2 \end{Bmatrix}$ *and vanishes at each corner of R for* $r = 0, 1, \ldots, \begin{Bmatrix} t_1 \\ t_2 \end{Bmatrix}$.

(ii) *the* $\left(\begin{Bmatrix} q_1 \\ q_2 \end{Bmatrix} + 1\right)$*-th derivatives in* $\begin{Bmatrix} x \\ y \end{Bmatrix}$ *are bounded in R and on its boundary and are continuous in R except for possibly at a finite set of points.*

Then (11.2.3) *is valid with* $s = \min\left\{\left[\frac{q_1 + 1}{2}\right], \left[\frac{q_2 + 1}{2}\right]\right\}$ *and for some* C_1 *and all m, n*

$$|b_{m,n}| \leqslant C_1(m^2 + n^2)^{-\delta/2}, \tag{11.2.4}$$

where

$$\delta = \min\{(q_1 + 1), (q_2 + 1), (s_1 + 1), (s_2 + 1)\},$$

and where

$$s_i = \begin{cases} t_i, & t_i \text{ even} \\ t_i + 1, & t_i \text{ odd} \end{cases} \qquad i = 1, 2.$$

We see that in order to achieve rapid convergence of the Fourier coefficients to zero it is necessary for a large number of low-order derivatives of f to vanish on the vertices of the square.

As it stands this method is rather limited in its application, although clearly it will yield results for other product trigonometric series such as $\{\cos mx \cos ny\}$. In the next section we introduce a rather more sophisticated scheme which can be derived from a generalisation of Green's theorem and is applicable whenever the expansion set is composed of products of trigonometrical or the classical orthogonal polynomials.

11.3 Generalised Green's theorem

The central theorem in this section is Green's theorem written in the trivial form:

Theorem 11.2. *Let $P(x, y)$ and $Q(x, y)$ be such that $P(x, y)$, $\partial P(x, y)/\partial y$, $Q(x, y)$ and $\partial Q(x, y)/\partial x$ are continuous and bounded inside a simply connected region R with smooth boundary C. Then*

$$\int_R \left\{\frac{\partial P}{\partial y} - \frac{\partial Q}{\partial x}\right\} \mathrm{d}x\,\mathrm{d}y = -\int_C \{P\,\mathrm{d}x + Q\,\mathrm{d}y\},$$

where the integrals are Cauchy-Riemann integrals.

We consider now product expansions of the form (11.1.4). Although the analysis does not require it, for simplicity of presentation we take

$$r_n(x) = s_n(x) = Q_{n,0}(x),$$

where the reason for the apparently redundant suffix in $Q_{n,0}$ will appear below.

Now the classical orthogonal functions on both finite and infinite intervals all satisfy relationships of the form

$$\frac{\mathrm{d}}{\mathrm{d}x}\left\{\rho_{i+1}(x)\frac{\mathrm{d}}{\mathrm{d}x}\right\} Q_{n,1}(x) = \lambda_{n,i}\rho_i(x)\,Q_{n,i}(x), \quad i = 0, 1, \ldots, n-1,$$

and

$$\gamma_{n,i}\frac{\mathrm{d}}{\mathrm{d}x}\{\rho_{i+1}(x)\,Q_{n,i+1}(x)\} = \lambda_{n,i}\rho_i(x)\,Q_{n,i}(x), \tag{11.3.1}$$

where $Q_{n,i}$ and the coefficients $\gamma_{n,i}$, ρ_i and $\lambda_{n,i}$ are given in Table 11.1. We note that the Legendre and Chebyshev polynomials are particular cases of the Jacobi polynomial ($\alpha = \beta = 0$ and $\alpha = \beta = -\frac{1}{2}$, respectively). Equation (11.1.6) can be written in the form

$$b_{m,n} = A_m A_n \int_R f(x, y)\,\rho_0(x)\,\rho_0(y)\,Q_{m,0}(x)\,Q_{n,0}(y)\,\mathrm{d}x\,\mathrm{d}y \tag{11.3.2}$$

where A_n is the usual normalisation constant.

Substitution of (11.3.1) into (11.3.2) yields

$$b_{m,n} = \frac{A_m A_n \gamma_{m,0}\gamma_{n,0}}{\lambda_{m,0}\lambda_{n,0}} \int_R f(x, y)\frac{\partial^2}{\partial x\,\partial y}\{\rho_1(x)\,\rho_1(y)\,Q_{m,1}(x)\,Q_{n,1}(y)\}\,\mathrm{d}x\,\mathrm{d}y. \tag{11.3.3}$$

TABLE 11.1

The coefficients γ, ρ, λ, Q of equations (11.3.1) and (11.3.2) for the classical orthogonal polynomials

Polynomial \ Factor	$Q_{n,i}(x)$	$\gamma_{n,i}$	$\rho_i(x)$	$\lambda_{n,i}$	A_n
Jacobi	$P_{n-i}^{(\alpha+i,\beta+i)}(x)$	$\frac{1}{2}(n+\alpha+\beta+i+1)$	$(1-x)^{\alpha+i}(1+x)^{\beta+i}$	$-(n-i)(n+\alpha+\beta+i+1)$	$\left\{\frac{(2n+\alpha+\beta+1)\,\Gamma(n+1)\,\Gamma(n+\alpha+\beta+1)}{2^{\alpha+\beta+1}\,\Gamma(n+\alpha+1)\,\Gamma(n+\beta+1)}\right\}^{1/2}$
Laguerre	$L_{n-i}^{(\alpha+i)}(x)$	-1	$e^{-x}x^{\alpha+i}$	$-n+i$	$\left\{\frac{n!}{\Gamma(n+\alpha+1)}\right\}^{1/2}$
Hermite	$H_{n-i}(x)$	$2(n-i)$	e^{-x^2}	$-2(n-i)$	$\{\pi^{1/2}2^n n!\}^{-1/2}$

We now choose

$$P = \gamma_{m,0}\gamma_{n,0} f \frac{\partial}{\partial x}\{\rho_1(x)\,\rho_1(y)\,Q_{m,1}(x)\,Q_{n,1}(y)\}$$

and

$$Q = \gamma_{m,0}\gamma_{n,0}\frac{\partial f}{\partial y}\rho_1(x)\,\rho_1(y)\,Q_{m,1}(x)\,Q_{n,1}(y),$$

and, supposing that P and Q satisfy the conditions of Theorem 11.2, we find

$$\begin{aligned} b_{m,n} = \frac{\gamma_{m,0}\gamma_{n,0}A_m A_n}{\lambda_{m,0}\lambda_{n,0}}\Bigg\{ & -\int_C f\frac{\partial}{\partial x}\{\rho_1(x)\,\rho_1(y)\,Q_{m,1}(x)\,Q_{n,1}(y)\}\,\mathrm{d}x \\ & -\int_C \frac{\partial f}{\partial y}\rho_1(x)\,\rho_1(y)\,Q_{m,1}(x)\,Q_{n,1}(y)\,\mathrm{d}y \\ & +\int_R \frac{\partial^2 f}{\partial x\partial y}\rho_1(x)\,\rho_1(y)\,Q_{m,1}(x)\,Q_{n,1}(y)\,\mathrm{d}x\,\mathrm{d}y\Bigg\} \end{aligned} \tag{11.3.4}$$

completing one step of integration by parts. The remainder term $\int_R$ in (11.3.4) has the same form as the right hand side in (11.3.2); under suitable conditions on f, we can therefore iterate this process. If we suppose, as before, that the operation may be performed a total of s times, we have the identity

$$\begin{aligned} b_{m,n} = A_m A_n\Bigg\{ & -\sum_{r=0}^{s-1}\left(\prod_{j=0}^{r}\frac{\gamma_{m,j}\gamma_{n,j}}{\lambda_{m,j}\lambda_{n,j}}\right) \\ & \times\left(\int_C \frac{\partial^{2r} f}{\partial x^r\partial y^r}\frac{\partial}{\partial x}\{\rho_{r+1}(x)\,\rho_{r+1}(y)\,Q_{m,r+1}(x)\,Q_{n,r+1}(y)\}\,\mathrm{d}x\right. \\ & \left.+\int_C \frac{\partial^{2r+1} f}{\partial x^r\partial y^{r+1}}\rho_{r+1}(x)\,\rho_{r+1}(y)\,Q_{m,r+1}(x)\,Q_{n,r+1}(y)\,\mathrm{d}y\right) \\ & +\prod_{j=0}^{s-1}\frac{\gamma_{m,j}\gamma_{n,j}}{\lambda_{m,j}\lambda_{n,j}}\int_R \frac{\partial^{2s} f}{\partial x^s\partial y^s}\rho_s(x)\,\rho_s(y)\,Q_{m,s}(x)Q_{n,s}(y)\,\mathrm{d}x\,\mathrm{d}y\Bigg\}. \end{aligned} \tag{11.3.5}$$

This equation is the counterpart of equation (11.2.2) of the previous section. For any given choice of the orthogonal function $Q_{n,0}$ we choose R to be the rectangle (possibly infinite) of orthogonality; then the contour C is piecewise linear and, as in Section 11.2, the results of Chapters 9 and 10 can be used to estimate the line integrals, the remainder term being bounded directly. We again omit details, which may be found in Bain (1974), but quote the results obtained for the normalised Jacobi polynomials:

$$Q_{n,0}(x) = A_n P_n^{(\alpha,\beta)}(x).$$

Following the definition of Section 9.4 we say that $f(x, y)$ belongs to the class $Cx^{(p_1, s_1)}_{(v_1, \mu_1)}(-1, 1)$ if for every fixed $y \in (-1, 1)$, $f(x, y) \in C^{(p_1, s_1)}_{(v_1, \mu_1)}(-1, 1)$. We then find the following:

Theorem 11.3. *Let*

$$f(x, y) \in Cx^{(p_1, s_1)}_{(v_1, \mu_1)}(-1, 1) \times Cy^{(p_2, s_2)}_{(v_2, \mu_2)}(-1, 1)$$

and set

$$p_0 = \min(p_1, p_2),$$

$$q_0 = \min[(\alpha - 2v_1) + 2s_1, (\beta - 2\mu_1) + 2s_1, (\alpha - 2v_2) + 2s_2, (\beta - 2\mu_2) + 2s_2].$$

Then for some K independent of m, n

$$|b_{m,n}| \leqslant K(mn)^{-l-3/2}, \quad \text{if } q_0 \neq p_0 - \tfrac{1}{2}, \tag{11.3.6}$$

where $l = \min(q_0, p_0 - \frac{1}{2})$.

In the special case $q_0 = p_0 - \frac{1}{2}$ a factor of $\log_e(m + n)$ appears in the right-hand side of the inequality.

A number of comments may be made about this result. Firstly, the analysis carries through even if the product expansion set were composed of products of different polynomials; for example, a half-range sine series in x and Legendre polynomials in y, or Jacobi polynomials with differing α, β. Secondly, we note that whereas the bound in (11.2.4) depends on products of $(m^2 + n^2)$, the bound in (11.3.6) depends on the product mn. This result does not reflect a difference between Fourier and polynomial expansions but is due entirely to the choice of second-order differential equation used in the analysis. A result of the general form (11.3.6) is obtained for a product sine expansion using the methods of this section if we identify

$$\begin{aligned} Q_{n,i} &= \sin nx, \quad i \text{ even}, \\ & \cos nx, \quad i \text{ odd}, \\ \rho_i(x) &= 1, \ \lambda_{n,i} = -n^2, \ \gamma_{n,i} = (-1)^i n, \end{aligned} \tag{11.3.7}$$

or, alternatively, using the technique of sect^con 11.2 and noting that the orthonormal functions

$$h_{m,n} = (2/\pi) \sin mx \sin ny$$

satisfy the differential equation

$$\frac{\partial^4}{\partial x^2 \partial y^2} h_{m,n} - m^2 n^2 h_{m,n} = 0.$$

Finally, we note that it should be possible to do better than (11.3.6), since intuitively it might be thought that a result of the sort $|b_{m,n}| \leqslant K m^{-l_1} n^{-l_2}$ would be obtainable, reflecting a lack of symmetry of $f(x, y)$ with respect to interchanging x, y. Equation (11.3.6) would then represent a weaker result with $l = \min(l_1, l_2)$. The final section describes a less elegant but more powerful method of analysis which yields characterisations of this type for product expansions.

11.4 A brute force method

For a product expansion, equation (11.1.6) can be separated in the form

$$b_{m,n} = \int_a^b w(x)\, r_m(x)\, F_n(x)\, dx \tag{11.4.1a}$$

where

$$F_n(x) = \int_A^B W(y)\, s_n(y)\, f(x, y)\, dy. \tag{11.4.1b}$$

Equations (11.4.1a, b) are formally one-dimensional so that the methods of Chapters 9 and 10 can be used to bound $F_n(x)$ for fixed x and then to bound $b_{m,n}$.

To compare the results which this method yields with those based on Green's theorem, we again consider a region R which is a square of side π and an expansion set $h_{m,n} = (2/\pi) \sin mx \sin ny$. Then

$$b_{m,n} = \frac{2}{\pi} \int_0^\pi \int_0^\pi f(x, y) \sin mx \sin ny\, dx\, dy,$$

and if $f \in D^{q_1, q_2}_{t_1, t_2}[R]$ (see Section 11.2) we may integrate

$$b_{m,n} = \int_0^\pi \sin mx g(x)\, dx,$$

where

$$g(x) = \frac{2}{\pi} \int_0^\pi \sin ny\, f(x, y)\, dy,$$

$(q_1 + 1)$ times to obtain

$$b_{m,n} = \frac{2}{\pi}\sum_{r=0}^{[q_1/2]} m^{-2r-1}\left\{\int_0^\pi \sin ny\left((-1)^{m+1}\frac{\partial^{2r} f}{\partial x^{2r}}\bigg|_{x=\pi} - \frac{\partial^{2r} f}{\partial x^{2r}}\bigg|_{x=0}\right)\mathrm{d}y\right\} + D,$$

where

$$D = \begin{cases} 2.\pi^{-1} m^{-q_1-1}(-1)^{q_1/2}\displaystyle\int_0^\pi\int_0^\pi \cos mx \sin ny \frac{\partial^{q_1+1} f}{\partial x^{q_1+1}}\,\mathrm{d}x\,\mathrm{d}y, & q_1 \text{ even},\\ 2.\pi^{-1} m^{-q_1-1}(-1)^{(q_1+1)/2}\displaystyle\int_0^\pi\int_0^\pi \sin mx \sin ny \frac{\partial^{q_1+1} f}{\partial x^{q_1+1}}\,\mathrm{d}x\,\mathrm{d}y, & q_1 \text{ odd}.\end{cases}$$

The order of integration in the remainder double integral is now reversed and it is integrated $q_2 + 1$ times with respect to y. For the case q_1 odd this yields:

$$\begin{aligned} b_{m,n} = \frac{2}{\pi}\Bigg\{&\sum_{r=0}^{(q_1/2]} m^{-2r-1}\left\{\int_0^\pi \sin ny\left((-1)^{m+1}\frac{\partial^{2r} f}{\partial x^{2r}}\bigg|_{x=\pi} - \frac{\partial^{2r} f}{\partial x^{2r}}\bigg|_{x=0}\right)\mathrm{d}y\right\}\\ &+ m^{-q_1-1}(-1)^{(q_1+1)/2}\sum_{s=0}^{[q_2/2]} n^{-2s-1}\\ &\times\left\{\int_0^\pi \sin mx\left((-1)^{n+1}\frac{\partial^{2s+q_1+1} f}{\partial x^{q_1+1}\partial y^{2s}}\bigg|_{y=\pi} - \frac{\partial^{2s+q_1+1} f}{\partial x^{q_1+1}\partial y^{2s}}\bigg|_{y=0}\right)\mathrm{d}x\right\}\\ &+ m^{-q_1-1} n^{-q_2-1}\int_0^\pi\int_0^\pi \sin mx\, E(y)\frac{\partial^{q_1+q_2+2}}{\partial x^{q_1+1}\partial y^{q_2+1}}\,\mathrm{d}x\,\mathrm{d}y\Bigg\},\end{aligned}$$

where

$$E(y) = \begin{cases}(-1)^{(q_1+q_2+1)/2} & \cos ny, \quad \text{if } q_2 \text{ even},\\ (-1)^{(q_1+q_2)/2+1} & \sin ny, \quad \text{if } q_2 \text{ odd}.\end{cases}$$

A similar result can be derived for q_1 even.

The remainder double integral can now be bounded directly and the single integrals bounded using the results of Chapter 9. This yields:

Theorem 11.4. *Let* $f \in D_{t_1,t_2}^{q_1,q_2}([0,\pi] \times [0,\pi])$*; then*

$$|b_{m,n}| \leqslant K(q_1, q_2)\, m^{-\delta_1} n^{-\delta_2}, \tag{11.4.2}$$

where $K(q_1, q_2)$ *is independent of* m, n *and*

$$\delta_1 = \min(q_1 + 1, s_1 + 1), \quad \delta_2 = \min(q_2 + 1, s_2 + 1),$$

where

$$s_i = \begin{cases} t_i, & t_i \text{ even},\\ t_i + 1, & t_i \text{ odd},\end{cases} \qquad i = 1, 2.$$

In terms of the subscript p (see (11.1.3)) this result implies

$$b_p = \mathcal{O}(p^{-\delta/2)}, \quad \text{where } \delta = \min(\delta_1, \delta_2),$$

in agreement with the results of Theorem 11.1.

As a second example, we apply this analysis to a product Jacobi expansion and obtain in a straightforward manner:

Theorem 11.5. *Under the conditions of Theorem* 11.3, *the bound* (11.3.6) *can be replaced by*

$$|b_{m,n}| \leqslant K_1 m^{-l_1-3/2} n^{-l_2-3/2}, \tag{11.4.3}$$

where $l_i = \min(q_i, p_i - \frac{1}{2})$, K_1 *is some constant, independent of m and n, and*

$$q_i = \min[(\alpha - 2v_i), (\beta - 2\mu_i)] + 2s_i, \quad i = 1, 2.$$

Clearly, this analysis will provide the results we require for product expansion sets whenever the one-dimensional results exist; moreover, the method generalises in a straightforward manner to higher dimensions.

The results we have listed are by no means as complete as those of Chapters 9 and 10 for one-dimensional expansions. However, in the chapters which follow, where use is made of the results, the *form* of the results as exemplified by (11.4.2) is more important than the value predicted for the exponents δ_1, δ_2. While it is mathematically pleasing, and even sometimes useful, to be able to characterise closely the conditions under which a given convergence rate is achieved, in practice users of expansion methods are unlikely to check beforehand on the conditions satisfied by their problems; and the programme will in any case *estimate* the parameters δ_1, δ_2 *a posteriori*. We do, however, make direct use of the functional form of (11.4.2), in analysing the matrix structure that results from multi-dimensional problems.

Section 3
Applications

12

The Galerkin Solution of Integral Equations

12.1 Introduction

In Chapter 4 we considered a particular integral equation, and showed explicitly that it led to a WAD system of equations. We now return to such problems, and show that the Galerkin solution of linear, second kind Fredholm or Volterra equations, with orthogonal expansion sets, leads quite generally to WAD systems.

We consider then the solution of the linear integral equation

$$f(x) = g(x) + \lambda \int_a^b K(x, y)\, f(y)\, \mathrm{d}y \tag{12.1.1}$$

which we write in the form

$$[I - \lambda \mathscr{K}]\, f = g. \tag{12.1.2}$$

A Ritz-Galerkin procedure for (12.1.2) chooses an expansion set $\{h_i\}$ and an inner product (,), and approximates f by f_N:

$$f = \sum_{i=1}^{\infty} b_i h_i \tag{12.1.3a}$$

$$f_N = \sum_{i=1}^{N} a_i^{(N)} h_i \tag{12.1.3b}$$

with the expansion coefficients $a_i^{(N)}$ determined from the algebraic equations

$$[\mathbf{M} - \lambda \mathbf{K}]\, \mathbf{a}^{(N)} = \mathbf{g} \tag{12.1.4}$$

where $\mathbf{M}$ is the Gram matrix and $\mathbf{K}$ the matrix of $\mathscr{K}$ in the representation

$\{h_i\}$:

$$M_{ij} = (h_i, h_j),\ K_{ij} = (h_i, \mathscr{K} h_j), \quad i, j = 1, \ldots, N, \tag{12.1.4a}$$

and

$$g_i = (h_i, g), \quad i = 1, \ldots, N. \tag{12.1.4b}$$

For simplicity, we assume that the interval $[a, b]$ is finite and that $\mathbf{g}$, $\mathbf{K}$ are real; and we consider (12.1.4) when the inner product is chosen to be

$$(f, g) = \int_a^b w(x)\, f(x)\, g(x)\, dx \tag{12.1.5}$$

where $w(x)$ is a positive weight function with respect to which the functions h_i are orthonormal:

$$\int_a^b h_i(x)\, h_j(x)\, w(x)\, dx = \delta_{ij}. \tag{12.1.5a}$$

Then the matrix $\mathbf{M}$ reduces to the unit matrix, and (12.1.4) takes the form

$$[\mathbf{I} - \lambda \mathbf{K}]\, \mathbf{a} = \mathbf{g} \tag{12.1.6}$$

$$\begin{aligned} g_i &= \int_a^b g(x)\, w(x)\, h_i(x)\, dx \\ K_{ij} &= \int_a^b w(x)\, h_i(x) \int_a^b h_j(y)\, K(x, y)\, dy\, dx. \end{aligned} \tag{12.1.7}$$

Under these conditions the matrix $\mathbf{I} - \lambda \mathbf{K}$ can be expected to be AD of type A, provided that the kernel $K(x, y)$ satisfies quite modest conditions. For we may rewrite (12.1.7) as

$$K_{ij} = \int_a^b \int_a^b w(x)\, w(y)\, h_i(x)\, h_j(y)\, \bar{K}(x, y)\, dx\, dy \tag{12.1.8}$$

where $\bar{K}(x, y) = K(x, y)/w(y)$, which identifies K_{ij} as the (i, j) expansion coefficient in the orthonormal expansion

$$\bar{K}(x, y) = \sum_{i, j = 1}^{\infty} K_{ij} h_i(x)\, h_j(y). \tag{12.1.9}$$

The discussion of Chapter 11 then indicates that provided the integral in (12.1.8) exists, K_{ij} will satisfy a bound of the form

$$|K_{ij}| \leqslant C i^{-p} j^{-q}. \tag{12.1.10}$$

Given (12.1.10), $\mathbf{L} = \mathbf{I} - \lambda \mathbf{K}$ is upper AD of type $A(q; C')$, and lower AD of type A(p; C'):

$$\frac{|L_{ij}|}{|L_{ii}|^{1/2}|L_{jj}|^{1/2}} \leqslant C' i^{-p}, \quad i > j, \qquad (12.1.11)$$

$$\leqslant C' j^{-q}, \quad j > i,$$

with

$$C' = \lambda C / \min_i |1 - \lambda K_{ii}|.$$

In the following sections we consider the conditions under which (12.1.10) holds for the two most important practical cases.

12.2 Expansions in Legendre polynomials

We set

$$w(x) = 1$$

$$h_i(x) = \left(\frac{2i-1}{b-a}\right)^{1/2} P_{i-1}(z) \qquad (12.2.1)$$

$$z = (2x - a - b)/(b - a).$$

Then $\bar{K}(x, y) = K(x, y)$ in (12.1.8). Given (12.1.7), we investigate the convergence and stability properties of the equations (12.1.6), and also the provision of practical error estimates.

12.2.1 Convergence

We show first that under suitable conditions the matrix $\mathbf{L} = \mathbf{I} - \lambda\mathbf{K}$ is AD.

Theorem 12.1. *Let*

$$K(x, y) \in C^{(p_x, s_x)}_{(v_x, \mu_x)}(a, b) \times C^{(p_y, s_y)}_{(v_y, \mu_y)}(a, b);$$

and let

$$g(x) \in C^{(p_g, s_g)}_{(v_g, \mu_g)}(a, b).$$

Further, define continuity indices

$$q_i = \min[p_i - \tfrac{1}{2}, 2(s_i - \mu_i), 2(s_i - v_i)] + \tfrac{3}{2}, \quad i = x, y, g, \qquad (12.2.2)$$

and let $\mathbf{K}$, $\mathbf{g}$ *satisfy* (12.1.4a, b) *with* h_i *given by* (12.2.1). *Then for some constants*

$C_k, C_g > 0$,

$$|K_{ij}| \leqslant C_k i^{-q_x} j^{-q_y}, \tag{12.2.3}$$

$$|g_i| \leqslant C_g i^{-q_g}. \tag{12.2.4}$$

Proof. (12.2.4) follows on setting $\alpha = \beta = 0$ in equation (9.4.10a), while (12.2.3) follows immediately from Theorem 11.5.

Corollary 12.1. Let $\min_i |1 - \lambda K_{ii}| = \alpha > 0$. Then $\mathbf{I} - \lambda \mathbf{K}$ satisfies (12.1.11) with $C' = \lambda(C_k/\alpha)$ and $p = q_x$, $q = q_y$.

The theory of Chapters 2 to 4 can now be used to characterise the convergence rates of the calculations:

Theorem 12.2. *Let* $\mathbf{K}, \mathbf{g}$ *satisfy the conditions of Theorem* 12.1 *with* $q_x, q_y, q_g > 1$. *Then for* λ *sufficiently small:*

(a) (12.1.4) *has a unique solution* $\mathbf{a}^{(N)}$ *for every* N;
(b) *For some constant* C_b *independent of* i,

$$|b_i| \leqslant C_b i^{-q} \quad \text{where } q = \min(q_g, q_x); \tag{12.2.5}$$

(c) *For some constant* C_b' *independent of* i, N,

$$|b_i - a_i^{(N)}| \leqslant C_b' N^{-(q_y + q) + 1}; \tag{12.2.6}$$

(d) $\|\mathbf{e}_N\|^2 = (f - f_N, f - f_N) \leqslant C_e N^{-2q+1}$,
where C_e *is given by* (12.2.8) *below.*

Proof. For λ sufficiently small, the constant C', equation (12.1.11) satisfies the conditions of Theorem 3.1 and part (a) follows.

Parts (b) and (c) then follow immediately from the unsymmetric versions of Theorems 2.2 and 2.3 (see Freeman and Delves, 1974), or from section 4.5.

Comment
Somewhat tighter bounds than (12.2.6) can be obtained by treating the matrix as WAD using the theory given in Chapter 3 with inequalities (12.1.10) rather than (12.1.11). However, this treatment leads to bounds on $|b_i|$ identical with (12.2.5), and the bounds (12.2.6) are sufficiently precise for our purposes.

Part (d): From (12.2.5) and (12.2.6) we derive the following error bounds in the natural norm:

$$\|e_N\|^2 = (f - f_N, f - f_N) = \sum_{i=1}^{N} (a_i^{(N)} - b_i)^2 + \sum_{i=N+1}^{\infty} b_i^2$$

$$\leqslant C_b'^2 N^{-2(q_y+q)+3} + \frac{C_b^2}{2q-1} N^{-2q+1} \tag{12.2.7}$$

$$\leqslant \left(C_b'^2 + \frac{C_b^2}{2q-1}\right) N^{-2q+1} = C_e N^{-2q+1}. \tag{12.2.8}$$

Equation (12.2.8) shows directly that for this class of problems and in this norm, the N-convergence rate of $\|e_N\|$ is that of approximation in the underlying space; the first term in (12.2.7) is dominated by the second, provided that $q_y > 1$.

12.2.2 Stability

Theorem 12.1 shows immediately that the system (12.1.6) with expansion set (12.2.1) is *relatively stable* in the sense of Definition 5.7. We therefore discuss here only those aspects required to provide the error estimates of the next section.

The computed matrix $\mathbf{L} = \mathbf{I} - \lambda\mathbf{K}$ and vector $\mathbf{g}$ contain round-off and quadrature errors; of these, the latter will almost certainly dominate, the former usually being negligible. We therefore solve the perturbed equations

$$(\mathbf{L} + \boldsymbol{\delta}\mathbf{L})(\mathbf{a} + \boldsymbol{\delta}\mathbf{a}) = \mathbf{g} + \boldsymbol{\delta}\mathbf{g} \tag{12.2.9}$$

and the error e'_N in the compound solution satisfies

$$\|e'_N\|^2 = \sum_{i=1}^{N} (a_i^{(N)} - b_i + \delta a_i)^2 + \sum_{i=N+1}^{\infty} b_i^2, \tag{12.2.10a}$$

$$\approx \sum_{i=1}^{N} (\delta a_i)^2 + \sum_{i=N+1}^{\infty} b_i^2, \tag{12.2.10b}$$

where the second estimate is valid in view of (12.2.5), (12.2.6), provided the quadrature errors are not smaller than the truncation errors; while, if they are, we may neglect $\boldsymbol{\delta}\mathbf{a}$ anyway.

For any N, we may bound $\|\boldsymbol{\delta}\mathbf{a}\|$ as in (5.3.1):

$$\|\boldsymbol{\delta}\mathbf{a}\| \leqslant \|\mathbf{L}^{-1}\|\{\|\boldsymbol{\delta}\mathbf{L}\|\,\|\mathbf{a}\| + \|\boldsymbol{\delta}\mathbf{g}\|\}/(1 - \|\mathbf{L}^{-1}\|\,\|\boldsymbol{\delta}\mathbf{L}\|) \tag{12.2.11}$$

provided that the denominator is positive.

But it follows from Corollary 12.1 and Lemma 5.1 that $\|\mathbf{L}^{(N)-1}\|$ is bounded uniformly in N; that is, for some K_L independent of N

$$\| \mathbf{L}^{(N)-1} \| \leqslant K_L. \tag{12.2.12}$$

Further, (12.2.5,6) above show that $\|\mathbf{a}\|$ is also uniformly bounded in N. Hence for $\|\boldsymbol{\delta}\mathbf{L}\|$ not too large there exist constants α, β such that

$$\|\boldsymbol{\delta}\mathbf{a}\| \leqslant \alpha \|\boldsymbol{\delta}\mathbf{L}\| + \beta \|\boldsymbol{\delta}\mathbf{a}\|. \tag{12.2.13}$$

and hence (see Definition 5.1) the calculation is *stable* in the strong sense of Mikhlin (1969).

12.2.3 Practical error estimates

The bound (12.2.8) of Theorem 12.2 ignores quadrature errors completely; we therefore use equation (12.2.10a), which yields a useful error estimate provided we can estimate numerically the various quantities which enter.

With little loss of efficiency we may disentangle the discretisation and quadrature errors:

$$\sum_{i=1}^{N} (a_i^{(N)} - b_i + \delta a_i)^2 \leqslant \left\{ \left[\sum_{i=1}^{N} (a_i^{(N)} - b_i)^2 \right]^{1/2} + \left[\sum_{i=1}^{N} \delta a_i^2 \right]^{1/2} \right\}^2.$$

We can then estimate the various contributions to (12.2.10a) as follows.

Truncation error
From (12.2.5) we extract the estimate

$$\sum_{i=N+1}^{\infty} b_i^2 \leqslant \frac{1}{2q-1} C_b^2 N^{-2q+1} \sim \frac{1}{2q-1} N b_N^2. \tag{12.2.14}$$

Now estimates of b_N and of q are of course available *a posteriori*; but to determine a suitable N we require an *a priori* estimate. We can obtain this from Theorem 12.2 and the bounds (12.2.3), (12.2.4):

$$|b_N| \sim G_b \max(|g_N|, |K_{N1}|). \tag{12.2.15}$$

(12.2.14), (12.2.15) can be used as estimators of the truncation error replacing (12.2.15) by the computed a_N *a posteriori* to improve the accuracy estimate. Here and below, G_i are constants which are difficult to estimate but which are (approximately) independent of N, and for which good estimates are therefore not really needed.

The discretisation error
This can also be estimated from (12.2.6):

$$\sum_{i=1}^{N} (a_i^{(N)} - b_i)^2 \leqslant C_b'^2 N^{-2(q_y+q)+3} \sim G_3 N^3 b_N^2 K_{1N}^2.$$

In practice, this term is negligible.

Quadrature errors

If the integrals involved in defining $\mathbf{K}$, $\mathbf{g}$ can be evaluated analytically, the "quadrature errors" come solely from round-off errors in evaluating the analytic forms, and the stability of the system usually ensures that these round-off errors are negligible. Normally, however, numerical quadrature is used; then the errors may be significant and depend strongly on the quadrature rule used. For a given choice of rule, it would be desirable to have available *a priori* estimates of $\|\delta\mathbf{K}\|$, $\|\delta\mathbf{g}\|$ to guide the choice of P, the number of points in the rule. Because the truncation error is a decreasing function of N, and the quadrature errors for fixed P increase with N, the optimal choice of P itself depends on N, and the functional form of this dependence is of particular interest.

Now the standard derivative form of the error estimates for a quadrature rule leads to the following (see, e.g. Haber, 1970, p. 489). Let $g(\mathbf{x})$ be a function in s dimensions defined over a hypercube $[-1, 1]^s$ and with partial derivatives of all orders up to r, such that

$$|g^{(r)}(\mathbf{x})| \leqslant M.$$

Then for a product rule Q of degree $q > r$ using P points per dimension:

$$E(g) = |(I - Q)\,g| \leqslant KMP^{-r}, \tag{12.2.16}$$

where Ig is the exact integral.

For the case in hand, we have one-dimensional integrals with integrands $g(x)\,h_i(x)$, $i = 1, \ldots, N$, and two-dimensional integrands $K(x, y)\,h_i(x)\,h_j(y)$, $i, j = 1, \ldots, N$. It is sufficient to consider the one-dimensional integrals. Suppose that $g(x)$ has r bounded derivatives on $[a, b]$; then so does $g(x)\,h_i(x)$. However, the size of these derivatives rises rapidly with i; it is straightforward to derive the bound for the set (12.2.1):

$$\left|\frac{\mathrm{d}^k}{\mathrm{d}x^k}\,h_n(x)\right| \leqslant K' n^{2k+1/2} \tag{12.2.17}$$

so that for some C

$$E(gh_N) \leqslant CN^{2r+1/2}\,P^{-r}.$$

This suggests that to ensure boundedness of the quadrature errors as $N \to \infty$ we require at least $P = \mathcal{O}(N^2)$. Unfortunately (or, perhaps, fortunately) the bound takes no cognizance of the special structure of the integrands, and is so pessimistic as to be useless. A somewhat better result may be obtained for a Gauss-Legendre rule by introducing the expansion of g. For any rule Q with weights λ_i and abscissae x_i we have

$$Q(gh_N) = \sum_{k=1}^{P} \lambda_k g(x_k)\,h_N(x_k) = \sum_{i=1}^{\infty} g_i \sum_{k=1}^{P} \lambda_k h_i(x_k)\,h_N(x_k).$$

But $h_i h_N$ is a polynomial of degree $N + i - 2$. Hence if the rule is of polynomial degree R

$$\sum_{k=1}^{P} \lambda_k h_i(x_k)\, h_N(x_k) = \int_a^b h_i(x)\, h_N(x)\, \mathrm{d}x = \delta_{iN},$$

provided that $N + i - 2 \leqslant R$.

Now the Gauss-Legendre rule has degree $R = 2P - 1$, and hence provided that $P \geqslant N$ we find

$$\delta g_N = Q(gh_N) - g_N = \sum_{i=2P-N+2}^{\infty} g_i \sum_{k=1}^{P} \lambda_k h_i(x_k)\, h_N(x_k)$$

whence

$$\left|Q(gh_N) - g_N\right| \leqslant K'^2 N^{1/2} \sum_{i=2P-N+2}^{\infty} i^{1/2} |g_i|,$$

where we have inserted the estimates (12.2.17) for $|h_q(x)|$ and the identity $\sum_{k=1}^{P} \lambda_k = (b - a)$. Inserting the bound (12.2.4) for $|g_i|$ we find finally

$$|\delta g_N| \leqslant \frac{K'^2 C_g N^{1/2}}{q_g - 3/2} (2P - N)^{-q_g + 3/2} (b - a). \tag{12.2.18}$$

Similar results can be obtained for the two-dimensional integrals. If we require $E(gh_N) \sim g_N$, (12.2.18) predicts that we should set $2P \sim N + N^\gamma$, $\gamma = 1 + 2/(q_g - 3/2)$. Alternatively, with $P \geqslant N - 1$ it predicts

$$|\delta g_N| \leqslant \frac{K'^2 C_g (b - a)}{q_g - 3/2} N^{-q_g + 2} \leqslant \frac{K'^2 (b - a)}{(q_g - 3/2)} N^2 |g_N| \tag{12.2.18a}$$

which shows that the choice $P = N$ is at least reasonable.

Equation (12.2.18) can be used as the basis of a (rather pessimistic) numerical estimate of $\|\boldsymbol{\delta}\mathbf{g}\|$.

Similar estimates can be given for $\|\boldsymbol{\delta}\mathbf{L}\|$. It then remains only to bound $\|\mathbf{L}^{(N)-1}\|$ in order to provide a numerical estimate of $\|\boldsymbol{\delta}\mathbf{a}\|$ from (12.2.11). A direct calculation of $\mathbf{L}^{(N)-1}$ has an operation count $\mathcal{O}(N^3)$, while the cost of solving for $\mathbf{a}^{(N)}$ using the iterative techniques described in Chapter 8 is only $\mathcal{O}(N^2)$. We avoid this cost by recalling that (see equation (12.2.12)) $\|\mathbf{L}^{(N)-1}\|$ is uniformly bounded in N; we therefore require only a crude estimate of this factor, and this suggests the use of the lower bound

$$\|\mathbf{L}^{(N)-1}\| \approx \|\mathbf{a}^{(N)}\| / \|\mathbf{g}^{(N)}\|. \tag{12.2.19}$$

This estimate has the merit of being very cheap. It is also in practice remarkably good as an estimate of the magnifying effect of the quadrature errors, rather than as an estimate of $\|\mathbf{L}^{(N)-1}\|$.

12.3 Expansions in Chebyshev polynomials

A similar analysis can be carried out when we use as expansion basis the Chebyshev polynomials $T_i(x)$, which are orthogonal on $[-1, 1]$ with weight $w(x) = (1 - x^2)^{-1/2}$, and we sketch only the differences which result. The choice $[a, b] = [-1, 1]$; $h_i(x) = (2/\pi)^{1/2} T_i(x)$, $i = 0,1,\ldots, N$ implies (see (12.1.8))

$$\bar{K}(x, y) = K(x, y)(1 - y^2)^{1/2}. \tag{12.3.1}$$

For smooth $K(x, y)$, $\bar{K}(x, y)$ is therefore not smooth, and expansion (12.1.9) converges only slowly. However, this does not imply that the calculation as a whole converges slowly. The convergence rate estimates (12.2.5)–(12.2.7) remain valid provided that in (12.2.3), (12.2.4) we insert the appropriate coefficients for Chebyshev expansions of $g(x)$ and of $\bar{K}(x, y)$. The relevant analysis is given in Section 9.4; if $g(x)$ is smooth we expect q_g large, and because the singularity in $\bar{K}(x, y)$ is a function of y only we expect q_x large (for smooth $K(x, y)$) but $q_y = 2$. However, q_y does not appear at all in (12.2.5 or 12.2.8), and the convergence rate implied by (12.2.6) remains high even for small q_y.

The singularity does have an effect on the numerical method used to produce the matrix **K** and vector **g** (see (12.1.7, 8)), which now have the defining forms

$$g_i = \left(\frac{2}{\pi}\right)^{1/2} \int_{-1}^{1} \frac{T_i(x)}{(1 - x^2)^{1/2}}\, g(x)\, dx,$$

$$K_{ij} = \frac{2}{\pi} \int_{-1}^{1} \frac{T_i(x)}{(1 - x^2)^{1/2}} \int_{-1}^{1} \frac{T_j(y)}{(1 - y^2)^{1/2}}\, K(x, y)(1 - y^2)^{1/2}\, dx\, dy. \tag{12.3.2}$$

The natural quadrature rule to use for g_i, and for the integral over x in (12.3.2) is the Gauss-Chebyshev rule:

$$\int_{-1}^{1} \frac{\phi(x)}{(1 - x^2)^{1/2}}\, dx = \frac{\pi}{P} \sum_{k=0}^{P}{}'' \phi\left(\cos\frac{\pi k}{P}\right) + E_P. \tag{12.3.3}$$

This rule is *not* appropriate for the y-integration in (12.3.2). We could use a Gauss-Legendre rule for this integral; but there are two reasons (other than a desire for elegance) for trying to retain the Gauss-Chebyshev rule.

Speed

Inserting (12.3.3) into the definition of g_i, and using the explicit form of the Chebyshev polynomials, we find

$$g_i = \frac{(2\pi)^{1/2}}{P} \sum_{k=0}^{P}{}'' g\left(\cos\frac{\pi k}{P}\right) \cos\frac{\pi i k}{P} + E_{Pi}, \quad i = 0,1,\ldots, N. \tag{12.3.4}$$

If we set $P = N$, (12.3.4) is in the form of a discrete cosine transform, and the g_i, $i = 0, \ldots, N$ can be accumulated using fast Fourier transform (FFT) techniques in $\mathcal{O}(N \ln N)$ operations rather than $\mathcal{O}(N^2)$. It would be usetul to arrange this for K_{ij} too.

Error estimates

The error E_P in the Gauss-Chebyshev rule has a very convenient expansion (Riess and Johnson, 1969), which with $P = N$ in (12.3.4) takes the form

$$E_{Pi} = \sum_{j=1}^{\infty} (g_{2Nj-i} + g_{2Nj+i}). \tag{12.3.5}$$

Equation (12.3.5) expresses the error in g_i as a sum over "missing" terms in the expansion of $g(x)$ itself, and if this expansion converges reasonably rapidly we can estimate E_{Pi} directly from the calculated terms:

$$|E_{Pi}| \approx |g_{2N-i}| \leqslant |g_N|, \quad i \leqslant N. \tag{12.3.5a}$$

This estimate is the Chebyshev counterpart of (12.2.18a).

To retain these advantages we may proceed as follows. Introduce

$$K_{0ij} = \frac{2}{\pi} \int_{-1}^{1} \frac{T_i(x)}{(1 - x^2)^{1/2}} \int_{-1}^{1} \frac{T_j(y)}{(1 - y^2)^{1/2}} K(x, y)\, dx\, dy, \quad i, j = 0, 1, \ldots, N, \tag{12.3.6}$$

which can be evaluated efficiently in $\mathcal{O}(N^2 \ln N)$ operations using FFT techniques and a product Gauss-Chebyshev rule. The K_{0ij} can be identified as the coefficients in the expansion:

$$K(x, y) = \sum_{i=0}^{\infty}{}' \sum_{j=0}^{\infty}{}' K_{0ij} T_i(x)\, T_j(x). \tag{12.3.7}$$

Now the function $(1 - y^2)^{1/2}$ has the Chebyshev expansion

$$(1 - y^2)^{1/2} = \frac{4}{\pi}\left[\frac{1}{2} - \sum_{r=1}^{\infty} \frac{T_{2r}(y)}{4r^2 - 1}\right] \tag{12.3.8}$$

and the coefficients K_{ij} can be computed by multiplying the Chebyshev series (12.3.7), (12.3.8), provided that attention is paid to the slow convergence of (12.3.8). A suitable algorithm, with operation count $\mathcal{O}(N^2 \ln N)$ and computable error estimate, is given in Delves, Abd-Elal and Hendry (1979). The error estimate given there is essentially that described in Section 12.2, with the inclusion of an error estimate for the quadrature errors based on (12.3.5a) and its extension to the two-fold integrals (12.3.6). The effectiveness of these estimates is illsurated in Chapter 1, Section 1.2.2.

TABLE 12.1

Solution of the integral equation (12.3.9) with parameter $\lambda = 1{\cdot}0$, $\beta = 1{\cdot}0$ using a Chebyshev expansion (Delves, 1977c). The actual error, $\| e \|_2$, and the error estimate sketched in Section 12.3, are shown

N	$\alpha = 4{\cdot}0$		$\alpha = 8{\cdot}0$		$\alpha = 12{\cdot}0$	
	Error	*Estimate*	*Error*	*Estimate*	*Error*	*Estimate*
4	$7{\cdot}9 \times 10^{-1}$	$5{\cdot}4 \times 10^{0}$	$1{\cdot}0 \times 10^{2}$	$4{\cdot}9 \times 10^{2}$		
6	3·5, −2	7·3, −2	1·4, +1	4·1, +1		
8	7·5, −4	1·3, −3	1·1, +0	3·0, +0	$8{\cdot}6 \times 10^{2}$	$2{\cdot}6 \times 10^{2}$
10	9·6, −6	1·6, −5	5·1, −2	1·4, −1	8·4, +1	2·4, +1
12	1·0, −7	2·2, −7	2·0, −3	4·3, −3	5·6, +0	2·0, +0
14	8·2, −10	1·2, −7	3·2, −5	1·0, −4	2·8, −1	7·1, −2
16			9·8, −7	7·5, −6	1·1, −2	4·4, −3
18					6·1, −4	1·3, −4

As a further illustration, we consider the family of problems

$$f(x) = g(x) + \lambda \int_0^1 e^{\beta xy} f(y)\,dy \tag{12.3.9}$$

$$g(x) = e^{\alpha x} - \lambda(e^{\alpha+\beta x} - 1)/(\alpha + \beta x)$$

with exact solution

$$f(x) = e^{\alpha x}. \tag{12.3.10}$$

Equation (1.2.8), with parameters specified by (1.2.17), is a member of this family, with $\alpha = \beta = \lambda = 1$.

Table 12.1 gives results obtained using the Chebyshev expansion method and error estimates sketched here. It is clear from these results that not only is convergence very rapid but also the error estimate reflects the behaviour of the actual error remarkably well, both when the error is large and when it is small.

12.4 Singular integral equations

Finally, we consider briefly the solution of second kind Fredholm equations with singular kernels or driving terms. Galerkin methods for singular equations yield two (related) problems:

(a) setting up the Galerkin equations accurately;
(b) obtaining rapidly convergent expansions for the solution.

The first problem can be solved in many cases when a Chebyshev expansion is used. We suppose that the singular kernel K can be factored into two parts:

$$K(x, y) = K_0(x, y)\, Q(x, y) \tag{12.4.1}$$

where $K_0(x, y)$ is smooth but $Q(x, y)$ is not. Then $K_0(x, y)$ can be expanded numerically as in (12.3.7); and if the Chebyshev expansion of the function $\bar{Q}(x, y) = Q(x, y)(1 - y^2)^{1/2}$ is available analytically, the two series can be multiplied together in $\mathcal{O}(N^2 \ln N)$ operations to yield an accurate Galerkin matrix.

This matrix will again be WAD, but in general with q_x, q_y (see (12.2.3)) small. Then (12.2.5) predicts a slowly convergent Chebyshev expansion for the solution. Normally this prediction will be borne out by the results. As an example, we consider the integral equation

$$f(x) = g(x) + \int_0^{\pi/2} x^{3/2} y^{3/2} f(y)\,dy$$
$$g(x) = x^{1/2} - (\pi^3/24)\, x^{3/2}. \tag{12.4.2}$$

The exact solution to this problem is

$$f(x) = x^{1/2}. \tag{12.4.3}$$

With a Chebyshev polynomial expansion, it follows on recalling the form of (12.3.2), suitably rescaling the interval $[0, \pi/2]$ to the interval $[-1, 1]$, and using (9.4.10a); or more immediately from the example (9.4.12), that the parameters q_x, q_y equation (12.2.3) and q_g, equation (12.2.4), have the values

$$q_g = 2, \quad q_x = 3, \quad q_y = 2. \tag{12.4.4}$$

Then from Theorem 12.2, the Chebyshev expansion coefficients b_i of the exact solution of (12.4.2), scaled to the interval $[-1, 1]$, have i-convergence rate q with

$$q \geqslant \min(2, 3) = 2 \tag{12.4.5}$$

and this bound is indeed attained by the exact solution (12.4.3). Theorem 12.2 also predicts the overall N-convergence rate

$$\|e_N\| \leqslant C_e^{1/2} N^{-(2q+1)/2} = C_e^{1/2} N^{-2\cdot 5}. \tag{12.4.6}$$

The numerical results obtained using the fast Galerkin algorithm of Delves (1977c) are shown in Table 12.2, together with the error estimate returned by the algorithm.

TABLE 12.2
Solution of the integral equation (12.4.2) using the Chebyshev expansion method of Delves (1977c). The error estimates are those returned by this method

N	$\|e_N\|_2$	*Estimate*
8	$2\cdot 4 \times 10^{-2}$	$7\cdot 4 \times 10^{-2}$
16	$3\cdot 7, -3$	$2\cdot 3, -2$
32	$1\cdot 3, -3$	$6\cdot 4, -3$

As predicted, the convergence is quite slow; a fit to these results yields the "observed" N-convergence rate

$$\|e_N\| \sim \mathcal{O}(N^{-2\cdot 1})$$

in reasonable agreement with the expected rate.

The close agreement between prediction and numerical results, which is reflected in the generally satisfactory error estimate, is very gratifying. However, it must be recalled that Theorem 12.2 supplies a uniform upper *bound* on the coefficients b_i, and hence a *lower bound* on the i-convergence rate. This bound is *not* necessarily *sharp*. In particular, it is possible for a singularity in the kernel to be matched by a singularity in the driving term

$g(x)$, the effects of these two singularities cancelling to yield a smooth solution. Such a cancellation *cannot* in general be reflected in theorems of the type 12·2, and Theorem 12.2 will in fact be pessimistic in such cases.

Cancellations of this sort occur particularly often in artificial examples, constructed to be awkward to solve but to have a known and usually simple solution. As an example, we consider the equation (Baker, 1977)

$$f(x) = g(x) + \int_0^1 \ln|x - y|\, f(y)\, \mathrm{d}y \tag{12.4.7}$$

$$g(x) = x - 0{\cdot}5\,[x^2 \ln x + (1 - x^2) \ln (1 - x) - (x + 0{\cdot}5)]$$

with exact solution

$$f(x) = x. \tag{12.4.8}$$

The result (9.4.10a) yields the estimate for q_g, (12.2.4):

$$q_g = 3{\cdot}5$$

and hence, without further consideration of the kernel, Theorem 12.2 predicts at best a convergence rate for the coefficients b_i and error $\|e_N\|$:

$$|b_i| = \mathcal{O}(i^{-3\cdot5}), \quad \text{and} \quad \|e_N\| = \mathcal{O}(N^{-3}) \tag{12.4.9}$$

whereas the exact expansion for $f(x)$ in fact terminates.

A suitably designed expansion method will reflect this type of cancellation when it occurs, and yield very rapid convergence. It is also possible to modify the error estimates based on Theorem 12.2 so that they, too, reflect the cancellation. We refer to Delves, Abd-Elal and Hendry (1979) for details, and merely give in Table 12.3 the results obtained for problem (12.4.7), together with those given by Baker (1977) using a modified Nystrom method, which takes account of the singular kernel but is unable to adapt to the "accidental" cancellation.

The Chebyshev method has achieved essentially machine accuracy with $N \geqslant 5$; more gratifying still, the error estimates reflect this.

TABLE 12·3

	Chebyshev			*Nystrom*
N	$\|\mathbf{e}_N\|_2$	*Estimate*	N	$\|\mathbf{e}_N\|_\infty$
3	$1{\cdot}7 \times 10^{-2}$	$7{\cdot}5 \times 10^{-2}$	8	$1{\cdot}0 \times 10^{-3}$
5	3·5, −12	2·1, −11	16	2·7, −4
7	1·1, −11	3·4, −11	32	7·1, −5
9	9·6, −12	4·8, −11	64	1·8, −5

Cancellation of singularities in the kernel are not shown only by artificial examples. There are two important classes of equations: those with Green's function and with Volterra kernels, which are very ill-behaved according to the criteria we have set down. The former have kernels with discontinuous first derivatives across the line $x = y$. The latter are those which come from equations of the form

$$f(x) = g(x) + \int_a^x K(x, y)\, f(y)\, \mathrm{d}y. \tag{12.4.10}$$

If (12.4.10) is treated as a Fredholm equation, the corresponding kernel is discontinuous across the line $x = y$. In neither case are the discontinuities reflected into the solution $f(x)$, and it is of interest to see why this should be. We consider a Volterra kernel, which the numerical procedure outlined above treats as a Fredholm kernel with a discontinuity along the line $x = y$; that is, it sets in (12.4.1)

$$\begin{aligned} Q(x, y) &= 1 \quad y < x, \\ &= 0 \quad y \geqslant x. \end{aligned} \tag{12.4.11}$$

The expansion of the corresponding function $\bar{Q}(x, y)$:

$$\bar{Q}(x, y) = \sum_{i=0}^{\infty}{}' \sum_{j=0}^{\infty}{}' q_{ij} T_i(x)\, T_j(y) \tag{12.4.12}$$

has coefficients

$$\begin{aligned} q_{0,0} &= 4/\pi, \quad q_{1,0} = 2/\pi, \\ q_{0,1} &= -1/\pi, \quad q_{2,1} = 1/2\pi, \\ q_{0,s} &= 4(-1)^{s-1}/(\pi(s^2 - 1)), \qquad s \geqslant 2, \\ q_{s-1,s} &= -1/(\pi(s-1)) \\ q_{s+1,s} &= 1/(\pi(s+1)) \end{aligned}$$

and all others zero; that is, its expansion has the form

$$\bar{Q}(x, y) = \sum_{l=0}^{\infty}{}' \left[a_l T_l(y) + b_l T_l(x)\, T_{l+1}(y) + c_l T_{l+1}(x)\, T_l(y)\right] \tag{12.4.13}$$

with bounds

$$\begin{aligned} |a_l| &\leqslant C_1 l^{-2}, \\ |b_l| &\leqslant C_2 l^{-1}, \\ |c_l| &\leqslant C_3 l^{-1}. \end{aligned} \tag{12.4.14}$$

Now suppose the kernel function $K_0(x, y)$, extended as required here over the whole square $[-1, 1] \times [-1, 1]$, has an expansion (12.3.7) with bounds

$$|K_{0ij}| \leqslant c_4 \hat{\imath}^{-p} \hat{\jmath}^{-q}, \tag{12.4.15}$$

(where $\hat{0} = 1; \hat{\imath} = i, i > 0$).

The Galerkin matrix $\mathbf{K}$ can be split into three terms

$$\mathbf{K} = \mathbf{K}_1 + \mathbf{K}_2 + \mathbf{K}_3$$

stemming from the three sums in (12.4.13).

The first term can be identified from (12.4.12) as the matrix of the function $K(x, y)(1 - y^2)^{1/2}$; the discussion of Section 12.3 then leads directly to the bound

$$|K_{1rs}| \leqslant D_1 \hat{r}^{-p} \hat{s}^{-2}. \tag{12.4.16}$$

The terms K_2, K_3 are very similar, and we treat only K_2. We require the expansion of the product

$$\left(\sum_{i,j=0}^{\infty}{}'' K_{0ij} T_i(x)\, T_j(y) \right) \left(\sum_{l=1}^{\infty} b_l T_{l-1}(x)\, T_l(y) \right).$$

The (r, s) term in this expression is

$$\begin{aligned} K_{2rs} = \sum_{l=1}^{\infty} \tfrac{1}{4} b_l [& K_{0|r+1-l|,|s-l|} + K_{0|r+1-l|,|s+l|} \\ & + K_{0|r+l-1|,|s-l|} + K_{0|r+l-1|,|s+l|}] \end{aligned} \tag{12.4.17}$$

and the bounds (12.4.14, 15) yield

$$|K_{2rs}| \leqslant C_4 \sum_{l=1}^{\infty} |r + \hat{1} - l|^{-p} |s \mathbin{\hat{-}} l|^{-q} \tag{12.4.18a}$$

where for simplicity we have noted that the first term in (12.4.17) is always dominant.

The sum (12.4.18a) can be bounded in a straightforward manner, splitting the range of l and using the lemmas of the Appendix. The result is

$$|K_{2rs}| \leqslant C_4 E |r - s|^{-t}, \quad r \neq s, \tag{12.4.18b}$$

where $t = \min(p, q)$ and the precise form of E is unimportant. The same result holds for K_3.

Finally, using the bound

$$(\hat{r}\hat{s})^{-1} \leqslant |r - s|^{-1}, \quad r \neq s, \tag{12.4.19}$$

we combine (12.4.16) (12.4.18b) to yield

$$\begin{aligned} |K_{rs}| &\leqslant F(r - s)^{-t}, \quad r > s, \\ &\leqslant G(s - r)^{-2}, \quad s > r. \end{aligned} \tag{12.4.20}$$

Thus, the Volterra matrix **K** is lower AD of type $B(0, t; F)$ and upper AD of type $B(0, 2; G)$. It then follows that the Galerkin calculation converges at a rate determined solely by the parameter t, the singularity across the line $x = y$ being "benign".

This analysis clearly applies also to "inverse Volterra" operators of the form

$$\int_x^b K(x, y) f(y)\,\mathrm{d}y \tag{12.4.21}$$

and hence to Green's function Fredholm equations:

$$f(x) = g(x) + \int_a^b K(x, y) f(y)\,\mathrm{d}y, \tag{12.4.22}$$

$$\begin{aligned} K(x, y) &= K_1(x, y), \quad x > y, \\ &= K_2(x, y), \quad x \leqslant y. \end{aligned}$$

As an example, we consider equation (12.4.22) with

$$\begin{gathered} [a, b] = [0, 1], \quad g(x) = x^2, \\ K_1(x, y) = y(x - 1), \quad K_2(x, y) = x(y - 1), \end{gathered} \tag{12.4.23}$$

and exact solution $f(x) = Ae^x + Be^{-x} - 2$,

$$A = (3 - e - 2)/(e^2 - 1); \quad B = e(2 - e - 3)/(e^2 - 1).$$

The functions $K_1(x, y)$ and $K_2(x, y)$ and $g(x)$ are all polynomials in this example; it follows that the Galerkin matrix **K** is lower AD of type $B(0, t; H)$ for all finite t, and that (12.2.4) is valid for all finite q_g. Then Theorem 4.3 predicts that $\|\mathbf{e}_N\|^2$ will converge to zero faster than any power of N. Numerical results are given in Table 12.4, for a Chebyshev basis.

TABLE 12.4
Solution of the integral equation problem (12.4.22), (12.4.23)

N:	2	3	4	5	6	7	8
$\|e_N\|$	$1{\cdot}2 \times 10^{-1}$	5·8, −4	1·2, −5	1·2, −6	1·8, −8	1·3, −9	1·1, −11
Est. error	$6{\cdot}0 \times 10^{-1}$	3·6, −4	2·3, −4	8·4, −7	3·9, −7	1·7, −9	1·1, −9

The convergence displayed is exponentially fast: $\|\mathbf{e}_N\| \sim CA^N$, $A \leqslant 0{\cdot}1$, in agreement with this prediction. The estimated errors in this table again come from a generalized version of the error analysis sketched in Sections 12.2 and 12.3, and are clearly very successful in reproducing the observed errors; see Delves, Abd-Elal and Hendry (1979) for details and further examples, and for the application to other types of singularity in both the kernel K and the driving term g.

13

Global Expansion Methods for Ordinary Differential Equations

13.1 Introduction

We now turn our attention to the study of expansion methods for the solution of (boundary value, second order) ordinary differential equations. As before, we restrict attention to those methods which use global rather than local expansion functions, and we show that these methods again lead to WAD matrices.

The study of differential equations is complicated (compared to that of integral equations) by the need to treat explicitly the boundary conditions of the problem. We sidestep this need by studying first only problems with homogeneous boundary conditions, and with expansion functions satisfying these conditions term by term.

We take as our standard problem the solution of the equation

$$\mathscr{L}f(x) \equiv \left[a_2(x)\frac{\mathrm{d}^2}{\mathrm{d}x^2} + a_1(x)\frac{\mathrm{d}}{\mathrm{d}x} + a_0(x) \right] f(x) = g(x) \qquad (13.1.1)$$

subject to suitable assumptions on the analyticity properties of the $a_i(x)$; and in Sections 13.2 and 13.3 we consider boundary conditions of the form

$$f(a) = f(b) = 0. \qquad (13.1.1a)$$

Given an expansion set $\{h_i\}$, and a weight function $w(x)$, the coefficients of the $N \times N$ Galerkin matrix $\mathbf{L}$ for this problem are

$$L_{nm} = \int_a^b w(x)\, h_n(x)\, \mathscr{L}h_m(x)\, \mathrm{d}x, \quad n, m = 1, 2, \ldots, N, \qquad (13.1.2)$$

and we shall be chiefly concerned with analysing the structure of $\mathbf{L}$. For boundary conditions (13.1.1a), and for expansion sets $\{h_i\}$ satisfying (13.1.1a) term by term, the Galerkin method leads to the approximations

$$f \approx f_N = \sum_{i=1}^{N} a_i h_i(x) \tag{13.1.3}$$

where

$$\mathbf{La} = \mathbf{g} \tag{13.1.4}$$

and

$$g_i = \int_a^b w(x)\, g(x)\, h_i(x)\, dx.$$

If $\{h_i(x)\}$ does not vanish at a, b, or if the boundary conditions are not homogeneous, equations (13.1.4) require either augmenting or modifying (Section 13.6). However, in either case the numerical properties of the method depend primarily on the structure of $\mathbf{L}$.

13.2 A simple example: the half-range sine expansion

As a simple example, we consider problem (13.1.1) subject to (13.1.1a). We take

$$a = 0, \quad b = \pi, \quad w(x) = 1, \quad h_i(x) = (2/\pi)^{1/2} \sin(ix), \quad i = 1, 2, \ldots, N, \tag{13.2.1}$$

and then we find

$$\begin{aligned} L_{nm} &= \frac{2}{\pi} \int_0^\pi \sin(nx)\, \mathscr{L}\, \sin(mx)\, dx \\ &= \frac{2}{\pi} \int_0^\pi \sin(nx) \{-m^2 a_2(x) \sin mx + m a_1(x) \cos mx + a_0(x) \sin mx\}\, dx. \end{aligned} \tag{13.2.2}$$

We expand the functions $a_i(x)$ in half-range series as follows:

$$\begin{aligned} a_i(x) &= \sum_{j=0}^{\infty} \left(\frac{2}{\pi}\right)^{1/2} b_j^{(i)} \cos jx, \quad i = 0, 2, \\ a_1(x) &= \sum_{j=1}^{\infty} \left(\frac{2}{\pi}\right)^{1/2} b_j^{(1)} \sin jx. \end{aligned} \tag{13.2.3}$$

Then using the product formulae

$$\sin nx \sin mx = \tfrac{1}{2}(\cos (n-m)x - \cos(n+m)x)$$

$$\sin nx \cos mx = \tfrac{1}{2}(\sin (n+m)x + \sin(n-m)x)$$

equation (13.2.2) becomes

$$L_{nm} = \frac{2^{1/2}}{\pi^{3/2}}\left\{-m^2\int_0^\pi (\cos(n-m)x - \cos(n+m)x)\sum_{j=0}^{\infty} b_j^{(2)} \cos jx\,dx\right.$$
$$+ m\int_0^\pi (\sin(n+m)x + \sin(n-m)x)\sum_{j=1}^{\infty} b_j^{(1)} \sin jx\,dx$$
$$\left.+\int_0^\pi (\cos(n-m)x - \cos(n+m)x)\sum_{j=0}^{\infty} b_j^{(0)} \cos jx\,dx\right\}.$$

But

$$\int_0^\pi \sin px \sin qx\,dx = (\pi/2)\,\delta_{pq}, \quad p \text{ and } q \text{ not both} = 0,$$

and

$$\int_0^\pi \cos px \cos qx\,dx = (\pi/2)\,\delta_{pq}, \quad p \text{ and } q \text{ not both} = 0,$$

whence we find:

$$L_{mm} = (2\pi)^{-1/2}\{-m^2(2b_0^{(2)} - b_{2m}^{(2)}) + mb_{2m}^{(1)} + 2b_0^{(0)} - b_{2m}^{(0)}\} \tag{13.2.4}$$

$$\begin{aligned} L_{nm} &= (2\pi)^{-1/2}\{-m^2(b_{n-m}^{(2)} - b_{n+m}^{(2)}) + m(b_{n+m}^{(1)} + b_{n-m}^{(1)}) \\ &\quad + b_{n-m}^{(0)} - b_{n+m}^{(0)}\}, \quad n > m, \\ &= (2\pi)^{-1/2}\{-m^2(b_{m-n}^{(2)} - b_{n+m}^{(2)}) + m(b_{n+m}^{(1)} - b_{m-n}^{(1)}) \\ &\quad + b_{m-n}^{(0)} - b_{n+m}^{(0)}\}, \quad n < m. \end{aligned} \tag{13.2.5}$$

Equations (13.2.4, 5) form a basis for an efficient numerical scheme for generating the matrix $\mathbf{L}$. The coefficients $b_j^{(i)}, j = 0, \ldots, N$ can be evaluated in $\mathcal{O}(N \ln N)$ operations, using fast Fourier transform techniques, and then the matrix $\mathbf{L}$ assembled from these in $\mathcal{O}(N^2)$ operations. This compares with an $\mathcal{O}(N^3)$ count if each element L_{nm} is approximated directly using numerical quadrature. Equations (13.2.4) and (13.2.5) also display the i- and j-convergence properties of the components L_{ij}. It follows from Chapter 9, Section 9.2, that the coefficients $b_j^{(i)}$ satisfy bounds of the form

$$|b_j^{(i)}| \leqslant c_i \hat{j}^{-p_i} \tag{13.2.6}$$

where the exponents p_i depend on the analytic structure of the functions $a_i(x)$; see Table 9.1. Now, inserting these bounds in (13.2.4), (13.2.5) and making the weak additional assumption

$$\min_m |L_{mm}| \geqslant \varepsilon > 0, \tag{13.2.7}$$

we find

$$|L_{nn}| \geqslant C_1 n^2, \tag{13.2.8}$$

and for $n \neq m$

$$|L_{nm}| \leqslant C_2 \max[m^2(|n-m|)^{-p_2}, m(|n-m|)^{-p_1}, (|n-m|)^{-p_0}]. \tag{13.2.9}$$

Hence introducing the normalised matrix $\mathbf{F}$

$$F_{nm} = \frac{|L_{nm}|}{\{|L_{nn}|.|L_{mm}|\}^{1/2}},$$

we find

$$F_{nm} \leqslant C_3 \max\begin{Bmatrix} n^{-1}m(|n-m|)^{-p_2} \\ n^{-1}(|n-m|)^{-p_1} \\ (nm)^{-1}(|n-m|)^{-p_0} \end{Bmatrix} \tag{13.2.10}$$

But $n^{-1}m < 1$ if $n > m$, and $n^{-1}m(m-n)^{-1} \leqslant 2$ if $m > n \geqslant 1$, and (13.2.10) therefore implies that the operator matrix $\mathbf{L}$ is (see Definition 4.2)

$$\text{ALD (AUD) of type B}(0, r_L(r_U); C_L(C_U)) \tag{13.2.11}$$

where

$$r_L = \min[p_2, p_1 + 1, p_0 + 1], \quad \text{and} \quad r_U = \min[p_2 - 1, p_1, p_0 + 1],$$

and C_L, C_U are constants independent of n, m.

We can distinguish two special cases:

(a) if $a_2(x)$ is constant then the terms in m^2 vanish from (13.2.5) and $\mathbf{L}$ is

$$\text{ALD (AUD) of type B}(1, r'_L(r'_U); C'_L(C'_U))$$

where

$$r'_L = \min[p_1, p_0], \quad \text{and} \quad r'_U = \min[p_1 - 1, p_0]; \tag{13.2.12}$$

(b) if in addition $a_1(x) = 0$ the terms in m vanish from (13.2.5) and $\mathbf{L}$ is then

$$\text{ALD (AUD) of type B}(2, p_0 - 1; C''_L(C''_U)).$$

13.3 Expansion by Chebyshev polynomials

Classical Fourier expansions were observed in Chapter 9 to converge rather slowly in most circumstances; better convergence properties are displayed by orthogonal polynomial expansions. We consider in the next two sections, Chebyshev and ultraspherical polynomial expansions, with the "natural" weight $w(x)$ with respect to which they are orthogonal. In each case we find that the operator matrix is AD of type $B(p, r; C)$, with parameters p, r which can be related to the properties of the coefficients $a_i(x)$.

We note that these polynomials do *not* satisfy homogeneous conditions at the boundary of the region; however, in Section 13.6 we show that the results obtained for Chebyshev expansions carry over rather easily to a modified polynomial basis which does vanish at the boundaries, and that the singularities which we calmly ignore in this section and in 13.4, do not then occur.

We choose, then:

$$[a, b] = [-1, 1]; \; w(x) = (1 - x^2)^{-1/2}; \tag{13.3.1}$$

$$h_i(x) = \left(\frac{2}{\pi}\right)^{1/2} T_i(x), \quad i = 0, 1, \ldots$$

Then

$$\begin{aligned} L_{nm} &= \frac{2}{\pi}\int_{-1}^{1} (1 - x^2)^{-1/2}\, T_n(x)\, \mathscr{L} T_m(x)\, \mathrm{d}x \\ &= \frac{2}{\pi}\left\{\int_{-1}^{1} (1 - x^2)^{-1/2}\, T_n(x)\left[a_2(x)\frac{\mathrm{d}^2}{\mathrm{d}x^2} T_m(x)\right.\right. \\ &\qquad \left.\left. + a_1(x)\frac{\mathrm{d}}{\mathrm{d}x} T_m(x) + a_0(x)\, T_m(x)\right]\mathrm{d}x\right\}. \end{aligned} \tag{13.3.2}$$

As with the sine expansions we find that the upper half elements of $\mathbf{L}(m > n)$ behave slightly differently from those of the lower half $(m < n)$. In the following, equations whose numbers are followed by (a) are used to estimate the upper half and those followed by (b) the lower half of $\mathbf{L}$.

The Chebyshev polynomials $T_m(x)$ satisfy the following relationships:

$$\left\{(1 - x^2)\frac{\mathrm{d}^2}{\mathrm{d}x^2} - x\frac{\mathrm{d}}{\mathrm{d}x} + m^2\right\} T_m(x) = 0 \tag{13.3.3}$$

$$(1 - x^2)\frac{\mathrm{d}}{\mathrm{d}x} T_m(x) = m[xT_m(x) - T_{m+1}(x)]. \tag{13.3.4a}$$

We use these to eliminate the first and second derivatives appearing in (13.3.2).

Then, defining $d_0(x) = a_0(x)$,

$$\begin{aligned} d_1(x) &= x(1 - x^2)^{-1} a_1(x) + x^2 a_2(x)(1 - x^2)^{-2}, \\ d_2(x) &= (1 - x^2)^{-1} a_2(x), \\ d_3(x) &= d_1(x)/x, \end{aligned} \tag{13.3.5}$$

and introducing the notation

$$(n|f|m) = \frac{2}{\pi} \int_{-1}^{1} (1 - x^2)^{-1/2} T_n(x) f(x) T_m(x) \, dx, \tag{13.3.6}$$

we find

$$L_{nm} = (n|d_0|m) + m(n|d_1|m) - m^2(n|d_2|m) - m(n|d_3|m+1) \tag{13.3.7a}$$

$$= (n|d_0|m) - m(n|d_1|m) - m^2(n|d_2|m) + m(n|d_3|m-1). \tag{13.3.7b}$$

Now proceeding formally (that is, ignoring the singularities in the functions $d_i(x)$) we expand the functions $d_i(x)$, $i = 0,1,2,3$ as a series of (normalised) Chebyshev polynomials:

$$d_i(x) = \left(\frac{2}{\pi}\right)^{1/2} \sum_{j=0}^{\infty}{}' b_j^{(i)} T_j(x) \tag{13.3.8}$$

and suppose that

$$|b_j^{(i)}| \leqslant C_4 \hat{j}^{-p_i} \quad j = 0,1,\ldots \quad i = 0,1,2,3. \tag{13.3.9}$$

Estimates for the exponents p_i are derived in Chapter 9, in terms of the analytic structure of the functions $d_i(x)$. Substitution of (13.3.8) into (13.3.7) leads to a series of integrals of the type

$$I_{nm}^{(i)} = \sum_{j=0}^{\infty}{}' b_j^{(i)} \int_{-1}^{1} (1 - x^2)^{-1/2} T_n(x) T_j(x) T_m(x) \, dx.$$

From the relationship

$$2T_n(x) T_m(x) = T_{n+m}(x) + T_{n-m}(x)$$

we obtain immediately

$$I_{nm}^{(i)} = \frac{\pi}{4}(b_{n+m}^{(i)} + b_{n-m}^{(i)}), \quad n \geqslant m.$$

And hence under the additional assumption (13.2.7) we find

$$|L_{nn}| \geqslant C_5 n^2,$$

and for $n \neq m$

$$|L_{nm}| \leqslant C_6 \max\{(|n-m|)^{-p_0}, m(|n-m|)^{-p_1}, m^2(|n-m|)^{-p_2}, m(|n-m|+1)^{-p_3}\},$$

and it follows as before that $\mathbf{L}$ is

$$\text{ALD (AUD) of type } \mathbf{B}(0, r_L(r_U); C_L(C_U)), \tag{13.3.10}$$

with

$$r_L = \min[p_0+1, p_1+1, p_2, p_3+1], \quad \text{and} \quad r_U = \min[p_0+1, p_1, p_2-1, p_3].$$

We can also distinguish two special cases:

(a) if in equation (13.3.2) $a_2(x) = C_1(1-x^2)$ where C_1 is a constant then the term in m^2 vanishes from (13.3.7) and we find that the matrix $\mathbf{L}$ is then

$$\text{ALD (AUD) of type } \mathbf{B}(1, r'_L(r'_U); C'_L(C'_U)),$$

where

$$r'_L = \min[p_0, p_1, p_3], \quad \text{and} \quad r'_U = \min[p_0, p_1-1, p_3-1]; \tag{13.3.11}$$

(b) if in addition $a_1(x) = -C_1 x$, so that the operator has the form

$$\mathscr{L} \equiv C_1\left\{(1-x^2)\frac{\mathrm{d}^2}{\mathrm{d}x^2} - x\frac{\mathrm{d}}{\mathrm{d}x}\right\} + a_0(x),$$

the terms in m also vanish from (13.3.7) and the operator matrix $\mathbf{L}$ is then

$$\text{ALD (AUD) of type } \mathbf{B}(2, p_0-1; C''_L(C''_U)). \tag{13.3.12}$$

We comment again that the validity of this analysis depends on the assumption that the $d_i(x)$ are regular enough for expansions (13.3.8) to exist. In practice, the singular factors which appear in the $d_i(x)$ can be removed by using a related set of polynomials and making only slight changes to the above analysis; see Section 13.5.

13.4. Expansions in ultraspherical harmonics

The Chebyshev polynomials form a limiting case of the (normalised) ultraspherical polynomials $B_n^{(\lambda)} P_n^{(\lambda)}(x)$ which, following the notation of Szego (1939), are orthonormal on $(-1, 1)$ with the weight

$$w(x) = (1-x^2)^{\lambda-1/2}, \quad \lambda > -\tfrac{1}{2}, \quad \lambda \neq 0,$$

with normalising factor given by

$$B_n^{(\lambda)} = \left\{ \frac{\Gamma(\lambda)^2 (n + \lambda)\, n!}{2^{1-2\lambda} \pi \Gamma(n + 2\lambda)} \right\}^{1/2}. \qquad (13.4.1)$$

The Galerkin matrix $\mathbf{L}$ has elements

$$L_{nm} = B_n^{(\lambda)} B_m^{(\lambda)} \int_{-1}^{1} (1 - x^2)^{\lambda - 1/2} P_n^{(\lambda)}(x) \left[a_2(x) \frac{d^2}{dx^2} P_m^{(\lambda)}(x) \right.$$
$$\left. + a_1(x) \frac{d}{dx} P_m^{(\lambda)}(x) + a_0(x) P_m^{(\lambda)}(x) \right] dx. \qquad (13.4.2)$$

The analysis of (13.4.2) follows along similar lines to the Chebyshev case. The ultraspherical polynomials $P_m^{(\lambda)}(x)$ satisfy the following relationships:

$$\left\{ (1 - x^2) \frac{d^2}{dx^2} - (2\lambda + 1)\, x \frac{d}{dx} + m(m + 2\lambda) \right\} P_m^{(\lambda)}(x) = 0, \quad (13.4.3)$$

$$(1 - x^2) \frac{d}{dx} P_m^{(\lambda)}(x) = (m + 2\lambda)\, x P_m^{(\lambda)}(x) - (m + 1)\, P_{m+1}^{(\lambda)}(x), \qquad (13.4.4a)$$

$$(1 - x^2) \frac{d}{dx} P_m^{(\lambda)}(x) = -mx P_m^{(\lambda)}(x) + (m + 2\lambda - 1)\, P_{m-1}^{(\lambda)}(x), \qquad (13.4.4b)$$

(see Szego, 1939, pp. 80, 83).

Using these and defining

$$\begin{aligned} d_0(x) &= a_0(x), \\ d_1(x) &= (1 - x^2)^{-2} [x^2(2\lambda + 1)\, a_2(x) + x(1 - x^2)\, a_1(x)], \\ d_2(x) &= (1 - x^2)^{-1} a_2(x), \\ d_3(x) &= d_1(x)/x, \end{aligned} \qquad (13.4.5)$$

and introducing the notation

$$(n|f|m) = \int_{-1}^{1} (1 - x^2)^{\lambda - 1/2} P_n^{(\lambda)}(x)\, f(x)\, P_m^{(\lambda)}(x)\, dx, \qquad (13.4.6)$$

we find

$$L_{nm} = B_n^{(\lambda)} B_m^{(\lambda)} \{ (n|d_0|m) + (m + 2\lambda)(n|d_1|m)$$
$$- m(m + 2\lambda)(n|d_2|m) - (m + 1)(n|d_3|m + 1) \}, \qquad (13.4.7a)$$

$$L_{nm} = B_n^{(\lambda)} B_m^{(\lambda)} \{ (n|d_0|m) - m(n|d_1|m)$$
$$- m(m + 2\lambda)(n|d_2|m) + (m + 2\lambda - 1)(n|d_3|m - 1) \}. \qquad (13.4.7b)$$

Assuming regularity of the d_i we introduce the expansions

$$d_i(x) = \sum_{j=0}^{\infty} B_j^{(\lambda)} b_j^{(i)} P_j^{(\lambda)}(x) \tag{13.4.8}$$

and again suppose that the $b_j^{(i)}$ satisfy (13.3.9).

Substitution of (13.4.8) into (13.4.7) leads to a series of integrals of the type

$$I_{nm}^{(i)} = \int_{-1}^{1} (1 - x^2)^{\lambda - 1/2} P_n^{(\lambda)}(x) \sum_{k=0}^{\infty} b_k^{(i)} B_k^{(\lambda)} P_k^{(\lambda)}(x) P_m^{(\lambda)}(x)\, \mathrm{d}x. \tag{13.4.9}$$

Lemma 13.1. (*see Hsien-yü Hsü* (1938))

$$\int_{-1}^{1} (1 - x^2)^{\lambda - 1/2} P_l^{(\lambda)}(x) P_m^{(\lambda)}(x) P_n^{(\lambda)}(x)\, \mathrm{d}x$$

$$= \frac{2^{1-2\lambda}\pi\Gamma(s + 2\lambda)}{\{\Gamma(\lambda)^2\}(s + \lambda)\Gamma(s + 1)}$$

$$\times \frac{\binom{s - l + \lambda - 1}{s - l}\binom{s - m + \lambda - 1}{s - m}\binom{s - n + \lambda - 1}{s - n}}{\binom{s + \lambda - 1}{s}}$$

or 0

where $\lambda > -\frac{1}{2}$, *l, m, n are non-negative integers, and* $s = (l + m + n)/2$.

The first of these values holds if $l + m + n$ is even, and a triangle exists with the sides l, m, n; the zero value holds in every other case.

Consider, for simplicity, the case $m + n$ even. Then, setting $k = m + n - 2j$ and using the triangle inequalities implied in Lemma 13.1, (13.4.9) can be rewritten as:

$$I_{nm}^{(i)} = \int_{-1}^{1} (1 - x^2)^{\lambda - 1/2} P_n^{(\lambda)}(x)$$

$$\times \sum_{j=0}^{\min(m,n)} b_{m+n-2j}^{(i)} B_{m+n-2j}^{(\lambda)} P_{m+n-2j}^{(\lambda)}(x) P_m^{(\lambda)}(x)\, \mathrm{d}x,$$

and again by Lemma 13.1 and (13.4.1) the above integral can be expressed as:

$$I_{nm}^{(i)} = \left\{\frac{2^{1-2\lambda}\pi}{[\Gamma(\lambda)]^6}\right\}^{1/2} \sum_{j=0}^{\min(m,n)} b_{m+n-2j}^{(i)} \left[\frac{(m + n - 2j + \lambda)(m + n - 2j)!}{\Gamma(m + n - 2j + 2\lambda)}\right]^{1/2}$$

$$\times A(m, n, j, \lambda), \tag{13.4.10}$$

where

$$A(m, n, j, \lambda) = \frac{\Gamma(m + n - j + 2\lambda)\,\Gamma(m - j + \lambda)\,\Gamma(j + \lambda) \times \Gamma(n - j + \lambda)\,(m + n - j)!}{(m + n - j + \lambda)\,(m + n - j)!\,(m - j)! \times (j)!\,(n - j)!\,\Gamma(m + n - j + \lambda)}. \quad (13.4.11)$$

Now for $\lambda \leqslant 1$, $A(m, n, j, \lambda)$ attains its maximum (in the range $j = 0(1) \min(m, n)$) when $j = \min(m, n)$. Taking this maximum value outside the sum, and using an integral bound for the remaining sum, we find eventually:

$$|I_{nm}^{(i)}| \leqslant \text{const.}\,(nm)^{\lambda - 1}|n - m|^{-p_i + 1}, \quad (13.4.12)$$

where p_i is defined by (13.3.9); and

$$|I_{nn}^{(i)}| = \mathcal{O}(n^{2\lambda - 2}). \quad (13.4.13)$$

From (13.4.12), (13.4.13) it follows, with the additional assumption (13.2.7), that **L** is

ALD (AUD) of type $B(0, r_L(r_U); C_L(C_U))$ with parameters

$$r_L = \min(p_0, p_1, p_2 - 1, p_3), \quad \text{and} \quad r_U = \min(p_0, p_1 - 1, p_2 - 2, p_3 - 1). \quad (13.4.14)$$

A comparison of (13.4.14) with (13.3.10) is possible if we let $\lambda \to 0$ in the case of the ultraspherical polynomial expansion. It is seen that the results obtained for r_L, r_U are poorer by one than those of (13.3.10); this is the result of having to bound the sum in (13.4.10) in this case, whereas for the Chebyshev polynomials it was possible simply to select the dominant term of two for the bound. Again we can distinguish two special cases:

(a) if in (13.4.2) $a_2(x) = C(1 - x^2)$ where C is some constant then **L** is

ALD (AUD) of type $B(1, r_L(r_U); C_L(C_U))$,

where

$$r_L = \min[(p_0 - 1), (p_1 - 1), (p_3 - 1)], \quad \text{and}$$

$$r_U = \min[(p_0 - 1), (p_1 - 2), (p_3 - 2)];$$

(b) if in addition $a_1(x) = Cx(2\lambda + 1)$, that is the operator $\mathscr{L}$ has the form

$$\mathscr{L} \equiv C\left\{(1 - x^2)\frac{\mathrm{d}^2}{\mathrm{d}x^2} - x(2\lambda + 1)\frac{\mathrm{d}}{\mathrm{d}x}\right\} + a_0(x)$$

then **L** is

ALD (AUD) of type $B(2, p_0 - 2; C_L(C_U))$.

13.5 The choice of weight function: polynomial expansions

So far we have allowed the choice of weight function $w(x)$ to be determined by the choice of the orthogonal basis used. But, for a *given* $w(x)$, the solution error, although not of course the structure of **L** or the convergence properties of the expansion coefficients, is independent of the particular form of polynomial basis used. In this section we consider again a problem with homogeneous boundary conditions, and ask: to what extent can the choice of weight function affect the convergence rate of the overall calculation? We take $[a, b] = [-1, 1]$ and consider the basis

$$h_n(x) = \left(\frac{2}{\pi}\right)^{1/2} (1 - x^2)\, T_n(x), \quad n = 0, 1, \ldots, \tag{13.5.1}$$

which satisfies the homogeneous conditions (13.1.1a), and the family of weight functions

$$w(x) = (1 - x^2)^{\delta - 1/2}, \quad \delta > -1. \tag{13.5.2}$$

We show below that the Galerkin matrix **L** is AD of type $B(0, r; C)$; the calculation in fact reduces to that of Section 13.3, but with functions $d_i(x)$ which are explicitly regular whenever the $a_i(x)$ are regular. The coefficient r depends on δ as well as on the analyticity properties of the $a_i(x)$ in (13.1.1). We then discuss the extent to which δ can be chosen to optimise the overall convergence rate of the calculation. This requires that we measure the N-convergence rate of $e_N = f - f_N$ in a fixed norm, which we take arbitrarily to be the L_2 norm with weight $(1 - x^2)^z$:

$$\| f \|_z^2 \equiv \int_{-1}^{1} |f(x)|^2 (1 - x^2)^z \, dx. \tag{13.5.3}$$

We evaluate in this norm

$$\| e_N \|_z = \| f - f_N \|_z.$$

Introducing the expansion of the exact solution:

$$f(x) = \sum_{i=0}^{\infty} b_i h_i(x) \tag{13.5.4a}$$

and defining

$$\begin{aligned} b_i' &= b_i, & i > N, \\ &= b_i - a_i^{(N)}, & i \leqslant N, \end{aligned} \tag{13.5.4b}$$

we find

$$\| e_N \|_z^2 = \sum_{\substack{i=0\\j=0}}^{\infty} b_i' b_j' \Omega_{ij}, \tag{13.5.5}$$

where

$$\begin{aligned}\Omega_{ij} &= \int_{-1}^{1} (1 - x^2)^z h_i h_j \, dx \\ &= \frac{2}{\pi} \int_{-1}^{1} (1 - x^2)^{-1/2} T_i(x) \{(1 - x^2)^{z + 5/2}\} T_j(x) \, dx. \end{aligned} \tag{13.5.6}$$

We can now proceed to estimate b_i' and Ω_{ij}; as a first step we show that the operator matrix **L** is AD of type B. Now,

$$L_{nm} = \frac{2}{\pi} \int_{-1}^{1} (1 - x^2)^{\delta - 1/2} \{(1 - x^2) T_n(x) \mathscr{L} (1 - x^2) T_m(x)\} \, dx,$$

which can be written in the alternative form

$$L_{nm} = \frac{2}{\pi} \int_{-1}^{1} (1 - x^2)^{-1/2} T_n(x) \tilde{\mathscr{L}} T_m(x) \, dx, \tag{13.5.7}$$

where

$$\tilde{\mathscr{L}} = c_2(x) \frac{d^2}{dx^2} + c_1(x) \frac{d}{dx} + c_0(x),$$

with

$$\begin{aligned} c_2(x) &= (1 - x^2)^{\delta + 2} a_2(x), \\ c_1(x) &= [(1 - x^2) a_1(x) - 4x a_2(x)] (1 - x^2)^{\delta + 1}, \\ c_0(x) &= [(1 - x^2) a_0(x) - 2x a_1(x) - 2a_2(x)] (1 - x^2)^{\delta + 1}. \end{aligned} \tag{13.5.8}$$

There is now a clear correspondence between equations (13.3.2) and (13.5.7), the functions $a_i(x)$ being replaced by $c_i(x)$ here. We may therefore proceed as in Section 13.3; and the structural results (13.3.10)–(13.3.12) all hold with the functions $d_i(x)$ given by (13.3.5), with $a_i(x)$ replaced by $c_i(x)$ or, alternatively, directly in terms of the $a_i(x)$ by:

$$\begin{aligned} d_0(x) &= [(1 - x^2) a_0(x) - 2x a_1(x) - 2a_2(x)] (1 - x^2)^{\delta + 1}, \\ d_1(x) &= [(1 - x^2) a_1(x) - 3x a_2(x)] \, x (1 - x^2)^{\delta}, \\ d_2(x) &= a_2(x) (1 - x^2)^{\delta + 1}, \\ d_3(x) &= d_1(x)/x. \end{aligned} \tag{13.5.9}$$

Now, to use the convergence theorems for type B matrices of Chapter 4 we must bound the right-hand side vector g, equation (13.1.4):

$$g_i = \left(\frac{2}{\pi}\right)^{1/2} \int_{-1}^{1} (1 - x^2)^{\delta - 1/2} g(x) (1 - x^2) T_j(x)\,dx$$

$$= \left(\frac{2}{\pi}\right)^{1/2} \int_{-1}^{1} (1 - x^2)^{-1/2} \tilde{g}(x) T_j(x)\,dx, \tag{13.5.10}$$

where

$$\tilde{g}(x) = (1 - x^2)^{\delta + 1} g(x). \tag{13.5.11}$$

The analysis of Chapter 9 then gives rise to an estimate of the form

$$|g_i| \leqslant \mathscr{C}\hat{j}^{-u}. \tag{13.5.12}$$

If $u \geqslant 1$ Section 4.5 yields, when converted to the normalisation appropriate to (13.5.4):

$$\begin{aligned} |b_i| &\leqslant A_1 i^{-t-1}, \\ |b_i - a_i^{(N)}| &\leqslant A_2 N^{-t}(N + 1 - i)^{-r_U + 1} i^{-1}, \quad i \leqslant N, \end{aligned} \tag{13.5.13}$$

where $t = \min\{u, r_L\}$ and r_L, r_U are given by (13.3.10).

Finally, to estimate $\|e_N\|_z$, equation (13.5.5), we need bounds on the elements Ω_{ij}. The defining integral in (13.5.6) may be analysed as before. We expand

$$(1 - x^2)^{z + 5/2} \equiv \sigma(x) = \sum_{i=0}^{\infty}{}' \, b_i^{(\sigma)} \left(\frac{2}{\pi}\right)^{1/2} T_i(x)$$

and applying relation (9.4.10a) we find the convergence rate estimate:

$$|b_n^{(\sigma)}| = \mathcal{O}(n^{-2(z+3)}), \quad n \neq 0,$$

provided $z + 3 > 0$. It then follows as in Section 13.3 that

$$\begin{aligned} |\Omega_{ij}| &\leqslant D_1(|i - j|)^{-2(z+3)}, \quad i \neq j, \\ |\Omega_{ii}| &\leqslant D_2, \end{aligned} \tag{13.5.14}$$

where D_1, D_2 are some positive constants.

Substitution of (13.5.14) and (13.5.13) into (13.5.5) finally yields, using the bounds of the Appendix, and assuming $r_U \geqslant 2$:

$$\|e_N\|_z \leqslant D_3 N^{-t}, \tag{13.5.15}$$

where

$$t = \min\{u, r_L\}, \quad \text{as in (13.5.13).}$$

We now relate u, r_L and hence t to the analytic structure of the coefficients $a_i(x)$ in the operator $\mathscr{L}$, the right-hand side function $g(x)$ and the weight function in the variational calculation.

In the notation defined in Chapter 9, suppose that

$$\begin{aligned} a_i(x) &\in C^{(p_i', s_i)}_{(\nu_i, \mu_i)}(-1, 1) \\ g(x) &\in C^{(p_g, s_g)}_{(\nu_g, \mu_g)}(-1, 1). \end{aligned} \tag{13.5.16}$$

Then from (13.5.11) and Chapter 9 it is evident that for integer δ

$$\tilde{g}(x) \in C^{(p_g, s_g + \delta + 1)}_{(\nu_g, \mu_g)}(-1, 1)$$

and hence from (9.4.10a) we may take

$$u = \min[p_g + 1, 2s_g + 2\delta + 3 - 2\theta_g], \tag{13.5.17}$$

where

$$\theta_g = \max(\nu_g, \mu_g).$$

Similarly for (13.3.10)

$$r_L = \min[p_0 + 1, p_1 + 1, p_2, p_3 + 1]$$

where from (13.5.16), (13.5.9) and (9.4.10a)

$$\begin{aligned} p_0 &= \min[p_0' + 1, p_1' + 1, p_2' + 1, 2s_0 + 2\delta + 5 - 2\theta_0, \\ &\qquad 2s_1 + 2\delta + 3 - 2\theta_1, 2s_2 + 2\delta + 3 - 2\theta_2], \\ p_1 = p_3 &= \min[p_1' + 1, p_2' + 1, 2s_1 + 2\delta + 3 - 2\theta_1, 2s_2 + 2\delta + 1 - 2\theta_2], \\ p_2 &= \min[p_2' + 1, 2s_2 + 2\delta + 3 - 2\theta_2], \end{aligned}$$

where

$$\theta_j = \max(\nu_j, \mu_j) \quad j = 0, 1, 2.$$

Hence

$$\begin{aligned} r_L = \min[&p_0' + 2, p_1' + 2, p_2' + 1, 2s_0 + 2\delta + 6 - 2\theta_0, \\ &2s_1 + 2\delta + 4 - 2\theta_1, 2s_2 + 2\delta + 2 - 2\theta_2], \end{aligned}$$

and thus from (13.5.15)

$$\begin{aligned} t = \min\{&p_g + 2, p_0' + 2, p_1' + 2, p_2' + 1, (2s_g + 2\delta + 3 - 2\theta_g), \\ &(2s_0 + 2\delta + 6 - 2\theta_0), (2s_1 + 2\delta + 4 - 2\theta_1), (2s_2 + 2\delta + 2 - 2\theta_2)\} \end{aligned}$$

provided that $z > -3$, $r_U \geqslant 2$.

From the above we may draw the following conclusions:

(i) The results obtained are independent of the particular weighted L_2 norm chosen to measure the N-convergence rate of e_N provided that the exponent $z > -3$. This restriction is equivalent to ensuring that the integral, which represents the inner products, is well defined.

(ii) When t is determined by p_g or by p_i', $i = 0,1,2$, that is, the "dominant" singularity of the functions $a_i(x)$, $g(x)$ lies in $(-1, 1)$, there is (not surprisingly) no choice of weight function from the class of weight functions $(1 - x^2)^\delta$ that will improve the error bound.

(iii) When it is a "boundary" singularity that characterises the N-convergence rate of e_N then the error bound may be improved (i.e. t increased) by increasing the (integer) weight function exponent δ. There is an obvious correspondence here between this result and that embodied in (9.4.10) where it is shown that when a function is expanded as a series of Jacobi polynomials then the effect on the convergence rate of singular behaviour of the function at the ends of the range can be arbitrarily reduced by increasing the relevant parameter α and/or β and hence increasing one or both of the exponents of the weight function $(1 - x)^\alpha (1 + x)^\beta$ in the integral representation of the Fourier coefficient.

13.6 Inhomogeneous boundary conditions

Inhomogeneous boundary conditions can be treated in a variety of ways; we discuss briefly the three most common, restricting the discussion for simplicity, but without essential loss of generality†, to Dirichlet conditions:

$$f(a) = \alpha, \quad f(b) = \beta. \tag{13.6.1}$$

13.6.1 Subtraction

We construct a function $f_0(x)$ satisfying (13.6.1); for example,

$$f_0(x) = \alpha + \frac{(x - a)}{b - a}(\beta - \alpha) \tag{13.6.2}$$

and write $f(x) = f_0(x) + f_1(x)$, where $f_1(x)$ satisfies

$$\begin{aligned} \mathscr{L}f_1 &= g - \mathscr{L}f_0, \\ f_1(a) &= f_1(b) = 0, \end{aligned} \tag{13.6.3}$$

thus converting the problem to one with homogeneous boundary conditions.

† For ordinary differential equations. For partial differential equations the "subtraction" technique is much more difficult when non-Dirichlet conditions are involved.

13.6.2 Least squares

We use a basis $\{h_i(x)\}$ which does not satisfy (13.6.1), and augment the Galerkin equations (13.1.4) by the boundary equations

$$f_N(a) = \sum_{i=1}^{N} a_i h_i(a) = \alpha,$$
$$f_N(b) = \sum_{i=1}^{N} a_i h_i(b) = \beta, \tag{13.6.4}$$

solving the resulting overdetermined set in the least squares sense. Provided that the Galerkin matrix $\mathbf{L}$ is WAD, this can be done using the technique described in Section 8.6 in $\mathcal{O}(N^2)$ operations.

13.6.3 Augmented functional

Instead of augmenting the equations, it is possible to augment the functional, the stationary point of which yields the Galerkin equations for a homogeneous problem, with additional terms which reflect the boundary conditions. There are many ways of doing this and we briefly discuss only one, the "global element" approach of Delves and Hall (1979).

For the formally self-adjoint problem

$$\left[-\frac{\mathrm{d}}{\mathrm{d}x} A(x) \frac{\mathrm{d}}{\mathrm{d}x} + B(x)\right] f(x) = g(x) \tag{13.6.5}$$

subject to Dirichlet conditions (13.6.1), this method leads (in the case that we do *not* subdivide the region $[a, b]$) to the functional

$$F(f) = \int_a^b [f'Af' + Bf^2 - 2gf]\,\mathrm{d}x - 2(\alpha - f(a))\,A(a)\,f'(a)$$
$$+ 2(\beta - f(b))\,A(b)\,f'(b). \tag{13.6.6}$$

For reasons which appear below we number the expansion functions used from -2:

$$f \simeq f_N = \sum_{i=-2}^{N} a_i h_i(x). \tag{13.6.7}$$

Then the approximate solution f_N is defined as the stationary point of F in the class of functions (13.6.7). This leads to the modified Galerkin equations for the expansion coefficients $\mathbf{a}$:

$$\mathbf{La} = [\mathbf{L}_0 + \mathbf{S}]\,\mathbf{a} = \mathbf{g} + \mathbf{k}, \tag{13.6.8}$$

where

$$(L_0)_{i,j} = \int_a^b [h_i' A h_j' + h_i B h_j] \, dx, \qquad i, j = -2, \ldots, N. \quad (13.6.9)$$

$$S_{ij} = A(a)\,[h_i(a)\,h_j'(a) + h_j(a)\,h_i'(a)] - A(b)\,[h_i(b)\,h_j'(b) + h_j(b)\,h_i'(b)], \qquad i, j = -2, \ldots, N,$$

$$g_i = \int_a^b h_i w g(x) \, dx, \qquad i = -2, \ldots, N,$$

$$k_i = [\alpha A(a)\,h_i'(a) - \beta A(b)\,h_i'(b)], \qquad i = -2, \ldots, N, \quad (13.6.10)$$

We now take $[a, b] = [-1, 1]$ and choose as expansion set the polynomial basis

$$h_{-2}(x) = 1; \; h_{-1}(x) = x;$$
$$h_i(x) = (1 - x^2)\,T_i(x), \quad i = 0, 1, \ldots, N. \quad (13.6.11)$$

Then the submatrix $\{\mathbf{L}_0 : i, j \geqslant 0\}$ is AD of type B(0, 2; C). This follows directly for the second term (the matrix of $B(x)$) in (13.6.9) from the analysis of Section 13.5; a proof for the derivative term is very similar. The low degree ($r = 2$) stems from the "missing" weight function $(1 - x^2)^{-1/2}$ in (13.6.6). The alternative choice

$$h_i(x) = (1 - x^2)\,P_i(x), \quad i = 0, 1, \ldots, N,$$

which is "natural" for the weight $w(x) = 1$, also leads to a type B submatrix but of high degree (assuming $A(x)$, $B(x)$ are smooth: set $\lambda = \frac{1}{2}$ in Section 13.4). The bands of $\mathbf{L}_0$ corresponding to $i, j = -2, -1$ also have decreasing elements; for example,

$$(L_0)_{-2,j} = \int_{-1}^{1} B(x)\,(1 - x^2)\,T_j(x) \, dx, \quad j > 0,$$

so that this row contains (apart from a constant factor) the Chebyshev expansion coefficients of the function $\bar{B}(x) = (1 - x^2)^{1/2} B(x)$; and hence certainly

$$|(L_0)_{-2,j}| < Cj^{-2}.$$

It therefore follows that the complete matrix $\mathbf{L}_0$ is AD of type B.

The convergence theory of Chapters 1–4 is not directly applicable to (13.6.8) because of the presence of the matrix $\mathbf{S}$. In practice, however, the method converges rapidly, and we can at least use the AD property above to yield an $\mathcal{O}(N^2)$ solution of the linear equation (13.6.8), by noting that the matrix $\mathbf{S}$ has rank 4, and that the rank modification technique of Section 8.6 is therefore directly applicable.

13.7 Least squares or Galerkin?

So far we have considered only Galerkin methods for (13.1.1). Restricting the discussion for simplicity to problems with homogeneous boundary conditions (or those which have been made homogeneous by boundary condition subtraction), an alternative approach is to minimise the residual norm:

$$\underset{\mathbf{a}}{\text{minimise}} \; \| \mathscr{L} f_N - g \|. \tag{13.7.1}$$

If we choose the continuous L_2 norm, the defining equations for **a** are then

$$\mathbf{Ma} = \bar{\mathbf{g}}, \tag{13.7.2}$$

where

$$M_{ij} = (\mathscr{L} h_i, \mathscr{L} h_j), \quad \text{and} \quad \bar{g}_i = (\mathscr{L} h_i, g).$$

We do not attempt to answer the question: which approach is better? Instead, we show that at least in the simplest case, both calculations will converge at the same rate. We consider the differential equation

$$\mathscr{L} f(x) = \frac{\mathrm{d}^2 f}{\mathrm{d}x^2} + a_0(x)\, f(x) = g(x) \tag{13.7.3}$$

with $[a, b] = [0, \pi]$; $f(0) = f(\pi) = 0$; and expansion set as in Section 13.2:

$$h_i(x) = \left(\frac{2}{\pi}\right)^{1/2} \sin ix, \quad i = 1, 2, \ldots. \tag{13.7.4}$$

From Chapter 9 and Section 13.2, we make the following regularity assumptions on $a_0(x)$, $g(x)$:

$$\begin{aligned} &\text{(a)}\ a_0(x) \in C^{(2m)}[0, \pi], \\ &\text{(b)}\ a_0^{(2i+1)}(0) = a_0^{(2i+1)}(\pi) = 0, \quad i = 0,1,2,\ldots, m-1, \\ &\text{(c)}\ g(x) \in C^{(2r)}[0, \pi], \\ &\text{(d)}\ g^{(2i)}(0) = g^{(2i)}(\pi), \quad i = 0,1,\ldots, r-1. \end{aligned} \tag{13.7.5}$$

These assumptions lead directly to the following results:

$$\begin{aligned} a_0(x) &= \sum_{i=1}^{\infty} b_i^{(0)} \left(\frac{2}{\pi}\right)^{1/2} \cos(i-1)\,x; \quad &|b_i^{(0)}| \leqslant C_1 i^{-(2m+2)}, \\ g(x) &= \sum_{i=1}^{\infty} g_i \left(\frac{2}{\pi}\right)^{1/2} \sin ix; &|g_i| \leqslant C_2 i^{-(2r+1)}. \end{aligned} \tag{13.7.6}$$

13.7.1 Galerkin

Setting

$$2m + 2 = p, \quad \text{and} \quad 2r + 1 = q, \tag{13.7.7}$$

we find immediately from (13.2.4), (13.2.5) that $|L_{ii}| = \mathcal{O}(i^2)$ and that $\mathbf{L}$ is AD of type $B(2, p-1; C_L(C_U))$. Then, taking due note of the normalisations involved, Section 4.2 yields the convergence rate

$$\|e_N\|_2^2 = \mathcal{O}(N^{-v}), \tag{13.7.8}$$

where $v = 2\min(p, q) + 3$.

13.7.2 Least squares

We may also estimate $\bar{g}_i$, M_{ij} directly. We find for $\bar{g}_i$:

$$\begin{aligned} \bar{g}_i &= -i^2 g_i + \tfrac{1}{2} \sum_{\substack{j=1\\k=1}}^{\infty} g_j b_k^{(0)} [\delta_{i+k-1,j} + \delta_{i+1-k,j} - \delta_{i+1-k,-j}] \\ &= \mathcal{O}(i^{-q+2} + i^{-p+1}) = \mathcal{O}(i^{-\bar{q}}), \end{aligned} \tag{13.7.9}$$

where $\bar{q} = \min(p, q-1) - 1$.

Similarly, we may show that

$$\begin{aligned} M_{ii} &= \mathcal{O}(i^4), \\ M_{ij} &= M_{ji} = \mathcal{O}(i^2(i-j)^{-p}), \quad i > j, \end{aligned} \tag{13.7.10}$$

and hence $\mathbf{M}$ is AD of type $B(2, p-2; C)$. for some constant C. Then Section 4.2 yields

$$\|e_N\|_2^2 = \mathcal{O}(N^{-\bar{v}})$$

with

$$\begin{aligned} \bar{v} &= 2\min(\bar{q} + 2, p) + 4 - 1 \\ &= 2\min(p + 1, q, p) + 3 = v. \end{aligned} \tag{13.7.11}$$

That is, the least squares and Galerkin calculations converge at the same rate.

14

Partial Differential Equations

14.1 Introduction

Finally, we consider briefly the numerical solution of elliptic second-order partial differential equations, and show that expansion methods based on product orthogonal function expansion sets (global variational method; global element method) lead to BWAD matrices, and hence can use the convergence theory of Chapter 7 and iterative solution techniques of the type discussed in Chapter 8.

The necessary background work for this analysis is given in Chapter 11; using the results of that chapter, the reduction to BWAD form follows closely the one-dimensional reduction of Chapter 13, and we therefore only sketch the major steps needed.

14.2 The model problem

We consider as a model problem the (formally) self-adjoint elliptic equation in two space variables x, y:

$$-\nabla A(x, y)\nabla f(x, y) + B(x, y)f(x, y) = g(x, y) \tag{14.2.1}$$

in a square region: $(x, y) \in [a, b] \times [a, b]$. We assume that $f(x, y)$ is approximated by the product expansion

$$f(x, y) \approx f_N(x, y) = \sum_{i=1}^{N} \sum_{j=1}^{N} a_{ij} h_i(x) h_j(y) \tag{14.2.2}$$

where the functions $\{h_i(x)\}$ are orthonormal with weight $w(x)$ on $[a, b]$. This model can serve as a fair prototype for methods based on the use of global orthogonal expansions; two such methods are discussed briefly in

Section 14.4. For simplicity, we omit irrelevant generalisations (such as the use of different expansion sets or different numbers of terms, in the two space variables); and we omit all reference to the boundary conditions, dealing only with the structure of the Galerkin matrix **L** with elements

$$L_{pq,\,rs} = \int_a^b w(x) \int_a^b w(y) h_p(x) h_q(y) [-\nabla A(x, y) \nabla + B(x, y)] h_r(x) h_s(y) \, dx \, dy \tag{14.2.3}$$

and the vector **g** with elements

$$g_{pq} = \int_a^b w(x) \int_a^b w(y) h_p(x) h_q(y) g(x, y) \, dx \, dy. \tag{14.2.4}$$

14.2.1 The vector g

It follows at once from the bounds of Section 11.4 that g_{pq} will satisfy a bounding relationship of the form

$$|g_{pq}| \leqslant C_g p^{-\alpha} q^{-\beta} \tag{14.2.5}$$

where α, β depend on the analyticity properties (smoothness) of $g(x, y)$ and on the choice of the set $\{h_i\}$.

14.2.2 The matrix L

We can split **L**, with an obvious notation, into two parts stemming from the terms $A(x, y)$ and $B(x, y)$ in (14.2.3):

$$L_{pq,\,rs} = A_{pq,\,rs} + B_{pq,\,rs}. \tag{14.2.6}$$

The matrix **B** can be computed from the expansion of $B(x, y)$:

$$B(x, y) = \sum_{i,\,j=1}^{\infty} b_{ij} h_i(x) h_j(y),$$

$$b_{rs} = \int_a^b w(x) \int_a^b w(y) B(x, y) h_r(x) h_s(y) \, dx \, dy, \tag{14.2.7}$$

in a way which is obviously a direct extension of that used in one dimension in Sections 12.2 and 12.3: we express products such as $h_p(x)h_q(x)$, in terms of the set $\{h_i(x)\}$, and then equate terms. For each of the sets discussed in Chapter 12 (Fourier, Chebyshev, ultraspherical polynomials) we find in this way a bound of the form

$$|B_{pq,\,rs}| \leqslant C_B |p \doteq r|^{-\gamma_B} |q \doteq s|^{-\delta_B} \tag{14.2.8}$$

where again γ_B, δ_B depend on $B(x, y)$ and $\{h_i\}$.

The structure of the matrix $A_{pq,\,rs}$ also follows in a straightforward way from the discussion of Sections 13.2 and 13.3. Expanding the first term in (14.2.3):

$$\nabla\delta(x,y)\nabla = \frac{\partial}{\partial x}A_{11}(x,y)\frac{\partial}{\partial x} + \frac{\partial}{\partial x}A_{12}(x,y)\frac{\partial}{\partial y} + \frac{\partial}{\partial y}A_{21}(x,y)\frac{\partial}{\partial x} + \frac{\partial}{\partial y}A_{22}(x,y)\frac{\partial}{\partial y}. \tag{14.2.9}$$

The simplest case to consider is that of the Laplace operator, with $A_{11} = A_{22} = 1$; $A_{21} = A_{12} = 0$; then the choice $h_i(x) = (2/\pi)^{1/2}\sin ix$, $0 \leqslant x \leqslant \pi$,

$$w(x) = 1,$$

yields

$$\nabla^2 h_r(x)h_s(y) = -(r^2 + s^2)h_r(x)h_s(y)$$

and hence we find for this case the explicit form

$$A_{pq,\,rs} = (r^2 + s^2)\delta_{pr}\delta_{qs} \leqslant (r+s)^2\delta_{pr}\delta_{qs},$$

while more generally we find bounds of the form

$$|A_{pq,\,rs}| \leqslant C_A(r+s)^2|p \mathrel{\hat{=}} r|^{-\gamma_A}|q \mathrel{\hat{=}} s|^{-\delta_A}. \tag{14.2.10}$$

Combining (14.2.8), (14.2.10) then yields the overall characterisation

$$|L_{pq,\,rs}| \leqslant C_L(r+s)^2|p \mathrel{\hat{=}} r|^{-\gamma}|q \mathrel{\hat{=}} s|^{-\delta}, \tag{14.2.11}$$

where $C_L = \max(C_A, C_B)$; $\gamma = \min(\gamma_A, \gamma_B)$; $\delta = \min(\delta_A, \delta_B)$.

14.3 Block matrix form

The relationship of (14.2.11) to the BWAD structure discussed in Chapter 7 depends on the way in which $\mathbf{L}$ is partitioned; we consider the commonest partitionings below.

14.3.1 *x*- or *y*- partitioning

These partitionings follow if we consider $\mathbf{L}$ as $N \times N$ block matrix $\mathbf{L}^{(x)}$ or $\mathbf{L}^{(y)}$, each of whose blocks is an $N \times N$ matrix:

$$\begin{aligned} L_{pq,\,rs} &= (\mathbf{L}^{(x)}_{pr})_{qs} \quad (x\text{-ordering}) \\ &= (\mathbf{L}^{(y)}_{qs})_{pr} \quad (y\text{-ordering}). \end{aligned} \tag{14.3.1}$$

We consider here an x-ordering, and show that $\mathbf{L}^{(x)}$ is BWAD:

(a) $$\|\mathbf{L}_{pr}\|_\infty = \max_{q=1}^{N} \sum_{s=1}^{N} |L^{(x)}_{pr,\,qs}| \leqslant C_L|p \doteq r|^{-\gamma} \max_{q=1}^{N} \sum_{s=1}^{N} (r+s)^2|q \doteq s|^{-\delta}$$
$$\leqslant C_L' r^2 |p \doteq r|^{-\gamma}, \qquad (14.3.2)$$

where C_L' is independent of N provided that $\delta \geqslant 2$.

(b) From (14.3.2), $\|\mathbf{L}^{(x)}_{pp}\|_\infty = \mathcal{O}(p^2)$. We assume that $\|\mathbf{L}_{pp}\| \neq 0, p = 1, \ldots, N$, and hence for some C_U

$$\|\mathbf{L}^{(x)}_{pp}\|_\infty \geqslant C_U p^2. \qquad (14.3.3)$$

Then

$$\frac{\|\mathbf{L}^{(x)}_{pr}\|_\infty}{[\|\mathbf{L}^{(x)}_{pp}\|_\infty \|\mathbf{L}^{(x)}_{pr}\|_\infty]^{1/2}} \leqslant C_L' C_U^{-1} r p^{-1} |p \doteq r|^{-\gamma}$$
$$\leqslant C_L' C_U^{-1} |p \doteq r|^{-\gamma}, \qquad p > r,$$
$$\leqslant 2C_L' C_U^{-1} |p \doteq r|^{-\gamma+1}, \quad r > p, \qquad (14.3.4)$$

where we have used the inequality $r/(p(r-p)) \leqslant 2$, which is valid for $r \geqslant p+1 \geqslant 2$.

Inequality (14.3.4) shows that $\mathbf{L}^{(x)}$ is BWAD, provided that the inverses of the diagonal blocks have norm uniformly bounded in N; that is, for some D independent of N

$$\|\mathbf{L}^{(x)-1}_{pp}\|_\infty \leqslant D, \quad \forall N.$$

The form of the bounds (14.3.4) is identical with those for a matrix AD of type B (see Section 4.5) and we refer to $\mathbf{L}$ as being BWAD of type B.

Now from (14.2.11);

$$|(\mathbf{L}^{(x)}_{pp})_{qs}| \leqslant C_L(p+s)^2|q \doteq s|^{-\delta}$$
$$= \mathcal{O}(p+s)^2, \quad q = s,$$

and hence, again barring "accidental" zero diagonal terms,

$$\frac{|(\mathbf{L}^{(x)}_{pp})_{qs}|}{|(\mathbf{L}^{(x)}_{pp})_{qq}(\mathbf{L}^{(x)}_{pp})_{ss}|^{1/2}} \leqslant C_L''(p+s)(p+q)^{-1}|q \doteq s|^{-\delta}$$
$$\leqslant C_L''|q \doteq s|^{-\delta}, \qquad q \geqslant s,$$
$$\leqslant 2C_L''|q \doteq s|^{-\delta+1}, \quad s > q, \qquad (14.3.5)$$

so that the diagonal blocks $\mathbf{L}^{(x)}_{pp}$, $p = 1, \ldots, N$ are themselves AD of type B, and hence provided that $\delta > 2$ (in general) have a bounded inverse; see Theorem 3.3.

14.3.2 "Modified speedometer" ordering

The expansion function ordering shown in Table 14.1 has the advantage that new functions can be added to increase the accuracy of a calculation without reordering those already present. It is referred to as a "speedometer ordering"; we could reverse the roles of p, q to obtain an alternative similar

TABLE 14.1
"Modified speedometer" ordering of the expansion set $h_p(x)h_q(y)$

p	q	$Q = p + q - 1$	i
1	1	1	1
2	1	2	1
1	2	2	2
3	1	3	1
2	2	3	2
1	3	3	3
4	1	4	1
3	2	4	2
2	3	4	3
1	4	4	4

ordering. The order of Table 14.1 corresponds to the mapping

$$\begin{aligned} p &= Q - i + 1, \\ q &= i, \end{aligned} \tag{14.3.6}$$

and with this ordering we obtain a partitioning $\mathbf{L}^{(S)}$ of $\mathbf{L}$ in which the block matrix $\mathbf{L}^{(S)}_{QQ'}$ is of order $Q \times Q'$ with elements

$$(\mathbf{L}^{(S)}_{QQ'})_{ij} = L_{Q-i+1,i,Q'-j+1,j} \tag{14.3.7}$$

Again, we can show that $\mathbf{L}^{(S)}$ is BWAD. For, from (14.2.11)

$$|(\mathbf{L}^{(S)}_{QQ'})_{ij}| \leqslant C_L(Q' + 1)^2 |(Q - Q') \hat{+} (j - 1)|^{-\gamma} |j \hat{-} i|^{-\delta}. \tag{14.3.8}$$

From (14.3.8) it follows immediately that the diagonal blocks $\mathbf{L}^{(S)}_{QQ}$ are AD of type B(0, $\gamma + \delta$; C_S). We can bound $||\mathbf{L}^{(S)}_{QQ'}||$ as follows. For any i, setting $l = j - i$:

$$\begin{aligned} \sum_{j=1}^{Q'} |(\mathbf{L}^{(S)}_{QQ'})_{ij}| &\leqslant C_L(Q' + 1)^2 \sum_{l=1-i}^{Q'-i} |\hat{l}|^{-\delta} |Q \hat{-} Q' + l|^{-\gamma} \\ &\leqslant C_L(Q' + 1)^2 \sum_{l=-\infty}^{\infty} |\hat{l}|^{-\delta} ||\widehat{Q - Q'}| - |l||^{-\gamma} \\ &\leqslant C_L(Q' + 1)^2 D_L |Q \hat{-} Q'|^{-t}, \end{aligned} \tag{14.3.9}$$

where $t = \min(\gamma, \delta) - 1$ and the result follows on noting that the summand is even in l, and splitting the range $0 \leqslant l < \infty$ into the subranges $([0, |Q - Q'| - 1]; |Q - Q'|; [|Q - Q'| + 1, \infty))$ and bounding the resulting series using the Appendix. The BWAD property follows at once from (14.3.9) and (14.3.8).

14.4 Particular methods

Finally, we mention briefly two (related) classes of solution techniques for elliptic problems which use expansions of the form (14.2.2).

(i) *The global variational method* of Yates (1975) maps the given region (assumed star-shaped) onto a circle, and introduces radial coordinates r, θ and a product expansion $h_i(r)k_j(\theta)$. The resulting matrix has elements of the general form (14.2.3), although the "natural" weights are not used. Boundary conditions with Yates' method are introduced using extra terms in the functional.

(ii) *The global element method* (Delves and Hall, 1977) splits the given region into a few (say, M) manageable pieces with three or four sides (global elements) and maps each onto a standard region: four-sided elements onto a square with cartesian coordinates, three-sided elements onto a semicircle with radial coordinates. The resulting equations have a matrix of form (14.2.3) for each element, together with additional boundary condition terms stemming from the functional used, and coupling terms from the continuity conditions between elements. The set of $MN^2 \times MN^2$ equations can be solved in $\mathcal{O}(MN^4)$ operations, using the BWAD structure of $\mathbf{L}$ and techniques for handling the boundary conditions and continuity terms which are direct generalisations of the rank-modification techniques described in Chapter 8 (see Delves and Phillips, 1980, for details).

Appendix

Some Series Bounds

We collect here for convenience a number of series bounds which are used in the body of the book to yield intermediate steps in a number of theorem proofs. Some of these lemmas are very trivial, but they are included for completeness. These lemmas mostly depend on the property of the integral bound on a sum. This property is as follows:

Let $f(x)$ be a real-valued, non-negative function on $[a, b]$, where $a < b$ are integers. Then:

(a) if $f(x)$ is increasing on $[a, b]$,

$$\sum_{i=a}^{b-1} f(i) \leqslant \int_a^b f(x)\,\mathrm{d}x; \tag{A.1}$$

(b) if $f(x)$ is decreasing on $[a, b]$,

$$\sum_{i=a+1}^{b} f(i) \leqslant \int_a^b f(x)\,\mathrm{d}x. \tag{A.2}$$

Lemma A.1. *For* $j + 1 \leqslant k \leqslant i - 1$

$$(k - j)(i - k) \geqslant (i - j)/2.$$

Proof.

$$\begin{aligned} 2(k - j)(i - k) - (i - j) &= 2(k - j)(i - k) - (i - k) - (k - j), \\ &= (i - k)(2(k - j) - 1) - (k - j), \\ &\geqslant 2(k - j) - 1 - (k - j), \quad \text{since } (i - k) \geqslant 1, \\ &\geqslant 0, \qquad\qquad\qquad\qquad\quad \text{since } (k - j) \geqslant 1. \end{aligned}$$

Hence

$$(k-j)(i-k) \geqslant (i-j)/2.$$

Lemma A.2

$$\sum_{k=1}^{j-1} (j-k)^{-r} \leqslant \frac{r}{r-1}, \quad \textit{provided } r > 1.$$

Proof

$$\begin{aligned}\sum_{k=1}^{j-1} (j-k)^{-r} &\leqslant 1 + \sum_{k=1}^{j-2} (j-k)^{-r} \\ &\leqslant 1 + \int_1^{j-1} (j-k)^{-r}\, \mathrm{d}k, \quad \text{using (A.1)}, \\ &\leqslant 1 + \left.\frac{(j-k)^{-r+1}}{r-1}\right]_1^{j-1}, \quad \text{provided } r \neq 1, \\ &\leqslant \frac{r}{r-1}, \quad \text{provided } r > 1.\end{aligned}$$

Lemma A.3

$$\sum_{k=j+1}^{i} k^{-p} \leqslant \frac{j^{-p+1}}{p-1}, \quad \textit{provided } p > 1.$$

Proof

$$\begin{aligned}\sum_{k=j+1}^{i} k^{-p} &\leqslant \int_j^i k^{-p}\, \mathrm{d}k, \quad \text{using (A.2)}, \\ &\leqslant \left.\frac{k^{-p+1}}{-p+1}\right]_j^i, \quad \text{provided } p \neq 1, \\ &\leqslant \frac{j^{-p+1}}{p-1}, \quad \text{provided } p > 1.\end{aligned}$$

Lemma A.4

$$\sum_{k=1}^{j-1} j^{-p}(j-k)^{-r} \leqslant j^{-p+1}, \quad \textit{provided } r \geqslant 0.$$

Proof

$$\begin{aligned}\sum_{k=1}^{j-1} j^{-p}(j-k)^{-r} &\leqslant \sum_{k=1}^{j-1} j^{-p}, \quad \text{provided } r \geqslant 0, \\ &\leqslant j^{-p+1}.\end{aligned}$$

Lemma A.5

$$\sum_{k=j+1}^{i} (k-j)^{-r} \leqslant \frac{r}{r-1}, \quad \textit{provided } r > 1.$$

Proof

$$\begin{aligned}
\sum_{k=j+1}^{i} (k-j)^{-r} &\leqslant 1 + \sum_{k=j+2}^{i} (k-j)^{-r} \\
&\leqslant 1 + \int_{j+1}^{i} (k-j)^{-r}\,\mathrm{d}k, \text{ using (A.2)}, \\
&\leqslant 1 + \left.\frac{(k-j)^{-r+1}}{-r+1}\right]_{j+1}^{i}, \quad \text{provided } r \neq 1, \\
&\leqslant \frac{r}{r-1}, \qquad\qquad \text{provided } r > 1.
\end{aligned}$$

Lemma A.6

$$\sum_{k=1}^{j} (i-k)^{-r} \leqslant (i-j)^{-r+1}\frac{r}{r-1}, \quad \textit{provided } r > 1.$$

Proof

$$\begin{aligned}
\sum_{k=1}^{j} (i-k)^{-r} &\leqslant (i-j)^{-r} + \sum_{k=1}^{j-1} (i-k)^{-r} \\
&\leqslant (i-j)^{-r} + \int_{1}^{j} (i-k)^{-r}\,\mathrm{d}k, \quad \text{using (A.1)}, \\
&\leqslant (i-j)^{-r} + \left.\frac{(i-k)^{-r+1}}{r-1}\right]_{1}^{j}, \quad \text{provided } r \neq 1, \\
&\leqslant (i-j)^{-r} + \frac{(i-j)^{-r+1}}{r-1}, \qquad \text{provided } r > 1, \\
&\leqslant (i-j)^{-r+1}\frac{r}{r-1}.
\end{aligned}$$

Lemma A.7

$$\sum_{k=j+2}^{i-2} (i-k)^{-r}(k-j)^{-r} \leqslant 2\left(\frac{r+1}{r-1}\right)(i-j)^{-r}, \quad \textit{provided } r > 1.$$

Proof. Firstly we consider $i + j$ even

$$\sum_{k=j+2}^{i-2} (i-k)^{-r}(k-j)^{-r} \leqslant 2\sum_{k=j+2}^{(i+j)/2} (i-k)^{-r}(k-j)^{-r}, \quad \text{since series is symmetric,}$$

$$\leqslant 2(i-j)^{-r}2^{r}\sum_{k=j+2}^{(i+j)/2} (k-j)^{-r},$$

$$\leqslant 2(i-j)^{-r}2^{r}\left\{2^{-r} + \sum_{k=j+3}^{(i+j)/2} (k-j)^{-r}\right\},$$

$$\leqslant 2(i-j)^{-r}2^{r}\left\{2^{-r} + \int_{j+2}^{(i+j)/2} (k-j)^{-r}\,\mathrm{d}k\right\}, \quad \text{using (A.2),}$$

$$\leqslant 2(i-j)^{-r}2^{r}\left\{2^{-r} + \left.\frac{(k-j)^{-r+1}}{-r+1}\right]_{j+2}^{(i+j)/2}\right\}, \quad \text{provided } r \neq 1,$$

$$\leqslant 2(i-j)^{-r}2^{r}\left\{2^{-r} + \frac{2^{-r+1}}{r-1}\right\}, \quad \text{provided } r > 1,$$

$$\leqslant 2(i-j)^{-r}\left(\frac{r+1}{r-1}\right).$$

Now, if $i + j$ is odd:

$$\sum_{k=j+2}^{i-2} (i-k)^{-r}(k-j)^{-r} \leqslant 2\sum_{k=j+2}^{(i+j-1)/2} (i-k)^{-r}(k-j)^{-r}, \quad \text{since series is symmetric,}$$

$$\leqslant 2(i-j)^{-r}2^{r}\sum_{k=j+2}^{(i+j-1)/2} (k-j)^{-r},$$

$$\leqslant 2(i-j)^{-r}\left(\frac{r+1}{r-1}\right), \quad \text{provided } r > 1.$$

Hence, for any i, j,

$$\sum_{k=j+2}^{i-2} (i-k)^{-r}(k-j)^{-r} \leqslant 2(i-j)^{-r}\left(\frac{r+1}{r-1}\right).$$

Lemma A.8

$$\sum_{j=N+1}^{\infty} (j-i)^{-r} \leqslant (N+1-i)^{-r+1}\frac{r}{r-1}, \quad \textit{provided } r > 1.$$

Proof

$$\begin{aligned}\sum_{j=N+1}^{\infty} (j-i)^{-r} &\leqslant (N+1-i)^{-r} + \sum_{j=N+2}^{\infty} (j-i)^{-r},\\ &\leqslant (N+1-i)^{-r} + \int_{N+1}^{\infty} (j-i)^{-r}\,\mathrm{d}j, \quad \text{using (A.2)},\\ &\leqslant (N+1-i)^{-r} + \left.\frac{(j-i)^{-r+1}}{-r+1}\right]_{N+1}^{\infty}, \quad \text{provided } r \neq 1,\\ &\leqslant (N+1-i)^{-r} + \frac{(N+1-i)^{-r+1}}{r-1}, \quad \text{provided } r > 1,\\ &\leqslant (N+1-i)^{-r+1}\frac{r}{r-1}.\end{aligned}$$

Lemma A.9

$$\sum_{j=1}^{i-1} (i-j)^{-r} j^{-q} \leqslant \frac{2^{t+1}t}{t-1} i^{-t}, \quad t = \min(q, r), \quad \textit{provided } t > 1.$$

Proof. We firstly consider $q > r$, i.e. $t = r$.

$$\begin{aligned}\sum_{j=1}^{i-1} (i-j)^{-r} j^{-q} &\leqslant \sum_{j=1}^{i-1} (i-j)^{-r} j^{-r} j^{-q+r},\\ &\leqslant \sum_{j=1}^{i-1} (i-j)^{-r} j^{-r},\\ &\leqslant 2\sum_{j=1}^{i/2} (i-j)^{-r} j^{-r}, \qquad i \text{ even},\\ &\leqslant 2i^{-r}2^{r} \sum_{j=1}^{i/2} j^{-r},\\ &\leqslant 2^{r+1} i^{-r}\left(1 + \sum_{j=2}^{i/2} j^{-r}\right),\\ &\leqslant 2^{r+1} i^{-r}\left(1 + \int_{1}^{i/2} j^{-r}\,\mathrm{d}j\right), \qquad \text{using (A.2)}\end{aligned}$$

$$\leqslant 2^{r+1}i^{-r}\left(1 + \left[\frac{j^{-r+1}}{-r+1}\right]_1^{i/2}\right), \quad \text{provided } r \neq 1,$$

$$\leqslant 2^{r+1}i^{-r}\frac{r}{r-1}, \quad \text{provided } r > 1,$$

$$\leqslant \frac{2^{t+1}t}{t-1}i^{-t}, \quad \text{since } r = t.$$

For i odd,

$$\sum_{j=1}^{i-1}(i-j)^{-r}j^{-q} \leqslant 2\sum_{j=1}^{(i-1)/2}(i-j)^{-r}j^{-r},$$

$$\leqslant 2(i+1)^{-r}2^r\sum_{j=1}^{(i-1)/2}j^{-r},$$

$$\leqslant 2^{r+1}i^{-r}\left(1 + \int_1^{(i-1)/2}j^{-r}\,\mathrm{d}j\right), \quad \text{using (A.2)}$$

$$\leqslant \frac{2^{t+1}t}{t-1}i^{-t}, \quad \text{provided } t > 1.$$

Secondly, we consider $q \leqslant r$, i.e. $t = q$.

$$\sum_{j=1}^{i-1}(i-j)^{-r}j^{-q} \leqslant \sum_{j=1}^{i-1}(i-j)^{-q}j^{-q}(i-j)^{-r+q}$$

$$\leqslant \sum_{j=1}^{i-1}(i-j)^{-q}j^{-q},$$

$$\leqslant 2^{q+1}\frac{i^{-q}q}{q-1}, \quad \text{provided } q > 1,$$

$$\leqslant \frac{2^{t+1}t}{t-1}i^{-t}, \quad \text{since } q = t.$$

Hence,

$$\sum_{j=1}^{i-1}(i-j)^{-r}j^{-q} \leqslant \frac{2^{t+1}t}{t-1}i^{-t}, \quad t = \min(r, q), \quad \text{provided } t > 1.$$

Lemma A.10

$$\sum_{j=i}^{N}j^{-q} \leqslant \frac{i^{-q+1}q}{q-1}, \quad \textit{provided } q > 1.$$

Proof

$$\begin{aligned}\sum_{j=i}^{N} j^{-q} &\leqslant i^{-q} + \sum_{j=i+1}^{N} j^{-q},\\ &\leqslant i^{-q} + \int_i^N j^{-q}\,\mathrm{d}j, \quad \text{using (A.2)},\\ &\leqslant i^{-q} + \left.\frac{j^{-q+1}}{-q+1}\right]_i^N,\\ &\leqslant i^{-q} + \frac{i^{-q+1}}{q-1},\\ &\leqslant \frac{i^{-q+1}q}{q-1}.\end{aligned}$$

Lemma A.11

$$\sum_{k=j}^{i} k^{b-a} \leqslant \begin{cases} i^{b-a+1}, & b \geqslant a,\\ ij^{b-a}, & b < a.\end{cases}$$

Proof

$$\begin{aligned}\sum_{k=j}^{i} k^{b-a} &\leqslant \sum_{k=j}^{i} 1 \begin{Bmatrix} i^{b-a}, & b \geqslant a\\ j^{b-a}, & b < a\end{Bmatrix},\\ &\leqslant \begin{cases} i^{b-a+1}, & b \geqslant a,\\ ij^{b-a}, & b < a.\end{cases}\end{aligned}$$

Lemma A.12

$$\sum_{k=j}^{i} k^{b-a} \leqslant \begin{cases} i^{b-a+1}, & b \geqslant a,\\ j^{b-a+1}\left(\dfrac{a-b}{a-b-1}\right), & b < a-1,\\ i^{b-a+1}\left(\dfrac{b-a+2}{b-a+1}\right), & a-1 < b < a,\\ ij^{-1}, & b = a-1.\end{cases}$$

Proof. When $b \geqslant a$,

$$\sum_{k=j}^{i} k^{b-a} \leqslant i^{b-a+1}.$$

When $b < a - 1$,

$$\begin{aligned}\sum_{k=j}^{i} k^{b-a} &\leqslant j^{b-a} + \sum_{k=j+1}^{i} k^{b-a},\\ &\leqslant j^{b-a} + \int_j^i k^{b-a}\,\mathrm{d}k, \quad \text{using (A.2)},\\ &\leqslant j^{b-a} + \left.\frac{k^{b-a+1}}{b-a+1}\right]_j^i, \quad \text{provided } b \neq a-1,\\ &\leqslant j^{b-a} + \frac{j^{b-a+1}}{a-b-1}, \quad \text{provided } b < a-1,\\ &\leqslant j^{b-a+1}\left(\frac{a-b}{a-b-1}\right).\end{aligned}$$

When $a - 1 < b < a$,

$$\begin{aligned}\sum_{k=j}^{i} k^{b-a} &\leqslant j^{b-a} + \sum_{k=j+1}^{i} k^{b-a}\\ &\leqslant j^{b-a} + \int_j^i k^{b-a}\,\mathrm{d}k, \quad \text{using (A.2)},\\ &\leqslant j^{b-a} + \left.\frac{k^{b-a+1}}{b-a+1}\right]_j^i, \quad \text{provided } b \neq a-1,\\ &\leqslant j^{b-a} + \frac{i^{b-a+1}}{b-a+1}, \quad \text{provided } b > a-1,\\ &\leqslant i^{b-a+1} + \frac{i^{b-a+1}}{b-a+1},\\ &\leqslant \left(\frac{b-a+2}{b-a+1}\right) i^{b-a+1}.\end{aligned}$$

When $b = a - 1$,

$$\begin{aligned}\sum_{k=j}^{i} k^{b-a} &\leqslant \sum_{k=j}^{i} k^{-1},\\ &\leqslant ij^{-1}.\end{aligned}$$

Lemma A.13

$$\sum_{j=1}^{i-1} (i-j)^{-r} j^{-v} \leqslant 2^{w_2+1}\frac{w_1}{(w_1-1)} i^{-u}, \quad \textit{provided } r > 1,$$

where $u = \min\{v, r\}$,

$$w_1 = \begin{cases} u; v > 1, \\ r; v \leqslant 1, \end{cases}$$

$$w_2 = \begin{cases} u; v > 1, \\ r; 0 < v \leqslant 1, \\ -1; v \leqslant 0. \end{cases}$$

Proof. When $v > 1$ the result immediately follows from Lemma A.9. When $0 < v \leqslant 1$, then

$$\begin{aligned} \sum_{j=1}^{i-1} (i-j)^{-r} j^{-v} &\leqslant \sum_{j=1}^{i-1} (i-j)^{-r} j^{-r} j^{r-v}, \\ &\leqslant i^{r-v} \sum_{j=1}^{i-1} (i-j)^{-r} j^{-r}, \\ &\leqslant i^{r-v} \frac{2^{r+1} r}{(r-1)} i^{-r}, \qquad \text{by Lemma A.9,} \\ &\leqslant i^{-v} \frac{2^{r+1} r}{(r-1)}. \end{aligned}$$

When $v \leqslant 0$,

$$\sum_{j=1}^{i-1} j^{-v} (i-j)^{-r} \leqslant i^{-v} \frac{r}{(r-1)}, \quad \text{by Lemma A.2.}$$

Hence the result follows immediately.

Bibliography

Ahlfors, L. V. (1953), "Complex Analysis". McGraw-Hill, New York.

Anderssen, R. S. and Omodei, B. J. (1974), On the stability of uniformly asymptotically diagonal systems. *Maths. Comput.*, **28**, 719–730.

Bain, M. (1974), Convergence of generalised Fourier series. Ph.D. Thesis, University of Liverpool.

Bain, M. (1978), On the uniform convergence of generalised Fourier series. *J. Inst. Math. and its Appl.*, **21**, 379–386.

Bain, M. and Delves, L. M. (1977), The convergence rates of expansions in Jacobi polynomials. *Num. Math.*, **27**, 219–225.

Bain, M., Delves, L. M. and Mead, K. O. (1978), On the rate of convergence of multi-dimensional orthogonal expansions. *J. Inst. Math. and its Appl.*, **22**, 225–234.

Baker, C. T. H. (1977), "The Numerical Treatment of Integral Equations". Oxford University Press, Oxford.

Crawford, C. R. (1973), Reduction of a band-symmetric generalised eigenvalue problem. *Commun. Ass. Comput. Mach.*, **16**, 41–44.

Davies, B. and Hendry, J. A. (1977), The indirect design of fluid channels using a global variational method with trial functions not satisfying the prescribed boundary conditions. *Int. J. Numer. Meth. Eng.*, **11**, 579–591.

Delves, L. M. (1968), Round-off errors in variational calculations. *J. Comput. Phys.*, **3**, 17–28.

Delves, L. M. (1973), A comparison of least squares and variational methods for the solution of differential equations. Research Report CSS/73/2/1, University of Liverpool.

Delves, L. M. (1974), An automatic Ritz–Galerkin procedure for the numerical solution of linear Fredholm integral equations of the second kind. Paper presented at the Conference on Mathematical Software, Purdue.

Delves, L. M. (1977a), The numerical solution of sets of linear equations arising from Ritz–Galerkin methods, *J. Inst. Math. and its Appl.*, **20**, 163–171

Delves, L. M. (1977b), A linear equation solver for Galerkin and least squares methods. *Comput. J.*, **20**, 371–374.

Delves, L. M. (1977c), A fast method for the solution of Fredholm integral equations. *J. Inst. Math. and its Appl.*, **20**, 173–182.

Delves, L. M. and Abd-Elal, L. F. (1977), The fast Galerkin algorithm for the solution of linear Fredholm equations. *Comput. J.*, **20**, 374–376.

Delves, L. M., Abd-Elal, L. F. and Hendry, J. A. (1979), A fast Galerkin alogorithm for singular integral equations. *J. Inst. Math. and its Appl.*, **23**, 139–166.
Delves, L. M. and Bain, M. (1977), On the optimal choice of weight functions in a class of variational calculations. *Num. Math.*, **27**, 209–218.
Delves, L. M. and Barrodale, I. (1979), A fast direct method for the least squares solution of slightly overdetermined sets of linear equations. *J. Inst. Math. and its Appl.*, **24**, 149–156.
Delves, L. M. and Freeman, T. L. (1975), On the convergence rates of variational methods: homogeneous systems. *Util. Math.*, **7**, 77–94.
Delves, L. M. and Hall, C. A. (1979), An implicit matching principle for global element calculations. *J. Inst. Math. and its Appl.*, **23**, 223–234.
Delves, L. M. and Mead, K. O. (1971), On the convergence rates of variational methods. I. Asymptotically diagonal systems. *Maths. Comput.*, **25**, 699–716.
Delves, L. M. and Musa, F. A. (1976), On a class of non-singular matrices. *Util. Math.*, **10**, 241–258.
Delves, L. M. and Phillips, C. (1980), A fast implementation of the global element method. *J. Inst. Math. and its Appl.*, **25**, 177–197.
Erdélyi, A., Magnus, W., Oberhettinger, F. and Tricomi, F. G. (1953), "Higher Transcendental Functions", Vol. II. McGraw–Hill, New York.
Flett, T. M. (1966), "Mathematical Analysis", McGraw–Hill, London.
Freeman, T. L. (1974), Convergence rates and stability of variational methods. Ph.D. Thesis, University of Liverpool.
Freeman, T. L. and Delves, L. M. (1975), Round-off errors and stability in variational calculations. *J. Inst. Math. and its Appl.*, **16**, 345–353.
Freeman, T. L., Delves, L. M. and Reid, J. K. (1974), On the convergence rates of variational methods. II. Systems of type B, C. *J. Inst. Math. and its Appl.*, **14**, 145–157.
Haber, S. (1970), Numerical evaluation of multiple integrals. *SIAM Rev.*, **12**, 481–526.
Hsien-yü Hsü (1938), Certain integrals and finite series involving ultraspherical polynomials and Bessel functions. *Duke Math. J.*, **4**, 374–383.
Mead, K. O. (1971), On the convergence of variational methods. Ph.D. Thesis, University of Liverpool.
Mead, K. O. and Delves, L. M. (1973), On the convergence rate of generalised Fourier expansions. *J. Inst. Math. and its Appl.*, **12**, 247–259.
Mikhlin, S. G. (1971), "The Numerical Performance of Variational Methods", (Moscow, 1966). English translation: Wolters–Noordhoff, Groningen.
Moler, C. B. and Stewart, G. W. (1973), An algorithm for generalised matrix eigenvalue problems. *SIAM J. Numer. Anal.*, **10**, 241–256.
Morse, P. M. and Feshbach, E. (1953), "Methods of Theoretical Physics". McGraw–Hill, New York.
Musa, F. A. (1973), A class of matrices in direct methods. Ph.D. Thesis, University of Liverpool.
Musa, F. A. and Delves, L. M. (1975), Block weakly asymptotically diagonal systems. Research Report CSS/75/1/1, University of Liverpool.
Natanson, I. P. (1965), "Constructive Function Theory", Vol. II. Ungar, New York.
Noble, B. (1969), "Applied Linear Algebra". Prentice-Hall, Englewood Cliffs, New Jersey.
Reid, J. K. (1971), A note on the stability of Gaussian Elimination. *J. Inst. Math. and its Appl.*, **8**, 374–375.

Richmond, N. (1981), Global Galerkin methods for linear parabolic partial differential equations. Ph.D. Thesis, University of Liverpool.

Riess, R. D. and Johnson, L. W. (1969), Estimating Gauss–Chebyshev quadrature errors. *SIAM J. Numer. Anal.*, **6**, 557–559.

Schultz, M. H. (1969), Rayleigh–Ritz–Galerkin methods for multi-dimensional problems. *SIAM J. Numer. Anal.*, **6**, 523–538.

Stewart, G. W. (1976), Simultaneous iteration for computing equivalence classes of non-Hermitian matrices. *Num. Math.*, **25**, 123–136.

Strang, G. J. and Fix, G. (1973), "An Analysis of the Finite Element Method". Prentice–Hall, Englewood Cliffs, New Jersey.

Szego, G. (1939), "Orthogonal Polynomials". American Mathematical Society Colloquium Publications, Vol. XXIII, Waverley Press.

Tapia, R. A. (1971), The Kantorovich theorem for Newton's method. *Am. Math. Mon.*, **78**, 389–392.

Titchmarsh, E. C. (1946), "Eigenfunction Expansions". Clarendon Press, Oxford.

Ursell, F. (1974), A problem in the theory of water waves. *In* "Numerical Solution of Integral Equations", L. M. Delves and J. Walsh (eds.), 291–299. Clarendon Press, Oxford.

Van Loan, C. (1975), A study of the matrix exponential. Numerical Analysis Report No. 10, Department of Mathematics, University of Manchester.

Wilkinson, J. H. (1965), "The Algebraic Eigenvalue Problem". Clarendon Press, Oxford.

Yates, D. F. (1975), A Rayleigh–Ritz–Galerkin method for the solution of one- and two-dimensional boundary value problems. Ph.D. Thesis, University of Liverpool.

Index